Lecture Notes in Computer Science 4726

Commenced Publication in 1973
Founding and Former Series Editors:
Gerhard Goos, Juris Hartmanis, and J   ven

T0216745

Nivio Ziviani   Ricardo Baeza-Yates (Eds.)

String Processing and Information Retrieval

14th International Symposium, SPIRE 2007
Santiago, Chile, October 29-31, 2007
Proceedings

 Springer

Volume Editors

Nivio Ziviani
Federal University of Minas Gerais
Department of Computer Science
Av. Antônio Carlos 6627, 31270-010 Belo Horizonte, MG, Brazil
E-mail: nivio@dcc.ufmg.br

Ricardo Baeza-Yates
Yahoo! Research Latin America
Blanco Encalada 2120, Santiago 6511224, Chile
E-mail: ricardo@baeza.cl

Library of Congress Control Number: 2007937296

CR Subject Classification (1998): H.3, H.2.8, I.2, E.1, E.5, F.2.2

LNCS Sublibrary: SL 1 – Theoretical Computer Science and General Issues

ISSN   0302-9743
ISBN-10   3-540-75529-2 Springer Berlin Heidelberg New York
ISBN-13   978-3-540-75529-6 Springer Berlin Heidelberg New York

Springer is a part of Springer Science+Business Media

springer.com

© Springer-Verlag Berlin Heidelberg 2007

Typesetting: Camera-ready by author, data conversion by Scientific Publishing Services, Chennai, India
Printed on acid-free paper   SPIN: 12171385   06/3180   5 4 3 2 1 0

Preface

This volume contains the papers presented at the 14th International Symposium on String Processing and Information Retrieval (SPIRE), held in Santiago, Chile, on October 29–31, 2007. SPIRE 2007 was organized in tandem with the 5th Latin American Web Congress (LA-WEB), with both conferences sharing a common day on Web Retrieval.

The papers in this volume were selected from 77 papers submitted from 25 different countries in response to the Call for Papers. Due to the high quality of the submissions, a total of 27 papers were accepted as full papers, yielding an acceptance rate of about 35%. SPIRE 2007 also featured three talks by invited speakers: Andrew Tomkins (Yahoo! Research, USA), Nivio Ziviani (Federal University of Minas Gerais, Brazil) and Justin Zobel (NICTA, Melbourne, Australia).

The SPIRE annual symposium provides an opportunity for researchers to present original contributions on areas such as *string processing* (dictionary algorithms, text searching, pattern matching, text compression, text mining, natural language processing, and automata based string processing), *information retrieval* (IR modeling, indexing, ranking and filtering, interface design, visualization, cross-lingual IR systems, multimedia IR, digital libraries, collaborative retrieval, and Web related applications), *interaction of biology and computation* (DNA sequencing and applications in molecular biology, evolution and phylogenetics, recognition of genes and regulatory elements, and sequence driven protein structure prediction), and *information retrieval languages and applications* (XML, SGML, information retrieval from semi-structured data, text mining, and generation of structured data from text).

Special thanks are due to the members of the Program Committee and the additional reviewers who worked very hard to ensure the timely review of all submitted manuscripts. Thanks are due to Fabiano Cupertino Botelho, a Ph.D. student volunteer who ran the OpenConf system during the reviewing process and helped with the editorial work for this volume. We also thank the local organizers for their support and organization of SPIRE, in particular Javier Velasco, Christian Middleton, and Sara Quiñones, as well as the local team of student volunteers, whose efforts ensured the smooth organization and running of the event.

We would like to thank the sponsoring institutions, the Millennium Nucleus Center for Web Research of the Dept. of Computer Science of the University of Chile, the Dept. of Computer Science of the Federal University of Minas Gerais and Yahoo! Research Latin America.

October 2007

Nivio Ziviani
Ricardo Baeza-Yates

SPIRE 2007 Organization

General Chair

Ricardo Baeza-Yates   Yahoo! Research (Spain & Chile) and
CWR/DCC, Universidad de Chile (Chile)

Program Committee Chair

Nivio Ziviani   Universidade Federal de Minas Gerais (Brazil)

Local Organization

Fabiano C. Botelho   Universidade Federal de Minas Gerais (Brazil)
Christian Middleton   Universitat Pompeu Fabra (Spain)
Sara Quiñones   Yahoo! Research Latin America (Chile)
Javier Velasco   CWR/DCC, Universidad de Chile (Chile)

Steering Committee

Alberto Apostolico   Università di Padova (Italy)
     and Georgia Tech (USA)
Ricardo Baeza-Yates   Yahoo! Research (Spain & Chile) and
     CWR/DCC, Universidad de Chile (Chile)
Mariano Consens   University of Toronto (Canada)
Fabio Crestani   Università della Svizzera Italiana (Switzerland)
Paolo Ferragina   Università di Pisa (Italy)
Massimo Melucci   Università di Padova (Italy)
Gonzalo Navarro   Universidad de Chile (Chile)
Berthier Ribeiro-Neto   Universidade Federal de Minas Gerais (Brazil)
Mark Sanderson   University of Sheffield (UK)
Nivio Ziviani   Universidade Federal de Minas Gerais (Brazil)

Program Committee Members

James Allan   University of Massachusetts Amherst (USA)
Amihood Amir   Bar-Ilan University (Israel)
Alberto Apostolico   University of Padova (Italy)
     and Georgia Tech (USA)
Chris Buckley   Sabir Research (USA)

Pável Pereira Calado	Instituto Superior Técnico/INESC-ID (Portugal)
Maxime Crochemore	University of Marne-la-Vallée (France)
Bruce Croft	University of Massachusetts Amherst (USA)
Martin Farach-Colton	Rutgers University (USA)
Edward Fox	Virginia Tech (USA)
Kimmo Fredriksson	University of Joensuu (Finland)
Raffaele Giancarlo	University of Palermo (Italy)
Marcos André Gonçalves	Federal University of Minas Gerais (Brazil)
Roberto Grossi	University of Pisa (Italy)
Heikki Hyyrö	University of Tampere (Finland)
Lucian Ilie	University of Western Ontario (Canada)
Costas Iliopoulos	University of London (UK)
Juha Kärkkäinen	University of Helsinki (Finland)
Mounia Lalmas	University of London (UK)
Tak-Wah Lam	University of Hong Kong (Hong Kong)
Gad Landau	University of Haifa (Israel)
Thierry Lecroq	University of Rouen (France)
Andrew MacFarlane	City University London (UK)
Veli Mäkinen	University of Helsinki (Finland)
Giovanni Manzini	University of Piemonte Orientale (Italy)
Massimo Melucci	University of Padua (Italy)
Alistair Moffat	University of Melbourne (Australia)
Edleno Silva de Moura	Federal University of Amazonas (Brazil)
Ian Munro	University of Waterloo (Canada)
Gonzalo Navarro	University of Chile (Chile)
Arlindo Oliveira	Inst. Superior Técnico/INESC-ID/IST (Portugal)
Sándor Pongor	Intl. Centre for Genetic Eng. and Biotechnology (Italy)
Bruno Pôssas	Google Inc.(Brazil)
Mathieu Raffinot	CNRS (France)
Kunihiko Sadakane	Kyushu University (Japan)
Marie-France Sagot	INRIA and University Claude Bernard, Lyon I (France)
João Setubal	Virginia Tech (USA)
Rahul Shah	Purdue University (USA)
Altigran Soares da Silva	Federal University of Amazonas (Brazil)
Fabrizio Silvestri	ISTI - CNR (Italy)
Wing-Kin Sung	National University of Singapore (Singapore)
Masayuki Takeda	Kyushu University (Japan)
Jorma Tarhio	Helsinki University of Technology (Finland)
Gabriel Valiente	Technical University of Catalonia (Spain)
Hugo Zaragoza	Yahoo! Research (Spain)
Justin Zobel	RMIT University (Australia)

Additional Reviewers

José Ramón Pérez Agüera
Guilherme Assis
Marie-Pierre Béal
Ido Dagan
Chiara Epifanio
Iman Hajirasouliha
Wing-Kai Hon
Tomi Klein
Jorma Laurikkala
Turkka Näppilä
Simon Puglisi
Leena Salmela
Tuomas Talvensaari
Jarkko Toivonen
Sebastiano Vigna

Diego Arroyuelo
Claudine Badue
Mathieu Constant
Gianna Del Corso
Ankur Gupta
Jan Holub
Petri Kalsi
Alberto H. F. Laender
Sabrina Mantaci
Hannu Peltola
Cenk Sahinalp
Borkur Sigurbjornsson
Andrew Turpin

Shuly Wintner

Previous Venues of SPIRE

The first four editions focused primarily on *string processing* and were held in Brazil and Chile. At that time SPIRE was called WSP (South American Workshop on String Processing). Starting in 1998, the focus of the workshop was broadened to include the area of *information retrieval*, due to the latter's increasing relevance and its inter-relationship with the area of string processing, and the name of the workshop was changed to the current one. In addition, since 2000, the symposium has been held alternately in Europe and Latin America, and has so far been held in Mexico, Spain, Chile, Portugal, Brazil, Italy, Argentina and the UK.

2006: Glasgow, UK
2005: Buenos Aires, Argentina
2004: Padova, Italy
2003: Manaus, Brazil
2002: Lisboa, Portugal
2001: Laguna San Rafael, Chile
2000: A Coruña, Spain
1999: Cancun, Mexico
1998: Santa Cruz de la Sierra, Bolivia
1997: Valparaiso, Chile
1996: Recife, Brazil
1995: Viña del Mar, Chile
1993: Belo Horizonte, Brazil

Table of Contents

A Chaining Algorithm for Mapping cDNA Sequences to Multiple Genomic Sequences

Mohamed Abouelhoda

Faculty of Engineering and Computer Science, Ulm University
D-89069 Ulm, Germany
mohamed.ibrahim@uni-ulm.de

Abstract. Given a set of matches between a cDNA sequence and *multiple* genomic sequences, we present a subquadratic chaining algorithm for computing an optimal chain of colinear matches, while allowing overlaps between the matches. Our algorithm improves upon the quadratic graph based solution, and extends the previous algorithms which are limited to matches between a cDNA sequence and a *single* genomic sequence. The matches of the resulting chain serve as anchors for computing a multiple alignment between the cDNA and the given sequences.

1 Introduction

A fundamental task of every genome annotation project is to locate each gene in the genome and to determine its structure. This knowledge serves as a basis for elucidating the gene function and studying the genome organization and evolution. One of the most successful methods for accomplishing this task is the mapping of cDNA sequences to the genomes they are transcribed from. A cDNA sequence is a complementary sequence to a mRNA. Because the introns are spliced out from a mRNA and just the exons remain, an alignment of a cDNA to the related genomic sequence locates the corresponding gene and directly reveals its exon-intron structure; see Figure 1 (a). The increasing number of full cDNA sequencing projects reflects the growing popularity of this method.

For high throughput mapping of cDNA sequences, standard dynamic programming algorithms are impractical due to their quadratic running time. Hence, heuristic algorithms have been developed; see e.g. [7,8,12] and the references therein. Most of these tools use an *anchor-based strategy* composed of three phases: (1) computation of fragments (regions in the sequences that are similar), (2) computation of an optimal chain of colinear fragments; these are the anchors that form the basis of the alignment, (3) alignment of the regions between the anchors considering the splice site signals.

The algorithm of Shibuya and Kurochkin [12] is superior to other ones because of two novel improvements: First, the fragments are of the type (rare) maximal exact match computed by means of the suffix tree of the genomic sequence in linear time and space. Second, in contrast to other algorithms, their chaining algorithm is geometry based and allows overlaps between the fragments.

N. Ziviani and R. Baeza-Yates (Eds.): SPIRE 2007, LNCS 4726, pp. 1–13, 2007.

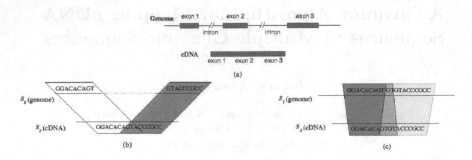

Fig. 1. (a): A cDNA mapped to a genomic sequence. The exons are separated by long introns in the genome. (b): Fragments (represented by parallelograms) overlap in the cDNA sequence only. (c) The overlap is in both the cDNA and the genome.

(The overlap lengths are taken into account, i.e., penalized, in the objective function.) Their chaining algorithm takes $O(m \log m)$ time and requires $O(m)$ space, where m is the number of the fragments. (Algorithms permitting *no* overlaps have been presented in [1,6,10,13].) Although this chaining algorithm is relatively complicated due to the combination of range maximum queries and the candidate list paradigm, it is an important improvement over the naive graph based solution that takes $O(m^2)$ time [12].

The rationale behind permitting overlaps is twofold: First, overlapping fragments were found to be very common in cDNA mapping [7,12], and they usually occur at the exon boundaries in the cDNA; see Figure 1 (b). Second, the amount of sequence covered by the chain will increase, which is crucial for both improving the sensitivity/specificity and for speeding-up the mapping task. Regarding the sensitivity/specificity, some fragments may be discarded as a result of permitting no overlap in the chain. This can reduce the chain coverage under the threshold defined by the user to filter out noisy chains, and consequently results in discarding the whole chain despite its potential significance. If one attempts to overcome this drawback by decreasing the threshold, many insignificant chains will not be filtered out and the specificity will decrease. Regarding the running time, the less the chain coverage, the higher the running time of the third phase in which an alignment on the character level is computed to finish the mapping task.

The genomes of very closely related species or the genomes of different strains of the same species share a very large sequence identity, and so does a cDNA sequence to the genome it stems from. Therefore, it is natural to extend the algorithm of Shibuya and Kurochkin to map a cDNA sequence to multiple genomes. Such an extension, in addition to the theoretical interest related to it, will help in both identifying the common genes among the genomes and determining the syntenic regions (regions of conserved gene-order) among the genomic sequences.

This extension, however, is not straightforward. While computing fragments from multiple genomic sequences can be easily achieved in linear time and space [2,5,11,9], the extension of the chaining algorithm of Shibuya and Kurochkin to chain fragments from k sequences while permitting overlaps is extremely complicated, if not infeasible. This is due to the difficulty of analyzing the overlaps, according to the objective function they suggested, and due to the difficulty of combining the range queries and candidate lists; Shibuya and Kurochkin noticed also these complications [12].

In this paper we handle the combinatorial chaining problem with overlap for mapping a cDNA sequence to multiple genomic sequences. We show in this paper that an efficient subquadratic chaining algorithm exists, if an objective function specific to the cDNA mapping task is used. We present this algorithm, and, moreover, address two special cases of practical interest: (1) the usage of rare *multi-MEMs*, and (2) constraining the amount of overlap. For these two cases, we show that the algorithm complexity can be further improved. Our algorithms are easy to implement, because they use solely range queries without any candidate lists. They are also so efficient that millions of fragments are processed in a few minutes.

In the following section, the definitions are stated. Section 3 introduces the chaining problem and the graph based solution. In Section 4, we present our geometry based algorithm. In Section 5, we focus on two special cases of the basic algorithm. Sections 6 and 7 contain experimental results and conclusions.

2   The Fragments

2.1   Definitions

For $1 \leq i \leq k$, S_i denotes a string of length $|S_i|$. In our application, S_i represents a cDNA or a genomic sequence. $S_i[h_1..h_2]$ is the substring of S_i starting at position h_1 and ending at position h_2, and $S_i[h_1]$ denotes the h_1^{th} character of S_i, $1 \leq h_1 \leq h_2 \leq |S_i|$. A *fragment* is a region of similarity among the given sequences. In this paper, we use fragments of the type *(rare) maximal multiple exact match*, denoted by *(rare) multi-MEM* and defined as follows.

A *multiple exact match* among k sequences $S_1, \ldots, S_k$ is a $(k+1)$-tuple $(l, p_1, .., p_k)$ such that $S_1[p_1..p_1+l-1] = \ldots = S_k[p_k..p_k+l-1]$; i.e., the l-character-long substrings of $S_1, \ldots, S_k$ starting at positions $p_1, \ldots, p_k$, respectively, are identical. A multiple exact match is *left maximal* if $S_i[p_i-1] \neq S_j[p_j-1]$ for any $1 \leq i \neq j \leq k$, and *right maximal* if $S_i[p_i+l] \neq S_j[p_j+l]$ for any $1 \leq i \neq j \leq k$, i.e., it cannot be extended to the left and to the right simultaneously in all the sequences. A *multi-MEM* is a left and right maximal multiple exact match.

A *multi-MEM* $(l, p_1, .., p_k)$ is called *rare*, if the substring $S_i[p_i..p_i+l-1]$ occurs at most r times in each S_i, $1 \leq i \leq k$. A *maximal multiple unique match* (*multi-MUM*) is a rare *multi-MEM* such that $r = 1$, i.e., $S_i[p_i..p_i+l-1]$ occurs exactly once in each S_i.

A hyper-rectangle in a k dimensional space ($\mathbb{R}^k$) can be represented by the k-tuple $([p_1..q_1], \ldots, [p_k..q_k])$, where $[p_i..q_i]$ is the interval on the coordinate axis x_i, $1 \leq i \leq k$. Equivalently, this hyper-rectangle can be denoted by $R(p, q)$, where $p = (p_1, .., p_k)$ and $q = (q_1, .., q_k)$ are its two extreme corner points. A fragment of the type (rare) *multi-MEM* $(l, p_1, .., p_k)$ can be represented by a hyper-rectangle in $\mathbb{R}^k$ with the two extreme corner points $(p_1, .., p_k)$ and

$(p_1 + l - 1, .., p_k + l - 1)$. In the following, we will denote these corner points by $beg(f) = (beg(f).x_1, .., beg(f).x_k)$ and $end(f) = (end(f).x_1, .., end(f).x_k)$, respectively. Furthermore, we define $f.length = l$ to denote the length of the *multi-MEM* corresponding to f.

Throughout this paper, the kth sequence is the cDNA sequence. For ease of presentation, we consider the point $0 = (0, \ldots, 0)$ (the origin) and the terminus $t = (|S_1| - 1, \ldots, |S_k| - 1)$ as fragments with length zero.

2.2 Computing the Fragments

Computing (rare) *multi-MEMs* between $k - 1$ genomic sequences and a cDNA database can be achieved in a linear time and space. One strategy is to proceed as follows: Construct the sequence S_k by appending unique characters to each cDNA and concatenating all of them. Construct the sequence $\hat{S}$ by appending unique characters to each genomic sequence and S_k and concatenating all of them. Then build the suffix tree (or the enhanced suffix array [2]) for $\hat{S}$. A *multi-MEM* $(l, p_1, .., p_k)$ is a match in $\hat{S}$ such that, $p_1 \in [1..(|S_1| + 1)]$, $p_2 \in [(|S_1| + 2)..(|S_1| + |S_2| + 2)], \ldots$ and $p_k \in [(|S_1| + \ldots + |S_{k-1}| + k)..(|S_1| + \ldots + |S_k| + k)]$. Computing *multi-MEMs* can be achieved by a bottom-up traversal of the suffix tree of S_k, as described in [2]. There it is also shown that the rareness constraint can be satisfied during the traversal without extra cost (the rareness value w.r.t. S_i in [2] is the value $C_P(S_i)$). For *multi-MUMs*, the algorithm in [5] requires a single scan of the enhanced suffix array, and it is easy to implement.

A more efficient strategy for computing (rare) *multi-MEMs* has recently been developed [11]. The idea is to construct the suffix tree (or the enhanced suffix array) for the shortest genomic sequence, say S_1. Then the remaining genomic sequences $S_2, \ldots S_{k-1}$ are sequentially matched against the suffix tree using the Chang-Lawler Algorithm [4]. During this matching, nodes of the suffix tree are annotated with match information. Only the nodes satisfying the rareness constraint are taken into account. Then the cDNA database is queried against the annotated suffix tree to further annotate more nodes. Finally, all (rare) *multi-MEMs* are reported through a bottom-up traversal of the suffix tree. The program `ramaco` is an implementation of the algorithm in [11]. The program M-GCAT [9], although no details are given, seems to use a similar approach for computing *multi-MUMs*.

3 Chaining Fragments with Overlaps

Definition 1. *Let f' and f be two fragments with $beg(f').x_i < beg(f).x_i$, for all $1 \leq i \leq k$. We say that f' overlaps with f in S_i iff (1) $end(f').x_i < end(f).x_i$ for all $1 \leq i \leq k$, and (2) $end(f').x_i \geq beg(f).x_i$, for any $1 \leq i \leq, k$.*

For $k = 2$, Figure 1 (b) shows two fragments overlapping in S_2 but not in S_1, while Figure 1 (c) shows two fragments overlapping in both S_1 and S_2.

Definition 2. *The relation $\ll$ on the set of fragments is defined as follows. $f' \ll f$ iff the following two conditions hold: $beg(f').x_i < beg(f).x_i$ and $end(f').x_i <$*

$end(f).x_i$, for all $1 \leq i \leq k$. If $f' \ll f$, then we say that f' precedes f. The fragments f' and f are colinear if either f' precedes f or f precedes f'.

Thus, two fragments are colinear if they appear in the same order in all sequences. Note that if we further have $end(f').x_i < beg(f).x_i$, for all $1 \leq i \leq k$, then f' and f are colinear and non-overlapping. A geometric representation of this relation for $k = 2$ is given in Figure 2 (a), where any fragment $f' \ll f$ must start in Region $A(f)$ and end in region $\{AB(f) \cup C(f)\}$; $A(f)$ is the rectangular region $R(0, beg(f))$, $AB(f)$ is the rectangle $([0..end(f).x_1 - 1], [0..beg(f).x_2 - 1])$, and $C(f)$ is the region $([0..end(f).x_1 - 1], [beg(f).x_2..end(f).x_2 - 1])$. For $k > 2$, $AB(f)$ and $C(f)$ are the hyper-rectangles $([0..end(f).x_1 - 1], .., [0..end(f).x_{k-1} - 1], [0..beg(f).x_k - 1])$, and $([0..end(f).x_1 - 1], .., [0..end(f).x_{k-1} - 1], [beg(f).x_k ..end(f).x_k - 1])$, respectively.

Definition 3. *For any two fragments f and f' from k sequences, where the kth sequence is the cDNA sequence, the amount of overlap in the cDNA sequence is*

$$overlap_k(f', f) = \begin{cases} end(f').x_k - beg(f).x_k + 1, & if\ beg(f).x_k \leq end(f').x_k \leq end(f).x_k \\ 0, & otherwise \end{cases}$$

Accordingly, the cDNA chaining problem can be formulated as follows.

Definition 4. *Given a set of m fragments, find a chain C of colinear fragments $f_1, f_2, .., f_t$ (i.e., $f_1 \ll f_2 \ll .. \ll f_t$) such that $score(C) = \sum_{i=1}^{t} f_i.length - \sum_{i=1}^{t-1} overlap_k(f_i, f_{i+1})$ is maximal.*

This objective function penalizes the overlaps and maximizes the amount of cDNA sequence mapped to the genomic sequence; which is the target of the cDNA mapping problem. It is easy to see that a perfect mapping has a score that equals the cDNA length. As we will show later in our geometry based solution, this objective function has the advantage that for each fragment f only two regions ($AB(f)$ and $C(f)$) are considered, independently of k, when constructing an optimal chain.

A straightforward solution to the cDNA chaining problem is to construct a weighted directed acyclic graph $G(V, E)$, where the set of vertices V is the set of fragments (including 0 and t), and the set of edges E is characterized as follows. For any two nodes $v' = f'$ and $v = f$, there is an edge $e(v' \rightarrow v) \in E$ with weight of $f.length - overlap(f', f)$, only if $f' \ll f$; see Figure 2 (b). An optimal chain corresponds to a path with maximum score from vertex 0 to vertex t in the graph. Because the graph is acyclic, such a path can be computed as follows. Let $f.score$ denote the maximum score of all chains ending with the fragment f. Clearly, $f.score$ can be computed by the recurrence

$$f.score = f.length + \max\{f'.score - overlap_k(f', f) | f' \ll f\} \qquad (1)$$

A dynamic programming algorithm based on this recurrence takes $O(m^2)$ time, where m is the number of fragments. However, this quadratic running time is a drawback for a large number of fragments. In the following section, we present a geometry based solution that runs in subquadratic time.

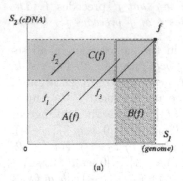 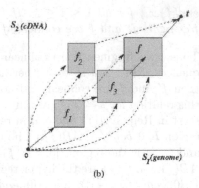

Fig. 2. (a) The fragments are represented by diagonal lines. Region $AB(f) = A(f) \cup B(f)$. The fragments f_1 and f_3 are colinear with fragment f, while f_2 is not. (b) A graph constructed over the set of fragments (drawn as 2D rectangles). An optimal chain is an optimal path (drawn as solid-arrows including $\{f_1, f_3, f\}$) in the graph.

4   Geometry-Based Solution

Because our algorithm is based on orthogonal range search for maximum, we have to recall two notions. Given a set of points in $\mathbb{R}^k$ with associated score, a *range maximum query* (RMQ) asks for a point of maximum score in $R(p, q)$. In the following, RMQ will also denote a procedure that takes two points p and q as input and returns a point of maximum score in the hyper-rectangle $R(p, q)$. Because all the points (representing the fragments) are given in advance, we use two semi-dynamic data structures D_1 and D_2 supporting RMQ with activation to manipulate the points. We write RMQ_{D_1} and RMQ_{D_2} to denote RMQs over D_1 and D_2, respectively. The points are initially inactive in these data structures.

We further use the line-sweep paradigm to construct an optimal chain. We sort the start points of the fragments w.r.t. their x_1 coordinate and process them in this order, which simulates a line (plane or hyper-plane in $\mathbb{R}^k$) that sweeps the points w.r.t. their x_1 coordinate. If a start point has already been scanned by the sweeping line, it is said to be *active* (and so does the respective fragment); otherwise it is said to be *inactive*. Each active fragment has score $f.score$ in D_1 and has score $f.score - end(f).x_k$ in D_2. While scanning the start point of a fragment f, we search for the fragment f' that maximizes Recurrence 1 among the active fragments by means of RMQs, considering both the overlap function $overlap(f', f)$ and the colinearity relation $f' \ll f$.

To take the overlap into account, we launch two range queries into two disjoint regions related to the fragment f: The first is RMQ_{D_1} into the region $AB(f)$, and the second is RMQ_{D_2} into the region $C(f)$. (Recall the definition of these regions from Section 3 and Figure 2 (a).) Fragments ending in $AB(f)$ do not overlap with f in the cDNA sequence, and those ending in $C(f)$ overlap with f in the cDNA sequence. As we will prove below, the scoring of each fragment in D_1 and D_2 takes the overlap into account. From the fragments retrieved by RMQ_{D_1} and RMQ_{D_2}, we choose the one that maximizes Recurrence 1.

To handle the constraint $f' \ll f$, each fragment f is represented by the $2k$-dimensional point $(beg(f).x_1, .., beg(f).x_k, end(f).x_1, .., end(f).x_k)$ in D_1 and D_2, and the range queries are formulated correspondingly. The first k components of this point will guarantee for any fragment $\hat{f}$ retrieved by the RMQs that $beg(\hat{f}) \in A(f)$ and the last k components will guarantee that $end(\hat{f}) \in \{AB(f) \cup C(f)\}$. That is, the search space becomes $2k$ dimensional to satisfy the colinearity constraint. The following Algorithm and proof formalize these ideas.

Algorithm 1

Sort all start points of the m fragments in ascending order w.r.t. their x_1 coordinate and store them in the array points.
For each fragment f, create the point $(beg(f).x_1, .., beg(f).x_k, end(f).x_1, .., end(f).x_k)$ and store it as inactive in the data structures D_1 and D_2.
for $1 \leq i \leq m$
  determine the fragment f with $beg(f).x_1 =$ points$[i]$
  $(b.x_1, .., b.x_k) := (beg(f).x_1, .., beg(f).x_k)$
  $(e.x_1, .., e.x_k) := (end(f).x_1, .., end(f).x_k)$
  $q_1 := RMQ_{D_1}([0..b.x_1 - 1], .., [0..b.x_k - 1], [0..e.x_1 - 1], .., [0..e.x_{k-1} - 1], [0..b.x_k - 1])$
  $q_2 := RMQ_{D_2}([0..b.x_1 - 1], .., [0..b.x_k - 1], [0..e.x_1 - 1], .., [0..e.x_{k-1} - 1], [b.x_k..e.x_k - 1])$
  determine the fragments f_1 and f_2 corresponding to q_1 and q_2, respectively
  $score_1 = f_1.score$
  $score_2 = f_2.score - (end(f_2).x_k - beg(f).x_k + 1)$
  $f.score = f.length + \max\{score_1, score_2\}$
  if $score_1 \geq score_2$ **then** *connect f_1 to f*
  else *connect f_2 to f*
  activate $(b.x_1, .., b.x_k, e.x_1, .., e.x_k)$ in D_1 with score $f.score$
  activate $(b.x_1, .., b.x_k, e.x_1, .., e.x_k)$ in D_2 with score $(f.score - end(f).x_k)$

It is clear that the RMQ_{D_1} in $AB(f)$ retrieves a highest scoring fragment in this regions. It remains to show that the RMQ_{D_2} in $C(f)$ retrieves a highest scoring fragment considering the overlap function. To this end, we introduce the following definition and lemma.

Definition 5. *The priority of a fragment $\hat{f}$, denoted by $\hat{f}.priority$, is defined as $\hat{f}.priority = \hat{f}.score - end(\hat{f}).x_k$, where the kth axis corresponds to the cDNA.*

Lemma 1. *Let f, f' and f'' be three fragments with $end(f') \in C(f)$ and $end(f'') \in C(f)$. We have $f''.priority < f'.priority$ if and only if $f''.score - overlap_k(f'', f) < f'.score - overlap_k(f', f)$.*

Proof.

$$f''.priority < f'.priority$$
$$\Leftrightarrow f''.score - end(f'').x_k < f'.score - end(f').x_k$$

by adding $beg(f).x_k$ to both sides, we obtain

$$f''.score - overlap_k(f'', f) < f'.score - overlap_k(f', f) \qquad \square$$

Thus, if f' is a fragment with highest priority in $C(f)$, then $f'.score - overlap_k$ (f', f) is maximal in $C(f)$. The priority of a fragment f' is independent of f. Hence, it can be computed in constant time when f' is scanned. This has the advantage that the overlaps between all fragments need not be computed in advance (note that this would yield a quadratic time algorithm).

The complexity of Algorithm 1 depends on the complexity of the RMQs with activation supported by the data structures D_1 and D_2. If D_1 and D_2 are implemented as range trees enhanced with priority queues as shown in [1], then the complexity of the algorithm is $O(m \log^{2k-1} m \log \log m)$ time and $O(m \log^{2k-1} m)$ space. (The range tree is built in $O(m \log^{2k-1} m)$ time and takes $O(m \log^{2k-1} m)$ space.) If the kd-tree [3] is used instead of the range tree, then the algorithm takes $O(m^{2-\frac{1}{2k}})$ time and requires $O(m)$ space. (The kd-tree is built in $O(m \log m)$ time and requires $O(m)$ space.)

Fortunately, it is possible to reduce the dimensionality of the RMQ by two, and accordingly improve the complexity of this algorithm. In fact, we can ignore the first range $[0..b(f).x_1]$ and the kth range $[0..b(f).x_k]$ of the RMQs. That is, it is sufficient to use the $(2k-2)$-dimensional queries $RMQ_{D_1}([0..b.x_2-1], .., [0..b.x_{k-1}-1], [0..e.x_1-1], .., [0..e.x_{k-1}-1], [0..b.x_k-1])$ and $RMQ_{D_2}([0..b.x_2-1], .., [0..b.x_{k-1}-1], [0..e.x_1 - 1], .., [0..e.x_{k-1} - 1], [b.x_k..e.x_k - 1])$ instead of the $2k$ dimensional ones, where b and e are $beg(f)$ and $end(f)$, respectively. (Accordingly, each fragment f is represented in both D_1 and D_2 by the $(2k - 2)$- dimensional point $(beg(f).x_2, \ldots, beg(f).x_{k-1}, end(f).x_1, \ldots, end(f).x_k)$.) The former range can be ignored because in the line sweep paradigm any activated fragment f' in D_1 and D_2 already satisfy $beg(f').x_1 < beg(f).x_1$. The latter one can be ignored because of the following: For RMQ_{D_1} in $AB(f)$, the final $((2k)^{th})$ range $[0..b(f).x_k]$, which restricts that the retrieved fragment f_1 satisfies $end(f_1).x_k \in [0..b(f).x_k]$, makes the inclusion of the kth range redundant. This is because $beg(f_1).x_k < end(f_1).x_k < beg(f).x_k$. For RMQ_{D_2} in $C(f)$, the kth range, which restricts that the retrieved fragment f_2 satisfies $beg(f_2).x_k \in [0..b(f).x_k]$, is dispensable if we examine the retrieved fragment f_2 as follows: If $beg(f_2).x_k < beg(f).x_k$, i.e., $beg(f_2) \in A(f)$, we take the retrieved fragment f_2 into further consideration, i.e., we compare its score to that of f_1 and proceed further in the algorithm. Otherwise, we ignore f_2. This ignorance will not affect the optimality of the chain because no fragment in region $C(f)$ in this case scores better than does the highest-scoring fragment in region $AB(f)$. The interested reader can find a proof of this point in Appendix I.

After this dimension reduction, the complexity of the algorithm becomes $O(m \log^{2k-3} m \log \log m)$ time and $O(m \log^{2k-3} m)$ space for the range tree, and $O(m^{2-\frac{1}{2k-2}})$ time and $O(m)$ space for the kd-tree.

5   Special Cases

In the previous algorithm, the RMQs are $(2k - 2)$-dimensional to guarantee that the retrieved fragment f' satisfies $beg(f') \in A(f)$ and $end(f') \in \{AB(f) \cup C(f)\}$. However, $(k - 1)$-dimensional RMQs suffice if $beg(f') \in A(f)$ implies $end(f') \in$

$\{AB(f) \cup C(f)\}$ and vice versa. There are two special cases of practical importance that meet this requirement. The first is the usage of *multi-MUMs* or rare *multi-MEMs* instead of *multi-MEMs*, and the second is the restriction to a certain amount of overlapping. The usage of rare *multi-MEMs* directly filters out repetitions, which saves using a repeat-masking program. The constraint on the amount of overlap is practically relevant, because overlaps are usually short.

5.1 Using *multi-MUMs* or Rare *multi-MEMs*

Suppose one uses *multi-MUMs* instead of *multi-MEMs*. When the sweep-line reaches the start point $beg(f)$ of fragment f, then it is sufficient to use the $(k-1)$-dimensional range queries $\text{RMQ}_{D_1}([0..e.x_2 - 1],.., [0..e.x_{k-1} - 1], [0..b.x_k - 1])$ and $\text{RMQ}_{D_2}([0..e.x_2 - 1], [0..e.x_{k-1} - 1], .., [b.x_k ..e.x_k - 1])$ instead of the $(2k - 2)$-dimensional RMQs of Algorithm 1, where b and e are $beg(f)$ and $end(f)$, respectively. This means that it is enough that D_1 and D_2 store the $k - 1$ dimensional point $(end(\hat{f}).x_2, .., end(\hat{f}).x_k)$ for each fragment $\hat{f}$. The correctness of this modification follows from the fact that no *multi-MUM* is enclosing or embedded in another *multi-MUM* in any sequence (dimension). For example, in Figure 3 fragment f_3 is enclosing f in S_1 and f_1 is embedded in f in S_2. These fragments cannot be *multi-MUMs*. (If such a *multi-MUM* existed, then the substring of the embedded *multi-MUM* would occur more than once in any S_i, which contradicts the definition of *multi-MUMs*). That is, it is guaranteed that the fragments f' retrieved by the RMQ satisfy $f' \ll f$. (Note that the component $[0..e.x_1 - 1]$ is ignored in the RMQs as a consequence of the line-sweep paradigm, because any fragment $\hat{f}$, where $end(\hat{f}).x_1 > end(f).x_1$ implying $beg(\hat{f}).x_1 > beg(f).x_1$, is not yet activated.) This dimension reduction improves the time and space complexity of the algorithm sketched above to $O(m \log^{k-2} m \log \log m)$ and $O(m \log^{k-2} m)$, respectively, using the range tree, and to $O(m^{2 - \frac{1}{k-1}})$ time and $O(m)$ space, using the kd-tree. (For 1D RMQs the range tree and kd-tree are equivalent and the algorithm takes $O(m \log m)$ time and $O(m)$ space.) For rare *multi-MEMs* with $r > 1$ (i.e., the string composing the match appears at most r times in any S_i), we have for each fragment at most $O(r - 1)$ fragments enclosing/embedded in f. Therefore, an $O(r)$ RMQs will retrieve the correct fragment to be connected. That is, the algorithm can be easily modified to deal with rare *multi-MEMs*.

5.2 Restricting the Amount of Overlapping

Suppose that the minimum fragment length is ℓ. If we tolerate overlapping of at most $\ell - 1$ characters between any two successive fragments f' and f in a chain (i.e., $end(f').x_i < beg(f).x_i + \ell$, $1 \leq i \leq k$), then it follows that $beg(f') \in AB(f)$ (i.e., for all $1 \leq i \leq k$, $beg(f').x_i < beg(f).x_i$). This property can be used to reduce the dimension of the RMQs to $k - 1$. To this end, D_1 and D_2 store the $k - 1$ dimensional point $(end(\hat{f}).x_2, .., end(\hat{f}).x_k)$ for each fragment $\hat{f}$. Then we attach to each fragment f the *virtual point* $v(f) = (beg(f).x_1 + \ell, .., beg(f).x_k + \ell)$. When the sweep-line reaches the point $v(f)$, we launch $\text{RMQ}_{D_1}([0, v(f).x_2 - 1], .., [0, v(f).x_{k-1} - 1], [0, beg(f).x_k])$ and $\text{RMQ}_{D_2}([0..v(f).$

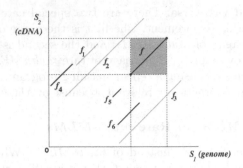

Fig. 3. Fragment f_1 is embedded in f in S_2, while f is embedded in f_2 in S_2 and in f_3 in S_1. Such embedding cannot occur if *multi-MUMs* are used instead of *multi-MEMs*.

$x_2 - 1], .., [0..v(f).x_{k-1} - 1], [beg(f).x_k..v(f).x_k - 1])$ to find the highest scoring fragments. Note that the processing starts when we scan $v(f)$, not $beg(f)$. In this case, each fragment is activated when its end point is scanned. This algorithm has the same time and space complexity as the one using *multi-MUMs*.

6   Experimental Results

We compared the graph based solution and our algorithms to one another. As an example dataset, we used the *D.melanogaster* full length cDNA database (na_cDNA.dros from BDGP, 11040 cDNAs of total length $\approx$ 23 Mbp). We mapped this database simultaneously to the *D.melanogaster* and *D.simulans* 2R chromosomes (Release 4 from UCSC Genome Browser). The length of the former chromosome is $\approx$ 21 Mbp, and the length of the latter is $\approx$ 20 Mbp. We used the program ramaco [11] to generate fragments of the type rare *multi-MEMs* of minimum length 15, and with different rareness values. We then ran three programs: (1) graph, which is an implementation of the recurrence in Section 3, i.e., graph-based (2) geom1, which is an implementation of Algorithm 1 of Section 4, and (3) geom2, which is an implementation of the algorithm in Subsection 5.2 that constrains the amount of overlap. The table in Figure 4 (left) shows for each rareness value the number of fragments and the time taken to process them using the above mentioned three algorithms. (The experiments were done using SUN-sparc machine, 1015 Mhz CPU, 6 GB RAM.) From the table it can be noted that the graph based solution is faster for small rareness values. This is due to the time taken to construct the kd-tree, which outweighs the speed-up achieved by the range queries. It is also noted that geom2 is always faster than geom1, which can be attributed to the dimensional reduction achieved in geom2.

To demonstrate that the determination of the syntenic regions is a byproduct of our chaining algorithm, Figure 4 (right) shows a projection of the chains w.r.t. the *D.melanogaster-D.simulans* 2R chromosomes. Each chain in this plot covers more than 30% of its cDNA length. From this plot, which represents a syntenic map of the two chromosomes, it is clear that the gene order is highly conserved.

r	frag. no.	graph	geom1	geom2
5	1656021	65	95	89
10	3876919	254	255	232
15	6883051	834	521	462
17	8580793	1254	671	592
18	9072474	1463	718	626

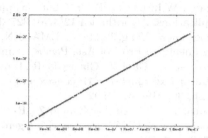

Fig. 4. Left: The experimental results for different rareness values, which are given in the first column titled with r. The second column contains the respective number of fragments. The other columns contain the running times in seconds for the programs graph, geom, and geom2. Right: Projection of the chains (shown as points) w.r.t. the *D.melanogaster* 2R chromosome (x-axis) and the *D.simulans* 2R chromosome (y-axis).

7   Conclusions

We have presented a subquadratic chaining algorithm that permits overlaps between the fragments of the type *multi-MEM*. The complexity of the algorithm is $O(m \log^{2k-3} m \log \log m)$ time and $O(m \log^{2k-3} m)$ space for the range tree, and $O(m^{2-\frac{1}{2k-2}})$ time and $O(m)$ space for the kd-tree. Furthermore, we addressed the use of rare *multi-MEMs* and the constraint on the amount of overlap. We have shown that the complexity of the algorithm significantly improves for these two cases: it takes $O(m \log^{k-2} m \log \log m)$ time and $O(m \log^{k-2} m)$ space for range trees, and $O(m^{2-\frac{1}{k-1}})$ time and $O(m)$ space for kd-trees.

We noted that the kd-tree is superior to the range tree in practice due to its reduced space consumption and its construction time. Although the range tree has theoretically better query time, querying the kd-tree is still faster, especially when the programming tricks of [3] are used.

The sensitivity of the approach in this paper can be further increased by using as short fragments as possible, increasing the rareness value, and by translating the sequences to the six reading frames (possibly combined by an alphabet transformation). This translation considers the phenomena that one amino acid maps to many genetic codes differing at the third (wobble) nucleotide.

References

1. Abouelhoda, M.I., Ohlebusch, E.: Chaining algorithms and applications in comparative genomics. J. Discrete Algorithms 3(2-4), 321–341 (2005)
2. Abouelhoda, M.I., Kurtz, S., Ohlebusch, E.: Enhanced suffix arrays and applicationss. In: Handbook of Computational Molecular Biology, CRC Press (2006)
3. Bently, J.L.: K-d trees for semidynamic point sets. In: Proc. of 6th Annual ACM Symposium on Computational Geometry, pp. 187–197. ACM Press, New York (1990)

4. Chang, W.I., Lawler, E.L.: Sublinear approximate string matching and biological applications. Algorithmica 12(4/5), 327–344 (1994)
5. Deogen, J.S., Yang, J., Ma, F.: EMAGEN: An efficient approach to multiple genome alignment. In: Proc. of Asia-Pacific Bioinf. Conf. pp. 113–122 (2004)
6. Eppstein, D., Galil, Z., Giancarlo, R., Italiano, G.F.: Sparse dynamic programming. I: linear cost functions; II: convex and concave cost functions. J. Assoc. Comput. Mach. 39, 519–567 (1992)
7. Florea, L., Hartzell, G., Zhang, Z., Rubin, G., Miller, W.: A computer program for aligning a cDNA sequence with a genomic DNA sequence. Genome Research 8, 967–974 (1998)
8. Kent, W.J.: BLAT—the BLAST-like alignment tool. Genome Research 12, 656–664 (2002)
9. Messeguer, X., Treangen, T.: M-GCAT: Interactively and efficiently constructing large-scale multiple genome comparison frameworks in closely related species. BMC Bioinformatics 7(433) (2006)
10. Myers, E.W., Miller, W.: Chaining multiple-alignment fragments in sub-quadratic time. In: Proc. of SODA, pp. 38–47 (1995)
11. Ohlebusch, E., Kurtz, S.: Space efficient computation of rare maximal exact matches between multiple sequences (submitted 2007)
12. Shibuya, S., Kurochkin, I.: Match chaining algorithms for cDNA mapping. In: Benson, G., Page, R.D.M. (eds.) WABI 2003. LNCS(LNBI), vol. 2812, pp. 462–475. Springer, Heidelberg (2003)
13. Zhang, Z., Raghavachari, B., Hardison, R.C., et al.: Chaining multiple-alignment blocks. J. Computional Biology 1, 51–64 (1994)

Appendix I

In Section 4, we mentioned that it is sufficient to use $2k - 2$-dimensional RMQs. Here, we prove that the kth range $[0..b(f).x_k]$ of RMQ_{D_2} in $C(f)$ in Algorithm 1 can be ignored provided that we check if $beg(f_2).x_k < beg(f).x_k$, where f_2 is the fragment retrieved by RMQ_{D_2}. That is, if this condition holds, we take the fragment into consideration, otherwise we ignore it. To this end, we show that this ignorance will not affect the optimality of the chain because in this case no fragment in region $C(f)$ scores better than does the highest-scoring fragment in region $AB(f)$.

Lemma 2. *Let C be a chain composed of the fragments $f_1, .., f_t$. For every index i, $1 \leq i \leq t - 1$, we have $f_{i+1}.priority \leq f_i.priority$.*

Proof.

$$
\begin{aligned}
f_{i+1}.priority &= f_{i+1}.score - end(f_{i+1}).x_k \\
&= f_i.score + f_{i+1}.length - overlap_k(f_i, f_{i+1}) - end(f_{i+1}).x_k \\
&= f_i.score - beg(f_{i+1}).x_k - overlap_k(f_i, f_{i+1}) + 1 \\
&\leq f_i.score - end(f_i).x_k \\
&\leq f_i.priority
\end{aligned}
$$

Note that if $overlap_k(f_i, f_{i+1}) = 0$, then $f_{i+1}.priority < f_i.priority$.

Now suppose that $beg(f_2).x_k \geq beg(f).x_k$, i.e., the start point of f_2 lies in $C(f)$. This implies that $overlap_k(f_2, f) \geq f_2.length$. According to Lemma 2, if there is a fragment f' connected to f_2 (i.e., f' is the predecessor of f_2 in a highest-scoring chain ending with f_2), then the end point $end(f')$ of f' must lie in $A(f)$. Hence, $overlap_k(f', f_2) = 0$ and we have

$$
\begin{aligned}
f'.score &= f'.score - overlap_k(f', f_2) \\
&= f_2.length + f'.score - overlap_k(f', f_2) - f_2.length \\
&= f_2.score - f_2.length \\
&\geq f_2.score - overlap_k(f_2, f)
\end{aligned}
$$

Recall from Lemma 1 that $f_2.score - overlap_k(f_2, f)$ is maximal in $C(f)$. Now it follows from $f'.score \geq f_2.score - overlap_k(f_2, f)$ in conjunction with $f'.score \leq f_1.score$ that f_2 can be safely ignored. (If f_2 has no predecessor, then f_2 can be also safely ignored by Lemma 1 because $f_2.score - overlap_k(f_2, f) = f_2.length - overlap_k(f_2, f) \leq 0$.)

Edge-Guided Natural Language Text Compression*

Joaquín Adiego, Miguel A. Martínez-Prieto, and Pablo de la Fuente

Depto. de Informática, Universidad de Valladolid, Valladolid, Spain
{jadiego,migumar2,pfuente}@infor.uva.es

Abstract. We describe a novel compression technique for natural language text collections which takes advantage of the information provided by edges when a graph is used to model the text. This technique is called *edge-guided* compression. We propose an algorithm that allows the text to be transformed in agreement with the edge-guided technique in conjunction with the spaceless words transformation. The result of these transformations is a PPM-friendly byte-stream that has to be codified with a PPM family encoder. The comparison with state-of-art compressors shows that our proposal is a competitive choice for medium and large natural language text collections.

Keywords: Text Compression, Natural Language Modelling, PPM.

1   Introduction

Information Retrieval Systems and Digital Libraries contain text documents and multimedia information. Although these environments can have a lot of multimedia data, the amount of textual data, predominant a few years ago, is not negligible. Besides, these systems are used in several environments such as networks or optical and magnetical media. On the other hand, in e-commerce and e-government environments almost all handled information is textual. Since processor speeds in the last few decades have increased much faster than disk transfer speeds, trading disk transfer times for processor decompression times has become a much better choice [17]. On the other hand, the use of compression techniques reduces transmission times and increases the efficiency using communication channels. These compression properties allow costs to be kept down.

In general, when a natural language text is modeled for compression either a dictionary or a sliding window technique is used. In both cases, the underlying model is a graph: each vertex represents a symbol and the edges connecting vertices are the transitions between symbols. A graph is indirectly reconstructed by a model while the symbols statistics are being obtained or the prediction

* This work was partially supported by the TIN2006-15071-C03-02 project from MCyT, Spain (first and third authors) and by the VA010B06 project from the C. Educación, JCyL, Spain (first author).

N. Ziviani and R. Baeza-Yates (Eds.): SPIRE 2007, LNCS 4726, pp. 14–25, 2007.
© Springer-Verlag Berlin Heidelberg 2007

tables constructed. Within the range of this model, information could be stored as much in the vertices as in the edges. Traditionally the information stored in vertices is used for codification, i.e. symbol frequency, and edges are used to predict the following symbols.

In this paper we explore the use of the information provided by the edges when a graph is used to model the text in order to improve the compression. The paper is organized as follows. In Section 2 a brief reminder of natural language text modelling is given. Section 3 covers related work on compression when k-order models are used. Section 4 explains the edge-guided text compression algorithm. Section 5 shows empirical results selecting some characteristics and comparing the compression of our proposal with other compression systems. We conclude in Section 6 with future work directions.

2   Modelling Natural Language Texts

Natural language is not only made up of words, considering a word as a maximal sequence of alphanumeric characters. There are also punctuation, separator, and other special characters. The sequence of characters between every pair of consecutive words is called a *separator*. Separators must also be considered as symbols of the source alphabet. There are even fewer different separators than different words, and their distribution is even more skewed. We will use the generic name *words* to refer to both text words and separators in this paper. With regard to compressing natural language texts the most successful techniques are based on models where the text words are taken as the source symbols [12], as opposed to the traditional models where the characters are the source symbols. Words reflect much better than characters the true entropy of the text [4]. For example, a semiadaptive Huffman coder, with the model that considers characters as symbols, typically obtains a compressed file whose size is around 60% of the original size, in natural language. A Huffman coder, when words are the symbols, obtains 25% [17].

Since the text is not only composed of words but also separators, a model must also be chosen for them. An obvious possibility is to consider the different inter-word separators as symbols too, and make a unique alphabet for words and separators. However, this idea does not take into account a fundamental *alternation* property: words and separators always follow one another. In [12,5] two different alphabets are used: one for words and one for separators. Once it is known that the text starts with a word or a separator, there is no confusion on which alphabet to use. This model is called **separate alphabets**.

In [14] a new idea for using the two alphabets is proposed, called **spaceless words**. An important fact that is not used in the method of separate alphabets is that a word is followed by a single space in most cases. In general, it is possible to emphasize that at least 70% of separators in text are single space [12]. Then, the spaceless words model takes a single space as a *default*. That is, if a word is followed by a single space, we just encode the word. If not, we encode the word and then the separator. At decoding time, we decode a word and assume that a space follows, except if the next symbol corresponds to a separator. Of course the alternation property does not hold anymore, so we have a single alphabet for words and separators (single space excluded). This variation achieves slightly better compression ratios in reported experiments.

3 K-th Order Models

PPM [8,15] is a statistical compressor that models the character frequencies according to the *context* given by the k characters preceding it in the text (this is called a k-th order model), as opposed to Huffman that does not consider the preceding characters. Moreover, PPM is adaptive, so the statistics are updated as the compression progresses. The larger k is, the more accurate the statistical model and the better the compression will be, but more memory and time will be necessary to compress and uncompress it. More precisely, PPM uses $k + 1$ models, of order 0 to k, in parallel. It usually compresses using the k-th order model, unless the character to be compressed has never been seen in that model. In this case it switches to a lower-order model until the character is found. The coding of each character is done with an arithmetic compressor, according to the computed statistics at that point.

The BWT [6] is a reversible permutation of the text, which puts together characters having the same k-th order context (for any k). Local optimization (for example, move-to-front followed by Huffman) over the permuted text obtains results similar to k-th order compression. PPM and BWT usually achieve better compression ratios than other families (around 20% on English texts), yet they are much slower to compress and decompress, and cannot uncompress arbitrary portions of the text collection. Well known representatives of this family are Seward's *bzip2*, based on the BWT, and Shkarin/Cheney's *ppmdi* and Bloom/Tarhio's *ppmz*, two PPM-based techniques.

In [13] the block-sorting algorithm of the BWT is extended to word-based models, including other transformations, like *spaceless words* mentioned above, in order to improve the compression. Experimental results show that the combination of word-based modelling, BWT and MTF-like transformations allows good compression effectiveness to be attained within reasonable resource costs.

A dictionary mapping on PPM modelling is described in [1]. This scheme is simple and efficient and allows words to be managed in PPM modelling when a natural language text file is being compressed. The main idea for managing words is to assign them codes to make them easier to manipulate and it also uses the *spaceless words* concept. Reported results show an excellent compression for natural language texts.

4 Edge-Guided Text Compression

4.1 Definitions

A graph describes a set of connections between objects. Each object is called a *vertex* or *node*. The connections themselves are called *edges*. Then, a graph consists of vertices and edges connecting certain pairs of vertices.

An *ordered pair* is a collection of two objects such that one can be distinguished as the first element and the other as the second element. An ordered pair with first element x and second element y is often written as (x, y) and defined as $(x, y) := \{\{\{x\}, \emptyset\}, \{\{y\}\}\}$. Therefore, ordering is significant in an ordered pair and consequently a pair of objects (x, y) is considered distinct from (y, x) for $x \neq y$. In graph theory, the first vertex in a directed edge is called the *source* and the second vertex is called the *destination*.

Let V be a set of vertices and A a set of ordered pairs of vertices, called *arcs* or *directed edges*. Then, a *directed graph* or *digraph*, short for directed graph, G is an ordered pair $G := (V, A)$ where V is the set that contains all the vertices that form G and A is the set that contains all the arcs that form G. The *degree* of a vertex in a graph is the number of edges connected to the vertex. If the graph is a directed graph the *in-degree* of $\nu \in V$ is the number of arcs where ν is the destination and the *out-degree* of $\nu \in V$ is the number of arcs where ν is the source.

4.2   Edge-Guided Compression Algorithm

In this section we describe the algorithm that allows the text to be transformed in agreement with the edge-guided technique in conjunction with the *spaceless words* transformation explained in Section 2. The result of these transformations is a byte-stream that has to be codified by some well-known encoder. Since this stream is PPM-friendly we selected a PPM family encoder.

Algorithm 1. (Edge-Guided Compression Algorithm)

```
V ← A ← ∅
current ← START_VERTEX
while (there are more words) do
        word ← get_word()
        if vertex_labeled(word) ∈ V
            then
                    destination ← V.vertex_labeled(word)
                    arc ← (current, destination)
                    if arc ∉ A then
                                    A ← A ∪ {arc}
                                    Encode(NEW_ARC(arc))
                            else
                                    Encode(FOLLOW(arc))
                    fi
            else
                    destination ← now vertex_labeled(word)
                    arc ← (current, destination)
                    V ← V ∪ {destination}
                    A ← A ∪ {arc}
                    Encode(NEW_VERTEX(destination))
        fi
        V.Update(α, destination)
        A.Update(arc)
```

  $current \leftarrow destination$
<u>od</u>

Algorithm 1 shows a generic scheme for compressing the text using the edge-guided compression. The algorithm codifies operations on a graph, within this range the algorithm transforms the text and codifies this transformed text. In order to codify the text a PPM encoder is used.

Let us suppose that we are using graph $G := (V, A)$ for compressing a natural language text with the algorithm. At a certain time $c \in V$ is the current node, when the next word, ω, is reached three possible situations can happen:

- ω appeared previously and, at some time, also preceded by the word represented by c. In this case, it is necessary to follow the arc connecting both vertices and a FOLLOW codification is carried out.
- ω appeared previously but in the past it was never preceded the word represented by c. In this case, it is necessary to create a new arc taking c as the source vertex and which destination is the vertex representing ω. A NEW_ARC codification is carried out, which represents the creation of this arc and to follow it to the destination vertex.
- ω is reached for the first time. In this case, it is necessary to create a new vertex, ν, representing ω and a new arc taking c as source vertex and ν as destination vertex. A NEW_VERTEX codification is carried out, which represents the creation of the new vertex, the creation of the arc and to follow it to the created vertex.

Natural language texts obey the Zipf and Heaps empirical laws: Zipf law observes that the relative frequency of the i-th most frequent word is $1/i^{\theta}$, for some $1 < \theta < 2$ [16,3]. Heaps law establishes that the number of different words in a text of n words is $O(n^{\beta})$, for some β between 0.4 and 0.6 [10,3], typically close to $O(\sqrt{n})$. Thus, the model size grows sublinearly with the collection size. In addition, many vertices will have in-degree and out-degree equal to 1 because they represent words that only appear a single time. On the other hand, the stopwords will be represented by vertices whose in-degree and out-degree will be very high but many arcs that arrive/leave these vertices leave/arrive to vertices with out-degree/in-degree equal to 1.

In order to save memory and improve the process time of the algorithm we propose several improvements that allow the graph to be dynamically handled according to the words that are reached. The idea behind these improvements is to fix a maximum value, α, for the out-degree of each vertex, that is, the model is limited restricting the number of arcs that leave each vertex. The idea behind this decision is to remove from the graph those arcs that will probably never be used in the future and once a word is represented in the graph it is codified by either its vertex identifier or by a dynamically changing arc identifier in the source vertex.

When the maximum value α is reached in a vertex $\nu \in V$ and a new arc is due to be inserted we must remove an arc of A whose source vertex is ν before

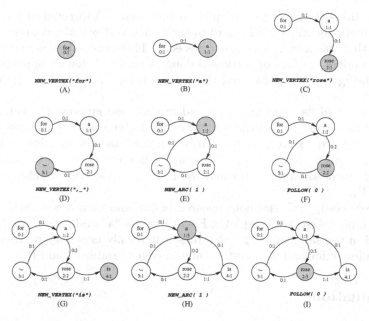

Fig. 1. Example graph generation and encoding when Algorithm 1 is used. Shaded nodes represent the current node.

inserting the new arc. The decision to select the arc that will be removed is not trivial which is why we propose two policies that allow this arc to be selected: the least-recently used arc and the least-frequently used arc. The least-recently used (LRU) policy selects the arc that has locally been more time without being followed. The least-frequently used (LFU) selects the arc that has locally been used less. Both policies reorganize the identifiers of the arcs and have well-known advantages and drawbacks and therefore the election of the policy, as well as the parameter α, will depend on the empirical results. In Algorithm 1 these actions are carried out in the *Update* sentences.

4.3   Example

Let us assume that we are compressing the sentence "for a rose, a rose is a rose" using Algorithm 1. Figure 4.3 shows graphically how the algorithm works step by step. For each step the generated graph and the codification type with real value (in italic) are shown. The process begins with an empty graph and then the first word, "for", is processed, a vertex is created. This vertex contains the word, a global numeric identifier (#0 for the first vertex) used in the encode process and a counter representing the number of times that this vertex has been reached (1 for the first time). The new vertex is inserted into the graph and becomes the current vertex (graph A).

Graph B is formed when next word, "a", is processed and this word is the first time it appears. In this situation it is necessary to create a new vertex and a new directed edge from current to created vertex. A directed edge contains a local numeric identifier with regard to the source vertex and a counter storing the times that the process has passed this edge. These two values are very useful when the update of edges or vertices is done. A similar behavior happens when both following words, "rose" and the separator ",_", are reached (graphs C and D).

Now, the word "a" appears again and in this case an edge is created from vertex #3 to vertex #1 and the counter of the latter vertex is updated (graph E). The next word, "rose", has previously appeared and also an edge exists from the current vertex to vertex #2. Therefore, it is only necessary to update the counter of the destination vertex and the counter of the edge it takes to reach it (graph F).

The next word, "is", has not appeared before and then a new vertex and a new edge must be created (graph G). Finally, words "a" and "rose" are reached again and in the same order, for this reason it is only necessary to update the corresponding information in vertices and in edges (graphs H and I).

5 Evaluation

Tests were carried out on the SuSE Linux 9.3 operating system, running on a computer with a Pentium IV processor at 1.2 GHz and 384 megabytes of RAM. We used a g++ compiler with full optimization. For the experiments we selected different size collections of WSJ, ZIFF and AP from TREC-3[1] [9]. In order to have heterogenous natural language texts we selected these three collections and we concatenated files so as to obtain approximately similar subcollection sizes from each collection, so the size in megabytes is approximate. With the purpose of testing the edge-guided compression itself we implemented a basic prototype, called *edppm*, and we use it to empirically analyze our technique and evaluate its performance. In this prototype we used the Shkarin/Cheney's *ppmdi* [15] encoder in order to obtain comparable results with the compressors mentioned below.

In Figure 2 we can see a comparison of the relationship between average compression and parameter α (the maximum out-degree value for each vertex) when least-recently used (LRU) and least-frequently used (LFU) policies are applied respectively. Both Figures are used to see graphically which is the best policy and which is the recommended value of α. When α is small a similar compression is obtained but as α increases policy LFU is better than policy LRU, improving compression on LRU by 8.5%. On the other hand, the LFU policy algorithm presents a higher asymptotical upper bound than the LRU algorithm and therefore LFU is about 30% slower than LRU but uses 1% less memory than LRU.

Then, let us focus on the LFU policy in order to obtain empirically the recommended value of α and compare it with other systems. In Figure 2 we can

---

[1] http://trec.nist.gov/

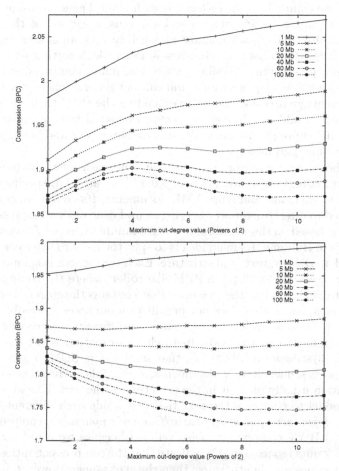

Fig. 2. Relationship between maximum out-degree value and average compression for LRU (above) and LFU (underneath) policies for each collection size. X-axis represents powers of two values.

observe in underneath graph that the compression is better for greater values of α and for values from 2^7 the compression remains roughly stable for medium collections. Therefore, in order to compare our prototype with other compressor systems we selected $\alpha = 2^7$ to look for a balance between time and memory requirements and compression since our algorithm is $O(\alpha n)$ and $\Omega(n \log n)$, where n is the number of vocabulary words.

Next, we compressed different size collections of WSJ, ZIFF and AP from TREC-3 in order to verify the behavior of the algorithms when managing medium and large collections. TREC-3 collections are formed by semistructured documents, this can harm the *edppm* prototype but allows us to compress documents with structure-aware compressors that obtain better compression than classical compressors.

Therefore, we compressed the collections with several general compressor systems: (1)*GNU gzip v.1.3.5*[2], which use LZ77 plus a variant of the Huffman algorithm (we also tried *zip* with almost identical results but slower processing); (2)*bzip2 v.1.0.2*[3], which uses the Burrows-Wheeler block sorting text compression algorithm, plus Huffman coding (where maximum compression is the default); (3)*mppm v.0.90*[4] [1], is a simple and efficient general scheme for compressing natural language text documents by extending the PPM to allow easy word handling using an additional layer. (4)*ppmdi*, extracted from *xmlppm v.0.98.2*, the same PPM compressor used in *xmlppm*, *scmppm*, *edppm* and in our prototype and with the same parameters.

On the other hand, we compressed the collections with other compression systems that exploit text structure: (1)*xmill v.0.8*[5] [11], an XML-specific compressor designed to exchange and store XML documents. Its compression approach is not intended to directly support querying or updating of the compressed documents. *xmill* is based on the *zlib* library, which combines Lempel-Ziv compression with a variant of Huffman. Its main idea is to split the file into three components: elements and attributes, text, and structure. Each component is compressed separately. (2)*xmlppm v.0.98.2*[6] [7], a PPM-like coder, where the context is given by the path from the root to the tree node that contains the current text. This is an adaptive compressor that does not permit random access to individual documents. The idea is an evolution over *xmill*, as different compressors are used for each component, and the XML hierarchy information is used to improve compression. (3)*scmppm v.0.93.3*[7] [2], that implements SCM, a generic model used to compress semistructured documents, which takes advantage of the context information usually implicit in the structure of the text. The idea is to use a separate model to compress the text that lies inside each different structure type. Like *xmlppm*, *scmppm* uses Shkarin/Cheney's *ppmdi* [15] compressors.

Word-based BWT compression was excluded because we could not find the software, yet results reported in [13] indicate that the compression ratios achieved for small collections are slightly worse than those of *mppm*. However, in order to be able to compare them, it is necessary to make more tests, mainly with files of greater size.

Table 1 shows the compression obtained with standard systems, structure-aware compressors and our *edppm* prototype for TREC-3 collections. Comparing with general compressor systems we can observe that the *gzip* obtained the worst compression ratios, not competitive in this experiment. It is followed by *bzip2* with a greater difference between it and *gzip*. The *ppmdi*, the base for the *mppm* and *edppm* compressors, obtains better compression but with compression ratios near to *bzip2*. The best general compressor is *mppm* with a very good

---

[2] http://www.gnu.org
[3] http://www.bzip.org
[4] http://www.infor.uva.es/~jadiego
[5] http://sourceforge.net/projects/xmill
[6] http://sourceforge.net/projects/xmlppm
[7] http://www.infor.uva.es/~jadiego

Table 1. Compression (bpc) for selected compressors for each TREC-3 collection

	Mbytes	gzip	bzip2	ppmdi	mppm	xmill	xmlppm	scmppm	edppm
	1	3.010	2.264	2.114	**1.955**	2.944	2.110	2.083	2.103
	5	3.006	2.193	2.057	**1.848**	2.910	2.052	2.000	1.928
	10	2.984	2.175	2.047	**1.823**	2.893	2.040	1.977	1.877
TREC-AP	20	2.970	2.168	2.041	**1.801**	2.877	2.036	1.963	1.825
	40	2.978	2.172	2.045	1.796	2.883	2.040	1.964	**1.788**
	60	2.983	2.174	2.046	1.795	2.888	2.044	1.964	**1.770**
	100	2.987	2.178	2.050	1.797	2.891	2.048	1.968	**1.757**
	Mbytes	gzip	bzip2	ppmdi	mppm	xmill	xmlppm	scmppm	edppm
	1	2.965	2.195	2.048	**1.932**	2.898	2.044	2.030	2.045
	5	2.970	2.148	2.034	**1.857**	2.878	2.029	1.984	1.901
	10	2.970	2.154	2.033	**1.832**	2.881	2.028	1.972	1.851
TREC-WSJ	20	2.973	2.153	2.035	1.820	2.882	2.030	1.971	**1.806**
	40	2.977	2.158	2.040	1.814	2.888	2.035	1.974	**1.768**
	60	2.983	2.160	2.043	1.814	2.891	2.038	1.975	**1.751**
	100	2.979	2.148	2.032	1.801	2.872	2.027	1.958	**1.727**
	Mbytes	gzip	bzip2	ppmdi	mppm	xmill	xmlppm	scmppm	edppm
	1	2.488	1.863	1.686	**1.652**	2.489	1.682	1.743	1.797
	5	2.604	1.965	1.803	**1.691**	2.596	1.799	1.782	1.794
	10	2.640	2.000	1.837	**1.708**	2.634	1.834	1.803	1.790
TREC-ZIFF	20	2.647	2.012	1.850	**1.710**	2.640	1.846	1.812	1.771
	40	2.649	2.013	1.851	**1.706**	2.639	1.847	1.808	1.738
	60	2.648	2.010	1.849	**1.701**	2.635	1.846	1.803	1.719
	100	2.654	2.016	1.853	**1.706**	2.640	1.849	1.807	**1.706**

compression. Our *edppm* prototype compressed significantly better than general compressors. It improves *gzip* by up to 72%, *bzip* by up to 24%, *ppmdi* by up to 17% and *mppm* by up to 4%.

Comparing with structure-aware compressors we can observe that the *xmill* obtains an average compression roughly constant in all cases because it uses *zlib* as its main compression machinery, and like *gzip*, its compression is not competitive in this experiment. On the other hand, *xmlppm* and *scmppm* obtain a good compression, both surpassing three of four general compressors. However, in this case, our *edppm* prototype also still obtains the best compression, reaching an improvement on *xmill* of up to 66%, on *xmlppm* of up to 17% and on *scmppm* of up to 13%.

A graphical representation of average compression is shown in Figure 3. In this graph we can observe that *edppm* prototype is better for medium and large collection than all compressor systems against which they have been compared. This can be due to the penalty of handling a dynamic graph to model the natural language which still does not have sufficient information to suitably model the source data. On average, *edppm* starts to be better from 40 Mbytes, and for 100 Mbytes it improves *mppm* by 6.4% and *ppmdi* by 15%.

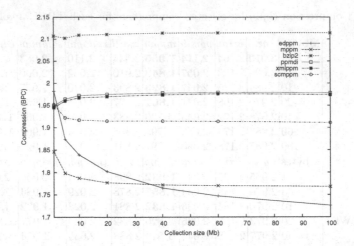

Fig. 3. Average compression for each TREC-3 collection size

In view of these results, we can conclude that *edppm* is an excellent alternative to compress natural language documents.

6   Conclusions and Future Work

We have proposed a new and efficient technique for compressing natural language text documents by encoding with PPM the transformed text using the edge-guided concept and the spaceless words transformation. When file size grows, our proposal improves compression up to 17% with respect to the character based PPM.

We have shown that the idea significantly improves compression and we have compared our prototype with standard and specific compressor systems, showing that our prototype obtains the best compression for medium and large collections, improving the compression when file size grows. An open possibility is to test our prototype with greater size collections in order to verify the real compression limit.

The prototype is a basic implementation and we are working on several improvements, which will make it more competitive. In order to improve the compression, we can tune our method including a new type of codification that allows several vertices to be followed at the same time. This is a simple matter if the involved vertices have out-degree equal to 1.

We are currently working in a semiadaptive version of the algorithm. In the first pass of the semiadaptive version, the graph is built. When the first pass is finished, the graph can be optimized in different ways, for instance by collapsing and pruning vertices. Finally, in the second pass, the text is codified using some well-known encoder.

References

1. Adiego, J., de la Fuente, P.: Mapping words into codewords on ppm. In: Crestani, F., Ferragina, P., Sanderson, M. (eds.) SPIRE 2006. LNCS, vol. 4209, pp. 181–192. Springer, Heidelberg (2006)
2. Adiego, J., de la Fuente, P., Navarro, G.: Merging prediction by partial matching with structural contexts model. In: DCC 2004, p. 522 (2004)
3. Baeza-Yates, R., Ribeiro-Neto, B.: Modern Information Retrieval. Addison-Wesley-Longman, Reading (1999)
4. Bell, T.C., Cleary, J.G., Witten, I.H.: Text Compression. Prentice-Hall, Englewood Cliffs, N.J (1990)
5. Bell, T.C., Moffat, A., Nevill-Manning, C., Witten, I.H., Zobel, J.: Data compression in full-text retrieval systems. Journal of the American Society for Information Science 44, 508–531 (1993)
6. Burrows, M., Wheeler, D.: A block sorting lossless data compression algorithm. Technical Report 124, Digital Equipment Corporation (1994)
7. Cheney, J.: Compressing XML with multiplexed hierarchical PPM models. In: DCC 2001, pp. 163–172 (2001)
8. Clearly, J.G., Witten, I.H.: Data compression using adaptive coding and partial string matching. IEEE Transactions on Communications COM-32(4), 396–402 (1984)
9. Harman, D.: Overview of the Third Text REtrieval Conference. In: Proc. Third Text REtrieval Conference (TREC-3), pp. 1–19. NIST Special Publication (1995) NIST Special Publication 500-207.
10. Heaps, H.S.: Information Retrieval - Computational and Theoretical Aspects. Academic Press, London (1978)
11. Liefke, H., Suciu, D.: XMill: an efficient compressor for XML data. In: Proc. ACM SIGMOD 2000, pp. 153–164. ACM Press, New York (2000)
12. Moffat, A.: Word-based text compression. Software - Practice and Experience 19(2), 185–198 (1989)
13. Moffat, A., Isal, R.Y.K.: Word-based text compression using the Burrows–Wheeler transform. Information Processing & Management 41(5), 1175–1192 (2005)
14. Moura, E., Navarro, G., Ziviani, N.: Indexing compressed text. In: Proceedings of the Fourth South American Workshop on String Processing, pp. 95–111 (1997)
15. Shkarin, D.: PPM: One step to practicality. In: DCC 2002, pp. 202–211 (2002)
16. Zipf, G.: Human Behaviour and the Principle of Least Effort. Addison–Wesley, Reading (1949)
17. Ziviani, N., Moura, E., Navarro, G., Baeza-Yates, R.: Compression: A key for next-generation text retrieval systems. IEEE Computer 33(11), 37–44 (2000)

Local Transpositions in Alignment of Polyphonic Musical Sequences

Julien Allali[1,*], Pascal Ferraro[1], Pierre Hanna[1], and Costas Iliopoulos[2,*]

[1] LaBRI - Université de Bordeaux 1, F-33405 Talence cedex, France
firstname.name@labri.fr
[2] Dept of Computer Science, King's College London, London WC2R 2LS
csi@dcs.kcl.ac.uk

Abstract. Music retrieval systems based on melodic similarity consider sequences of notes. Adaptations of edit-based algorithms, mainly applied in bioinformatic applications, to the musical domain lead to promising results. However, new problems are raised when considering polyphonic music. Existing representations of notes do not allow retrieval systems to be transposition invariant. In this article, we propose a new dynamic programming algorithm that permits to take into account multiple local transpositions. Experiments with MIREX collections have been performed to evaluate the improvements induced by this algorithm. The results clearly show the contribution of this algorithm: it is confirmed as the most accurate solution for a music retrieval system based on alignment algorithm to be transposition invariant.

1 Introduction

The number of audio documents available on the Internet is considerably increasing. New methods for browsing, retrieving or classifying have to be proposed to users. The growing Music Information Retrieval research community identifies and explicates the problems induced by these new methods. One of the key problems of this research area is the estimation of the musical similarity between symbolic music data.

Measuring similarity between sequences is a well-known problem in computer science which has applications in many fields such as computational biology, text processing, optical character recognition, image and signal processing, error correction, pattern recognition, etc [1,2]. However, musical sequences are characterized by specific properties. That is why developing efficient and accurate algorithms requires information about sound perception and music theory.

The notion of similarity is very difficult to precisely define, and music similarity remains one of the most complex problems in the field of music information retrieval. It may depend on the musical culture, on personal opinion, on mood, etc. For example, two musical pieces can be evaluated as very similar, if the composer, the performer, the style, the instrument, ... are the same. Since research

---

* This research was partially supported by the CCS9169 Royal Society UK-France collaborative grant.

N. Ziviani and R. Baeza-Yates (Eds.): SPIRE 2007, LNCS 4726, pp. 26–38, 2007.
© Springer-Verlag Berlin Heidelberg 2007

works in the domain of the musical information retrieval generally concern Western music, we focus on the melodic characteristics of musical pieces because it is one of the main characteristics of this kind of music.

Several techniques for evaluating music similarities have been introduced during the last few years. Geometric algorithms consider geometric representations of melodies and compute the distance between objects. Some systems [3] are closely linked to the well-known piano-roll representation. Other geometric systems represent notes by weighted points [4]. The weight is related to the duration of the note. Distances between such geometric representations are calculated according to the Earth Mover's Distance.

Another algorithm adapted from string matching domain has been proposed in [5]. N-grams techniques involve counting the distinct terms that the query and a potential answer have in common. This approach is very simple but appears to be very efficient [6]. Nevertheless, the algorithm applied for the computation of the similarity measure (counting the matching subsequences) does not take into account the musical properties of the music. For example, only two cases are assumed, the subsequence does match or not. This limitation has significant consequences on the quality of the retrieval system based on N-gram techniques.

Last but not least, other existing systems are based on atomic edit operations. As our work uses this approach, it is presented later. In this article, we propose a new dynamic progamming algorithm dealing with polyphonic musical sequences, and allowing multiple local transpositions. In the next sections, we present the main existing representations of musical pieces and the principle of edit operations and alignment. Section 2 is devoted to the problem of the transposition invariance in the polyphonic context. The algorithm proposed is detailed in section 3. Results of experiments on real databases are exposed in section 4.

1.1 Representation of Musical Pieces

Following Mongeau and Sankoff's model (90), any monophonic score can be represented as a sequence of ordered pairs (pitch, duration) of each note. Thus, the sequence **(B4 B4 r4 C4 G4 E2 A2 G8)** represents the example illustrated in Figure 1.

Several alphabets of characters and set of numbers have been proposed to represent pitches and durations [6]. The absolute pitch simply indicates the exact pitch (MIDI notation). In order to reduce the vocabulary, this exact pitch can be represented by their modulo-12 values. The melodic contour can also be taken into account by using positive values when the melody moves up and negative

Fig. 1. Example of monophonic musical score

values when it moves down. The *directed modulo*−12 *absolute pitch* sequence corresponding to the melody represented by fig. 1 is **11,11,+0,-7,+4,-9,-7**.

The *exact interval* representation is simply the number of half-tones between two successive notes. This representation can also be limited with modulo-12. Information about melodic direction can also be indicated: **0,+1,-5,+9,-7,-2**.

The *key relative* representations indicate the difference in half-tones between notes and the key of the melody. In the case of figure 1, the key signature corresponds to C major. This representation can also be limited according to modulo−12 and the information about melodic contour can be indicated: **1,1,+0,-7,+4,-9,-7**. The limitations of the *key relative* representation is closely linked to the correct choice of the key.

Concerning note durations, the same representations are possible: the absolute representation simply indicates the length of the note (in sixteenth notes for example), whereas the difference of durations between successive notes can be expressed as duration subtraction or duration ratio [7].

1.2 Alignment, Best-Fit and Edit Distance

Measuring similarity between sequences is a well-known problem in computer science which has applications in many fields [1,2]. In the early seventies, Needleman and Wunsch [8] and then Wagner and Fisher [9] proposed algorithms to compute an alignment of two strings. Given two strings q and t of lengths m and n on alphabet Σ, an alignment of these sequences is defined as a couple of sequences (q', t') of same length, on $\Sigma \cup \{-\}$. If we remove the character $'-'$ from q', resp. t', we obtain q, resp. t. Also, q' and t' cannot have the letter $'-'$ at the same position. For each position, if one of the two letters is a $'-'$ we have a gap. If the letters are equal we have a match. If they are different we have a mismatch.

Now, we have a function that gives a score to each couple of letters. The score of the alignment is equal to the sum of the score on each position in q' and t'. The goal is to find the alignment with the best score (the highest one), this can be done in $O(nm)$ with a dynamic programming algorithm.

A variant of this problem is to find the factor f of t such that the alignment score between q and f is maximal. This problem is called "best-fit" alignment and is computed with the same algorithm as the "global" alignment; only the initialisation step of the algorithm changes.

Finally, the edit distance defined in [10] is equivalent to alignment. In edit distance, we have a set of three edit operations: the substitution, the deletion and the insertion of a letter. We assign a cost to each of these operations and the *edit distance* of q and t is defined as the minimum cost of edit operations that transform q into t.

2 Transposition Invariance

The main applications of systems that estimate the similarity between audio musical pieces are the Query-by-Humming/Singing/Whistling systems. The main

Fig. 2. Example of a polyphonic musical score (top), a monophonic query not transposed (middle) and a monophonic query characterized by two local transpositions (bottom)

idea of these applications is to help users to retrieve musical pieces from a single melody line (whistled, hummed or sung).

Such applications generally consider a monophonic query and a polyphonic database. That is the reason why retrieval systems have to be able to evaluate the similarity between a monophonic musical piece and a polyphonic one. Monophonic music is assumed to be composed of only one dominant melody. In a stricter sense, it implies that no more than one note is sounded at any given time. In the polyphonic context, more than one note can sound at a given time. Fig 2 shows an example of a polyphonic musical piece and two monophonic queries. Query-by-humming systems have to retrieve the polyphonic piece from any of these queries.

Furthermore, some properties of the melodic retrieval system are expected. For instance, since a query can be transposed (one or several times), without degrading the melody, retrieval systems have to be transposition invariant.

2.1 Limitations of the Representation of Notes in Polyphony

In the monophonic context, a few representations enable systems to be transposition invariant. For example, representing pitches by the difference between successive pitches (interval representation), or, in the case of tonal music, by the difference between the pitch and the key of the musical piece (key-relative representation) allows the query to be transposed.

Concerning polyphonic music, the representation of pitches raises new problems. As for monophonic music, the *absolute* representation cannot be applied since it does not allow transposition invariance. It is also obvious that the *interval* and *contour* representations cannot be applied in the polyphonic context, since several notes may sound at the same time. Moreover, when considering the key-relative representation, the condition of transposition invariance is the before-hand knowledge of the key of the musical pieces studied. A false estimation of the key, or a bad choice leads to high errors in the similarity measurement.

One can conclude that no pitch representation correctly enables retrieval polyphonic music systems to be transposition invariant. A new approach has to be proposed, independent from the note representation chosen. We give an original algorithm which permits retrieval systems based on alignment to deal with transposed polyphonic music.

2.2 Multiple Local Transpositions

Some applications relying on music retrieval systems consider queries that are produced by human beings. These queries can not only be totally transposed, but also can be composed of several parts that are independently transposed. For example, if the original musical piece searched is composed of different harmonic voices, the user may sing different successive parts with different keys. Pieces of popular music are sometimes composed of different choruses sung based on different tonic. A sung query may imitate these characteristics.

Moreover, errors in singing or humming may occur, especially for common users that are not used to perfectly controlling their voice like professionnal singers. From a musical point of view, sudden tonal changings are disturbing. But, if these changes sound during a long time, they may not disturb listeners.

Fig 2 shows an example of query that is characterized by two local transpositions. This piece sounds very similar to the polyphonic piece, whereas the two resulting sequences are very different whatever the representation of notes chosen. That are the reasons why we propose an algorithm that allows edit based algorithms to take into account successive local transpositions.

3 Algorithm

We start with a few basic definitions and then we give a formal definition of the problem. We then proceed with the description of a dynamic programing algorithm, followed by an improvement and a demonstrative example.

3.1 Definitions and Problem

For the rest of the article, let Σ be a set of sets of integers. Its size is $|\Sigma|$. Let q and t be two texts over Σ. Thus q and t can be polyphonic, since each letter of Σ can represent a note (set with single value) or a chord.

For the sake of simplicity, we do not include here additional information such as note or chord duration, sound level etc.

We denote by $|s|$ the *length* of any text s. The $i+1^{th}$ letter of $s = s_0 s_1 \ldots s_{|s|-1}$ is s_i for all $i \in [0..|s|-1]$, The subword $s_i s_{i+1} \ldots s_j$ of s is denoted by $s_{i...j}$. For any set $\Gamma \subseteq \Sigma$ and integer v, $\Gamma + v$ is the set $\{\gamma \in \Gamma : \gamma + v\}$.

A *local transposition* h over a text s is represented by triplet $h = (b, v, e)$ such that, $\{b, e\} \in [0, |s|-1]$ and v is an integer. Applying h to s consists in building a new string s' of length $|s|$ such that $s'_i = s_i$ for all $i \notin [b, e]$ and $s'_i = s_i + v$ for all $i \in [b, e]$. Thus h represents a change of the pitch level of consecutive chords in s.

Definition 1 (Covering Transposition). *A series of local transpositions $T = \{(b_i, v_i, e_i)\}$ (order by b_i) is said to be a covering transposition of a given text s if $b_0 = 0$, $b_i = e_{i-1} + 1$, $\forall\, i > 0$ and $e_{|T|-1} = |s| - 1$. That is, the local* *transpositions in T are consecutive and cover all the positions of s. We denote $s^{\to T}$ the text obtained by successively applying each transposition of T to s.*

Last, we make use of a function called $alignment(q, t)$ that computes the alignment score between q and t. We choose the alignment concept but the algorithm can be easily adapted to work with the edit distance.

Now, given a query q and a text t, our aim is to find a covering transposition T on q such that, the alignment score between t and $q^{\to T}$ is optimal.

Problem 1 (Alignment with local transpositions). Given a query q and a text t over Σ, an alignment function $alignment(u, v)$ and a number $M \leq 0$, the problem is to find alignment of q and t with local transpositions:

$$LTalignment(q, t) = \max_{T}\{alignment(q^{\to T}, t) + (|T| - 1) * M\}$$

The value M is a malus used to control the gain provided by a transposition. Note that if $M = -\infty$ the solution consists in computing the alignment with q entirely transposed by the same value. If M is equal to zero, the best solution consists in transposing each of the letters of q to match letters of t.

Finally, the problem is defined for both "global" alignment and "best fit" variants of the problem. It can be easily adapted to the local alignment problem using the same technique presented in [11].

3.2 Algorithm

We now present a dynamic programming algorithm to answer this problem. Basically, it consists in computing simultaneously the alignment matrices between t and all possible transpositions of q. At each position in each matrix, we consider, if it is better to continue with the current transposition or to start a new transposition from another matrix. Let $score(\alpha, \beta)$ be the scoring function used for the alignment.

Beforehand we have to compute all interesting transposition values. This consists in finding all values v such that there exist letters α, β occuring in q, t such that $score(\alpha + v, \beta) \geq 0$. We call Δ_T a sorted array containing these values.

Now, we fill $|\Delta_T|$ matrices M^i using a dynamic programming algorithm. Each matrix M^i corresponds to the *alignment with local transpositions* of a prefix of the query with a prefix of the text such that the last transposition applied to the query is about $\Delta_T[i]$. More formally, the value $M^i[r][c]$ corresponds to the best alignment with local transposition between $t_{0...c}$ and $q_{0...r}$ with the restriction that the last transposition applied to $q_{0...r}$ is $\Delta_T[i]$. The size of the matrices is $(|q| + 1) \times (|t| + 1)$ as we place the query vertically and the text horizontally.

Following is the recurrence formula used to fill the matrices:

$$M^i[r][c] = \max \begin{cases} M^i[r - 1][c] + score(q_{r-1} + \Delta_T[i], \epsilon) \\ M^i[r][c - 1] + score(\epsilon, t_{c-1}) \\ M^i[r - 1][c - 1] + score(q_{r-1} + \Delta_T[i], t_{c-1}) \\ M^j[r - 1][c - 1] + score(q_{r-1} + \Delta_T[i], t_{c-1}) + M, \quad for\ j \neq i \end{cases}$$

This formula is illustrated by fig. 3. The first and second cases correspond to a gap in the query or the text. The third corresponds to a substitution between the letter $q_{r-1} + \Delta_T[i]$ and t_{c-1}. Finally, we look for starting a new transposition from one of the other matrices. Depending on the scoring scheme that we will show later, this can be optimised. The proof of correctness is not given here as it is similar to Dan Gusfield's proof in [2] page 218.

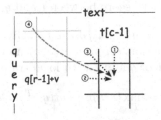

Fig. 3. Illustration of the recurrence formulae to fill the case $M^i[r][c]$: $\boxed{1}$ a gap in the query or $\boxed{2}$ in the text, $\boxed{3}$ substitution between $q_{r-1} + v$ and t_{c-1} ($v = \Delta_T[i]$), $\boxed{4}$ starting a new transposition with value v by coming from another matrix.

The initialisation of the matrices described on algorithm 1 depends on the variants of the problem. If we are computing a "global" alignment, then the line 6 is used. For the "best-fit" variant of the problem, the line 7 is used.

Algorithm 1. Initialisation of matrix M^i
1: $M^i[0][0] = 0$
2: **for** r from 1 to $|q|$ **do**
3: $\quad M^i[r][0] = M^i[r-1][0] + score(q[r-1] + \Delta_T[i], \epsilon)$
4: **end for**
5: **for** c from 1 to $|t|$ **do**
6: $\quad M^i[0][c] = M^i[0][t-1] + score(\epsilon, t[c-1])$ // global alignment case
7: $\quad M^i[0][c] = 0$ // best fit case
8: **end for**

Algorithm 2 gives the pseudo-code used to fill the matrices. The value of the alignment with local transpositions between q and t is $\max_i(M^i[r][c])$ (bottom right corner) for the global alignment and $\max_{i,k}(M^i[r][k])$ (bottom row) for the best fit problem.

The complexity of this algorithm is $O(|\Delta_T|^2 * |q| * |t|)$ in time and $O(|\Delta_T| * |q|)$ in space. It is possible to significantly improve the running time by skipping the loop at line 8 to 10 that is, the part where we look for a new local transposition.

3.3   Improvement

Assume that at line 6, the substitution score between the letter in the query transposed and the letter in the text is the worst possible. Normally we should start a new transposition by coming from matrix j. But there is no reason is

Algorithm 2. Fill the matrices
1: **for** r from 1 to $|q|$ **do**
2:    **for** c from 1 to $|t|$ **do**
3:       **for** i from 0 to $|\Delta_T|$ **do**
4:          $del = M^i[r-1][c] + score(q_{r-1} + \Delta_T[i], \epsilon)$
5:          $ins = M^i[r][c-1] + score(\epsilon, t_{c-1})$
6:          $subscore = score(q_{r-1} + \Delta_T[i], t_{c-1})$
7:          $sub = M^i[r-1][c-1] + subscore$
8:          **for** j from 0 to $|\Delta_T|$; $j \neq i$ **do**
9:             $sub = \max(M^j[r-1][c-1] + subscore + M; sub)$
10:          **end for**
11:          $M^i[r][c] = \max(del, ins, sub)$
12:       **end for**
13:    **end for**
14: **end for**

to start this new transposition as the substitution played in matrix j will lead to a score equal or better. If the scoring scheme is constant for the substitution (fixed value for a match and fixed value for a mismatch), then it is equivalent to state that each local transposition must start by a match.

To implement this improvement, for each letter in t, we compute in array named WS, the worst score with all possible letters of q including the ones obtained by transposition. This can be computed during the same time Δ_T is built. Now the line 8 to 10 are replaced by:

   if $subscore > WS[t_{c-1}]$ **then**
     for j from 0 to $|\Delta_T|$; $j \neq i$ **do**
      $sub = \max(M^j[r-1][c-1] + subscore + M; sub)$
     end for
   end if

In the constant scoring scheme and if the query and the text are monophonic the time complexity becomes $O(|\Delta_T| * |q| * |t|)$.

3.4 Demonstrative Example

We present a simple example to illustrate the algorithm. For the query we take the second and third bars of the monophonic version of "brother john" and for the text we take the same bars of the polyphonic version of "brother john".

We represent all notes by their pitch level modulo 12, that is C is 0 and B is 11. Thus the size of Δ_T is bounded by 12. So the query q is represented by the sequence $\{9\}$ $\{11\}$ $\{1\}$ $\{9\}$ $\{4\}$ $\{5\}$ $\{7\}$ and the text t is $\{5\}$ $\{0, 7\}$ $\{5, 9\}$ $\{0, 5\}$ $\{5, 9\}$ $\{0, 7, 10\}$ $\{5, 9, 0\}$.

The scoring function we use is basic, we consider that we have a match if the note of the query occurs in the chord of the text:

$$score(\alpha, \beta) = \begin{cases} +2 & \text{if } \alpha \in \beta \\ -1 & \text{if } \alpha \notin \beta \text{ or } \alpha = \epsilon \text{ or } \beta = \epsilon \end{cases}$$

For the practical implementations, we used a more sophisticated scoring scheme, presented in the next section.

For this example, the algorithm builds 12 matrices one for each of the twelve possible transposition (from 0 to 11). We represent below the matrices for the transposition value 8 and 5, the malus M is -3.

The score of the global alignment with local transpositions is 11 and the covering transposition is $T = \{(0,8,3),(4,5,6)\}$. This corresponds to the score of the alignment of $q^{\to T} = \{5\}\ \{7\}\ \{9\}\ \{5\}$-$\{9\}\ \{10\}\ \{0\}$ with t which is 14 minus 3 (one malus). On these matrices, we put in bold the values, where we look for the other matrices to start a new local transpositions. The backtrace is shown by the surrounded values.

$\Delta[i] = 8$			7	9	5	9	7 (10)	5 (9)
	ϵ	5	0	5	0	5	0	0
ϵ	**0**	-1	-2	-3	-4	-5	-6	-7
9+8=5	-1	**2**	1	0	-1	-2	-3	-4
11+8=7	-2	1	**4**	3	2	1	0	-1
1+8=9	-3	0	3	**6**	5	4	3	2
9+8=5	-4	**-1**	2	5	**8**	7	6	5
4+8=12	-5	-2	1	4	7	7	6	5
5+8=1	-6	-3	0	3	6	6	6	5
7+8=3	-7	-4	-1	2	5	5	5	4

$\Delta[i] = 5$			7	9	5	9	7 (10)	5 (9)
	ϵ	5	0	5	0	5	0	0
ϵ	0	-1	-2	-3	-4	-5	-6	-7
9+5=2	-1	-1	-2	-3	-4	-5	-6	-7
11+5=4	-2	-2	-2	-3	-4	-5	-6	-7
1+5=6	-3	-3	-3	-3	-4	-5	-6	-7
9+5=2	-4	-4	-4	-4	-4	-5	-6	-7
4+5=9	-5	-5	-5	1	0	**7**	6	5
5+5=10	-6	-6	-6	0	0	6	**9**	8
7+5=0	-7	-7	-4	-1	**2**	5	8	**11**

4   Experiments

One of the main problems in the music information retrieval domain is the problem of the evaluation of the system proposed. The first Music Information Retrieval Evaluation eXchange (MIREX) [12] is a contest whose goal is to compare state-of-the-art algorithms and systems relevant for Music Information Retrieval. During this first contest, an evaluation topic about symbolic melodic similarity has been performed. Participants have discussed the process of evaluation and proposed an evaluation procedure. The experiments presented in this paper are based on the MIREX procedures.

4.1   Scoring Scheme

We used the same scoring scheme as the ones we introduce in [13]. Mainly, the score between two notes takes into account the pitch, the duration and the consonance of notes. For example, the fifth (7 semitones) and the third major or minor (3 or 4 semitones) are the most consonant intervals in Western music. By consequence, these intervals are encouraged.

As we limit our experiments to monophonic queries, the score between a note and a chord is equal to the best score between the note and each note of the chord. The development of pertinent scoring functions for systems considering polyphonic queries remain an open problem.

4.2   Monophonic Music

The MIREX 2005 database is the RISM A/II (International inventory of musical sources) collection, which is composed of one half-million notated real world compositions. The incipits are symbolically encoded music. They are monophonic and contain between 10 and 40 notes. 11 incipits have been randomly chosen. A ground truth has been established [14] by combining ranked lists that were created by 35 music experts. A specific measure has been proposed: the Average Dynamic Recall (ADR) [15]. It takes into account the ranked groups of the ground truth by indicating how many of the documents that should have appeared before or at a given position in the result list actually have appeared. The higher the ADR measure is, the more accurate the tested system is.

The goal of the first experiments we have performed is to verify that the algorithm proposed in this paper improves the quality of music retrieval systems for monophonic musical pieces. The results of the symbolic melodic similarity contest that was proposed during MIREX 2005 are presented in Tab. 1 and can be compared to the ones obtained by our retrieval system considering *interval* and *key relative* representations for pitch, and by applying the local transposition algorithm described in this paper.

Table 1. Results of the evaluation of retrieval systems based on alignment during MIREX 2005 compared to the results obtained by a edit-distance based retrieval system with or without the algorithmic improvement proposed

Algorithm	average ADR
Local transposition algorithm	77.1
key relative representation	60.6
interval representation	76.8
Edit distance I/R (Grachten)	66
Edit distance (Lemstrm)	54

At first sight, results seem disappointing because tab. 1 does not show any better result when applying the local transposition algorithm. In this case, the average ADR is 77, whereas it is respectively 61 and 77 when considering the *key relative* and *interval* representations, and without applying the algorithm presented. But it is important to note that the application of the local transposition algorithm does not require any knowledge about the musical pieces tested, at the contrary of the *key relative* representation. One of the conclusion of these experiments is the accuracy of the local transposition algorithm, since it obtains at least the same results as systems considering *interval* representations obtain. Moreover, one of the improvements of our algorithm is that it allows multiple local transpositions. However, the MIREX 2005 queries are too short to permit to highlight these improvements: for short queries, two local transpositions imply a too important perceptual dissimilarity.

4.3   Polyphonic Music

During MIREX 2006[1], the second task of the symbolic melodic similarity contest consisted in retrieving the most similar pieces from mostly polyphonic collections given a monophonic query. Two collections was considered, and 11 queries (hummed or whistled) was proposed. The *mixed* collection is composed of 10000 randomly picked MIDI files that were harvested from the Web and which include different genres. The *karaoke* collection is composed of about 1000 `.kar` files (Karaoke MIDI files) with mostly Western popular music. Tab. 2 presents the results obtained with these two collections. Since only a few algorithms have participated to the MIREX 2006 contest, we think that the ground truth established is not significant. That's why we propose to evaluate our algorithm according to two measures: the *average precision*, and the *precision at N documents* (N is the number of relevant documents).

Table 2. Average Precision (AP) and Precision at N Documents (PND) obtained by edit-distance based retrieval systems with or without the algorithmic improvement proposed for MIREX 2006 databases and queries (absolute representation, key relative representation, absolute representation with local tranposition algorithm and algorithm submitted by Uitdenbogerd during MIREX)

Collection		Absolute	Key rel.	**Local Transp.**	Uitdenbogerd
Karaoke	AP	0.20	0.35	**0.78**	0.36
	PND	0.20	0.40	**0.83**	0.33
Mixed	AP	0.21	0.16	**0.67**	0.52
	PND	0.20	0.17	**0.66**	0.55

Results presented in Tab. 2 clearly show that the algorithm proposed allowing multiple transpositions improves retrieval systems. Concerning the *karaoke* collection, the average precision obtained is near 0.80 whereas it is only 0.20 when considering the *absolute* representation without the algorithm proposed, and only 0.35 when considering the *key-relative* representation. This significant difference is also observed for the *mixed* collection. The average precision is 0.67 instead of 0.21 or 0.0. The main justification is the lack of reliable representation for pitch in polyphonic music. As explained previously, errors in key estimation imply limitations for the *key relative* representation. That is why existing systems based on alignment remain limited. For example, the algorithm submitted to the MIREX 2006 by Uitdenbogerd [16] respectively obtains 0.52 and 0.36. The algorithm we proposed computes the transposition that allows the best match between musical pieces. It seems to be a great solution to this problem of representation of notes for polyphonic music. For now, to our knowledge, no other solution as accurate as this one exists.

---

[1] http://www.music-ir.org/mirex2006/index.php/MIREX2006_Results

5   Discussion and Conclusion

In this paper, we proposed a new algorithm which allows music similarity systems based on alignment to take into account multiple local transpositions. The experiments presented show that this improvement increases the accuracy of retrieval systems in both monophonic and polyphonic contexts. It is important to note that time computation added by the algorithm is significant and may be justified only for long queries in the monophonic context.

Also, we think it is possible to improve the running time of the algorithm by filtering the set of interesting transpositions Δ_T. For example, on the alignment given in section 3.4 it would be possible to not compute matrix for value 7 because this transposition value brings only one match.

Finally, the system proposed for comparing polyphonic musical sequences has to be improved. The algorithm submitted by Typke [17] during MIREX 2006 obtains better results with different techniques. We aim at improving our system in order to reach the same accuracy.

References

1. Sankoff, D., Kruskal, J.B., eds.: Time Wraps, Strings Edits, and Macromolecules: the Theory and Practice of Sequence Comparison. Addison-Wesley Publishing Company Inc, University of Montreal, Quebec, Canada (1983)
2. Gusfield, D.: Algorithms on Strings, Trees and Sequences - Computer Science and Computational Biology. Cambridge University Press, Cambridge (1997)
3. Ukkonen, E., Lemstrm, K., Mkinen, V.: Geometric Algorithms for Transposition Invariant Content-Based Music Retrieval. In: Proc. of the 4th International Conference on Music Information Retrieval, Baltimore, USA (2003) 193–199
4. Typke, R., Veltkamp, R.C., Wiering, F.: Searching Notated Polyphonic Music Using Transportation Distances. In: Proceedings of the ACM Multimedia Conference, New-York, USA (October 2004) 128–135
5. Doraisamy, S., Rger, S.: Robust Polyphonic Music Retrieval with N-grams. Journal of Intelligent Information Systems 21(1) (2003) 53–70
6. Uitdenbogerd, A.L.: Music Information Retrieval Technology. PhD thesis, RMIT University, Melbourne, Victoria, Australia (July 2002)
7. Ferraro, P., Hanna, P.: Optimizations of Local Edition for Evaluating Similarity Between Monophonic Musical Sequences. In: Proc. of the 8th International Conference on Information Retrieval (RIAO), Pittsburgh, PA, USA. (may 2007)
8. Needleman, S., Wunsch, C.: A General Method Applicable to the Search for Similarities in the Amino Acid Sequences of Two Proteins. Journal of Molecular Biology 48 (1970) 443–453
9. Wagner, R.A., Fisher, M.J.: The String-to-String Correction Problem. Journal of the association for computing machinery 21 (1974) 168–173
10. Levenshtein, V.I.: Binary codes capable of correcting deletions, insertions and reversals. Sov. Phys. Dokl. 6 (1966) 707–710
11. Smith, T., Waterman, M.: Identification of Common Molecular Subsequences. Journal of Molecular Biology 147 (1981) 195–197
12. Downie, J.S., West, K., Ehmann, A.F., Vincent, E.: The 2005 Music Information retrieval Evaluation Exchange (MIREX 2005): Preliminary Overview. In: ISMIR. (2005) 320–323

13. Hanna, P., Ferraro, P.: Polyphonic Music Retrieval by Local Edition of Quotiented Sequences. In: Proceedings of the 5th International Workshop on Content-Based Multimedia Indexing (CBMI'07), Bordeaux, France (June 2007)
14. Typke, R., den Hoed, M., de Nooijer, J., Wiering, F., Veltkamp, R.C.: A Ground Truth For Half A Million Musical Incipits. Journal of Digital Information Management **3**(1) (2005) 34–39
15. Typke, R., Veltkamp, R.C., Wiering, F.: A Measure for Evaluating Retrieval Techniques Based on Partially Ordered Ground Truth Lists. In: Proc. of the International Conference on Multimedia and Expo, Toronto, Canada (2006) 128–135
16. Uitdenbogerd, A.: Variations on Local Alignment for Specific Query Types. In: 2nd Music Information Retrieval Evaluation eXchange, Victoria, Canada (2006)
17. Typke, R., Veltkamp, R.C., Wiering, F.: MIREX Symbolic Melodic Similarity and Query by Singing/Humming. In: 2nd Music Information Retrieval Evaluation eXchange, Victoria, Canada (2006)

Efficient Computations of ℓ_1 and ℓ_∞ Rearrangement Distances

Amihood Amir[1,2,*], Yonatan Aumann[1], Piotr Indyk[3], Avivit Levy[1,**],
and Ely Porat[1]

[1] Department of Computer Science, Bar Ilan University, Ramat Gan 52900, Israel
{amir,aumann,levyav2,porately}@cs.biu.ac.il
[2] Department of Computer Science, Johns Hopkins University, Baltimore, MD 21218
[3] Department of Computter Science, MIT, Cambridge, Massachusetts 02139
indyk@theory.lcs.mit.edu

Abstract. Recently, a new pattern matching paradigm was proposed, *pattern matching with address errors*. In this paradigm approximate string matching problems are studied, where the content is unaltered and only the locations of the different entries may change. Specifically, a broad class of problems in this new paradigm was defined – the class of *rearrangement errors*. In this type of errors the pattern is transformed through a sequence of *rearrangement operations*, each with an associated *cost*. The natural ℓ_1 and ℓ_2 rearrangement systems were considered. A variant of the ℓ_1-rearrangement distance problem seems more difficult – where the pattern is a general string that may have repeating symbols. The best algorithm presented for the general case is $O(nm)$. In this paper, we show that even for general strings the problem can be approximated in linear time! This paper also considers another natural rearrangement system – the ℓ_∞ *rearrangement distance*. For this new rearrangement system we provide efficient exact solutions for different variants of the problem, as well as a faster approximation.

1 Introduction

The historical challenge of approximate pattern matching was coping with errors in the data. The traditional *Hamming distance* problem assumes that some elements in the pattern are erroneous, and one seeks the text locations where the number of errors is sufficiently small [1,2,3], or efficiently calculating the Hamming distance at every text location [4,1,5]. The *edit distance* problem adds the possibility that some elements of the text are deleted, or that noise is added at some text locations [6,7]. Indexing and dictionary matching under these errors has also been considered [8,9,10,11]. The implicit assumption in all these problems is that there may indeed be errors in the *content* of the data, but the

---

* Partly supported by ISF grant 35/05.
** Partly supported by a BIU President Fellowship. This work is part of A. Levy's Ph.D. thesis.

N. Ziviani and R. Baeza-Yates (Eds.): SPIRE 2007, LNCS 4726, pp. 39–49, 2007.

order of the data is inviolate. Data may be lost or noise may appear, however, the order of the data was assumed to be ironclad.

Nevertheless, some non-conforming problems have been gnawing at the walls of this assumption. The *swap* error, motivated by the common typing error where two adjacent symbols are exchanged [12], does not assume error in the content of the data, but rather, in the order. However, here too the general order was assumed accurate, with the difference being at most one location away. The advent of computational biology has added several problems wherein the "error" is in the order, rather than the content. During the course of evolution, whole areas of genome may translocate, shifting from one location in genome to another. Alternatively, two pieces of genome may exchange places. These cases represent a situation where the content of the individual entries does not change, but rather their locations may be different. Several works have considered specific versions of this biological setting, primarily focusing on the sorting problem (*sorting by reversals* [13,14], *sorting by transpositions* [15], and *sorting by block interchanges* [16]).

Motivated by these questions, a new pattern matching paradigm was proposed, *pattern matching with address errors* [17]. In this paradigm approximate string matching problems are studied, where the content is unaltered and only the locations of the different entries may change. This new approximate matching model opens a whole new world of problems and introduces new techniques to the field of pattern matching (see [18], [19]). Specifically, [17] defined a broad class of problems in this new paradigm – the class of *rearrangement errors*. In this type of errors the pattern is transformed through a sequence of *rearrangement operations*, each with an associated *cost*. The cost induces a distance measure between the strings, defined as the total cost to convert one string to the other. Given a pattern and a text, the problem is to find the subsequence of the text closest to the pattern. [17] consider several natural distance measures, including the analogues of the ℓ_1 and ℓ_2 distances, as well as two *interchange measures*. For these, they provide efficient algorithms for different variants of the associated string matching problems.

A variant of the ℓ_1-rearrangement distance problem seems more difficult – where the pattern is a general string (i.e. strings that may have repeating symbols). [17] present a linear time algorithm for the problem where the pattern is a string with distinct letters, however, the techniques used fail in the general pattern case. The best algorithm presented for the general case is $O(nm)$. In this paper, we show that even for general strings the problem can be approximated in linear time! Our solution utilizes properties of p-stable distributions to derive a very efficient approximation to this problem.

In addition, another natural rearrangement system – the ℓ_∞ *rearrangement system* – is considered in this paper for the first time. For this new rearrangement system we provide efficient exact solutions for different variants of the problem, as well as a faster approximation algorithm. Formal definitions of the rearrangement pattern matching problems are given below.

Rearrangement Distances. Consider a set A and let x and y be two m-tuples over A. [17] formally defined the process of converting x to y through a sequence of *rearrangement* operations in the following way. A *rearrangement operator* π is a function $\pi : [0..m-1] \to [0..m-1]$, with the intuitive meaning being that for each i, π moves the element currently at location i to location $\pi(i)$. Let $s = (\pi_1, \pi_2, \ldots, \pi_k)$ be a sequence of rearrangement operators, and let $\pi_s = \pi_1 \circ \pi_2 \circ \cdots \circ \pi_k$ be the composition of the π_j's. s *converts* x *into* y if for any $i \in [0..m-1]$, $x_i = y_{\pi_s(i)}$. That is, y is obtained from x by moving elements according to the designated sequence of rearrangement operations.

Let Π be a set of rearrangement operators, Π *can convert* x *to* y, if there exists a sequence s of operators from Π that converts x to y. Given a set Π of rearrangement operators, a non-negative *cost* is associated with each sequence from Π, $cost : \Pi^* \to R^+$. The pair $(\Pi, cost)$ is called a *rearrangement system*. Consider two vectors $x, y \in A^m$ and a rearrangement system $\mathcal{R} = (\Pi, cost)$, the distance from x to y under $\mathcal{R}$ is defined to be:

$$d_\mathcal{R}(x, y) = \min\{cost(s) | s \text{ from } \mathcal{R} \text{ converts } x \text{ to } y \}$$

If there is no sequence that converts x to y then the distance is ∞.

The String Matching Problem. Let $\mathcal{R}$ be a rearrangement system and let $d_\mathcal{R}$ be the induced distance function. Consider a text $T = T[0], \ldots, T[n-1]$ and pattern $P = P[0], \ldots, P[m-1]$ $(m \le n)$. For $0 \le i \le n-m$ denote by $T^{(i)}$ the m-long substring of T starting at location i. Given a text T and pattern P, we wish to find the i such that $d_\mathcal{R}(P, T^{(i)})$ is minimal.

The ℓ_1 and ℓ_∞ Rearrangement Distances. The simplest set of rearrangement operations allows any element to be inserted at any other location. Under the ℓ_1 *Rearrangement System*, the cost of such a rearrangement is the sum of the distances the individual elements have been moved. Formally, let x and y be strings of length m. A rearrangement under the ℓ_1 operators is a permutation $\pi : [0..m-1] \to [0..m-1]$, where the cost is $cost(\pi) = \sum_{j=0}^{m-1} |j - \pi(j)|$. [17] call the resulting distance the ℓ_1 *Rearrangement Distance.*

We define the ℓ_∞ Rearrangement System in which we use the same set of operators, with the cost being the maximum of the distances the individual elements have been moved. Formally, let x and y be strings of length m. A rearrangement under the ℓ_∞ operators is a permutation $\pi : [0..m-1] \to [0..m-1]$, where the cost is $cost(\pi) = \max_{j \in \{0,\ldots,m-1\}} |j - \pi(j)|$. We call the resulting distance the ℓ_∞ Rearrangement Distance. We prove:

Theorem 1. *For T and P of sizes n and m respectively $(m \le n)$, the ℓ_1 Rearrangement Distance can be approximated to a constant $\epsilon > 0$ in time $O(n)$.*

Theorem 2. *For T and P of sizes n and m respectively $(m \le n)$ the ℓ_∞ Rearrangement Distance can be computed in time $O(m(n-m+1))$. If all entries of P are distinct then the ℓ_∞ Rearrangement Distance can be computed in time $O(n \log m)$.*

Theorem 3. *For T and P of sizes n and m respectively ($m \leq n$), the ℓ_∞ Rearrangement Distance can be approximated to a factor of 2 in time $O(n \log m)$.*

2 Approximating the ℓ_1 Rearrangement Distance

In this section we show how the ℓ_1 rearrangement distance can be approximated in linear time for general strings.

2.1 Preliminaries

Stable Distributions. A distribution $\mathcal{D}$ over $\mathbb{R}$ is called *p-stable*, if there exists $p \geq 0$ such that for any m real numbers $a_1, \ldots, a_m$ and i.i.d variables $X_1, \ldots, X_m$ with distribution $\mathcal{D}$, the random variable $\sum_j a_j X_j$ has the same distribution as the variable $(\sum_j |a_j|^p)^{1/p} X$, where X is a random variable with distribution $\mathcal{D}$.

It is known [20] that stable distributions exist for any $p \in (0, 2]$. In particular, a *Cauchy* distribution (denoted $\mathcal{D}_C$), defined by the density function $C(x) = \frac{1}{\pi} \frac{1}{1+x^2}$, is 1-stable. Specifically, we use the following lemma.

Lemma 1. *[Indyk] [21] Let $X_0, \ldots, X_{m-1}$ be random variables drawn from $\mathcal{D}_C$ distribution, and let $X_1, \ldots, X_l$ be independent samples of $X_0, \ldots, X_{m-1}$, where $l = c/\epsilon^2 \log 1/\delta$ and c is a suitable constant, then the probability that*

$$\text{median}(|\sum a_j x_{j,1}|, |\sum a_j x_{j,2}|, \ldots, |\sum a_j x_{j,l}|) \in [(1-\epsilon)\sum |a_j|, (1+\epsilon)\sum |a_j|]$$

is greater than $1 - \delta$.

Remark. In [21] it is shown that Lemma 1 can be applied even when a bounded precision of $O(\log m)$ bits is used.

The ℓ_1 Pairing Lemma. The main difficulty in the case of general strings is that repeating symbols have multiple choices for their desired destination. Let x and y be strings of length m. Our goal is to pair the locations in x to destination locations in y, so that repeating symbols can be labelled in x and y to get strings with the same m distinct letters (permutation strings). Such a labelling can be viewed as a permutation of the indices of x. Clearly, if x contains distinct elements then only one labelling permutation can convert x to y. However, there can be many labelling permutations if x contains multiple occurrences of elements. Trying all labelling permutations π to choose the one that gives the minimal distance between the permutation strings resulting from the labelling according to π, is impractical. Fortunately, we can characterize a labelling permutation of indices that gives the minimum ℓ_1-distance. This will be enough to derive a polynomial algorithm for the ℓ_1-distance in the general strings case as well.

Lemma 2. *[Amir et al.] [17] Let $x, y \in \Sigma^m$ be two strings such that $d_{\ell_1}(x, y) < \infty$. Let π_o be the permutation that for any a and k, moves the k-th a in x to the location of the k-th a in y. Then, $d_{\ell_1}(x, y) = cost(\pi_o)$.*

2.2   The Approximation Algorithm

By Lemma 2 in order to find the ℓ_1 distance of P and $T^{(i)}$ we need to compute $\sum_{j=0}^{m-1} |j - \pi_o^i(j)|$. By Lemma 1, it is enough to compute $|\sum_{j=0}^{m-1}(j - \pi_o^i(j)) \cdot x_{j,k}|$ for $k = 1, \ldots, l$ samples of $X_0, \ldots, X_{m-1}$ drown from Cauchy distribution, and take the median value. In the sequel we explain how to compute $|\sum_{j=0}^{m-1}(j - \pi_o^i(j)) \cdot x_{j,k}|$ for all text locations in time $O(n)$.

Note that the above sum is the difference of the following two sums: $\sum_{j=0}^{m-1} j \cdot x_{j,k}$ and $\sum_{j=0}^{m-1} \pi_o^i(j) \cdot x_{j,k}$. The first sum can be easily be computed in $O(m)$ time, so we only have to explain how to compute $\sum_{j=0}^{m-1} \pi_o^i(j) \cdot x_{j,k}$ for all text locations in $O(n)$ time.

First note that by simple counting we can easily find all locations for which the distance is ∞, i.e. the locations for which there is no way to convert the one string to the another. Thus, we need only regard the substrings $T^{(i)}$ which are a permutation of P. Consider an element $a \in A$, and let $occ_a(x)$ be the number of occurrences of a in x. For these substrings, $occ_a(P) = occ_a(T^{(i)})$ for all $a \in P$. Consider a symbol a, and let $\psi_a(P)$ and $\psi_a(T)$ be the lists of locations of a in P and T, respectively. Note that these two lists need not be of the same length. Similarly, let $\psi_a(T^{(i)})$ be the list of locations of a in $T^{(i)}$. Then, for any $T^{(i)}$ (which is a permutation of P):

$$\sum_{j=0}^{m-1} \pi_o^i(j) \cdot x_{j,k} = \sum_{a \in P} \sum_{j=0}^{occ_a(P)-1} \psi_a(T^{(i)})[j] \cdot x_{j,k} \tag{1}$$

We now wish to express the above sum using $\psi_a(T)$ instead of the individual $\psi_a(T^{(i)})$'s. Note that all the a's referred to in $\psi_a(T^{(i)})$ are also referred to in $\psi_a(T)$. However, $\psi_a(T)$ gives the locations with regards to the beginning of T, whereas $\psi_a(T^{(i)})$ gives the locations with regards to the beginning of $T^{(i)}$ - which is i positions ahead.

For each i and a, let $match_a(i)$ be the index of the smallest entry in $\psi_a(T)$ with value at least i. Then, $match_a(i)$ is the first entry in $\psi_a(T)$ also referenced by $\psi_a(T^{(i)})$. Then, for any a, i and $j \leq occ_a(P)$:

$$\psi_a(T^{(i)})[j] = \psi_a(T)[match_a(i) + j] - i.$$

Thus, (1) can now be rewritten as:

$$\sum_{j=0}^{m-1} \pi_o^i(j) \cdot x_{j,k} = \sum_{a \in P} \sum_{j=0}^{occ_a(P)-1} (\psi_a(T)[match_a(i) + j] - i) \cdot x_{j,k} \tag{2}$$

Finally, (2) can be rewritten as:

$$\sum_{j=0}^{m-1} \pi_o^i(j) \cdot x_{j,k} = \sum_{a \in P} \sum_{j=0}^{occ_a(P)-1} \psi_a(T)[match_a(i) + j] \cdot x_{j,k} - i \cdot \sum_{a \in P} \sum_{j=0}^{occ_a(P)-1} x_{j,k} \tag{3}$$

Using (3) we can compute separately $\sum_{a \in P} \sum_{j=0}^{occ_a(P)-1} x_{j,k}$ (in $O(m)$ time) and

$$\sum_{a \in P} \sum_{j=0}^{occ_a(P)-1} \psi_a(T)[match_a(i)+j] \cdot x_{j,k}$$

for all i, then in $O(n)$ time get the ℓ_1 distance for every location, and choose the minimum value. It remains to explain how to compute:

$$\sum_{a \in P} \sum_{j=0}^{occ_a(P)-1} \psi_a(T)[match_a(i)+j] \cdot x_{j,k}$$

for all i in time $O(n)$. The important observation here is that the values $x_{j,k}$ are all random values drown from Cauchy distribution, so it does not matter which of the values is currently multiplied as long as we have $occ_a(P)$ different sample values for each text location. Thus, the above sum can be simply computed in $O(occ_a(P))$ time for the first location in $\psi_a(T)$, and in $O(1)$ for every other location using a sliding window of size $occ_a(P)$. Doing this for every $a \in A$ and summing the results gives the requested sum. We have proved Theorem 1.

3   The ℓ_∞ Rearrangement Distance

3.1   Exact ℓ_∞ Rearrangement Distance

Let x and y be strings of length m. Clearly, if x contains distinct elements then only one permutation can convert x to y. However, there can be many such permutations if x contains multiple elements. Computing the cost for each of them in order to find the distance between x and y might be too expensive. Fortunately, similar to the ℓ_1 case, we can characterize a minimal cost permutation converting x to y.

Lemma 3. *Let $x, y \in A^m$ be two strings such that $d_{\ell_\infty}(x, y) < \infty$. Let π_o be the permutation that for all $a \in A$ and k, moves the k-th a in x to the location of the k-th a in y. Then,*

$$d_{\ell_\infty}(x, y) = cost(\pi_o).$$

Proof. For a permutation π, and $i < j$ such that $x[i] = x[j]$, say that π *reverses* i *and* j if $\pi(j) > \pi(i)$. Note that π_o is characterized by having no *reversals*. Now we show that it has the least cost. Let τ be a permutation converting x to y of minimal cost that has the minimal number of reversals. If there are no reversals in τ, then there is nothing to prove, since it is exactly the permutation π_o. Otherwise, suppose τ reverses j and k ($j < k$). Let τ' be the permutation which is identical to τ, except that $\tau'(j) = \tau(k)$ and $\tau'(k) = \tau(j)$. Then, clearly τ' also converts x to y. We show that $cost(\tau') \leq cost(\tau)$. Consider two cases:

Case 1: $\tau(j) \geq k \geq \tau(k) \geq j$ or $\tau(k) \leq j \leq \tau(j) \leq k$. Consider the case $\tau(j) \geq k \geq \tau(k) \geq j$. We get:

$cost(\tau) - cost(\tau') =$

$= \max\{|\tau(j) - j|, |\tau(k) - k|\} - \max\{|\tau'(j) - j|, |\tau'(k) - k|\}$

$= \max\{|\tau(j) - j|, |\tau(k) - k|\} - \max\{|\tau(k) - j|, |\tau(j) - k|\}$

$= (\tau(j) - j) - \max\{|\tau(k) - j|, (\tau(j) - k)\} \geq 0.$

The argument for $\tau(k) \leq j \leq \tau(j) \leq k$ is symmetrical.

Case 2: $j < \tau(k) < \tau(j) < k$. Then,

$cost(\tau) - cost(\tau') =$

$= \max\{|\tau(j) - j|, |\tau(k) - k|\} - \max\{|\tau'(j) - j|, |\tau'(k) - k|\}$

$= \max\{|\tau(j) - j|, |\tau(k) - k|\} - \max\{|\tau(k) - j|, |\tau(j) - k|\} > 0.$

Since, $|\tau(j) - j| > |\tau(k) - j|$ and $|\tau(k) - k| > |\tau(j) - k|$.

Thus, the cost of τ' is at most that of τ, and there is one less reversal in τ', in contradiction.

Thus, in order to compute the ℓ_∞ distance of x and y, we create for each symbol a two lists, $\psi_a(x)$ and $\psi_a(y)$, the first being the list of locations of a in x, and the other - the locations of a in y. Both lists are sorted. These lists can be created in linear time. Clearly, if there exists an a for which the lists are of different lengths then $d_{\ell_\infty}(x, y) = \infty$. Otherwise, for each a, compute the differences between the corresponding elements in the lists, and take the maximum over all a's. This provides a linear time algorithm for strings of identical lengths, and an $O(m(n - m + 1))$ algorithm for the general case. This proves the first part of Theorem 2.

Patterns with Distinct Letters. We now show that if all entries of P are distinct, then the problem can be solved in $O(n \log m)$. In this case, w.l.o.g. we may assume that the pattern is simply the string $0, 1, \ldots, m - 1$. The basic idea is first to compute the distance for the first text location, as described above, while keeping information in appropriate data structures (to be explained). Then inductively compute the distance for the next text location, based on the information from previous location, making proper adjustments. Consider a text location i such that $d_{\ell_\infty}(P, T^{(i)}) < \infty$. Then, since all entries of P are distinct, for each $j \in P$ there is exactly one *matching* entry in $T^{(i)}$. As we move from one text location to the next, the matching symbols all move one location to the left – relative to the pattern, except for the leftmost — which falls out; and the rightmost — which is added. For all symbols that are further to the right in the text than in the pattern, this movement decreases their difference by 1. Thus, their relative order is unchanged, and can be kept in a priority queue R-*PriorityQueue* (by their original difference) so that we can keep track of their maximum. For all symbols that are further to the left in the text than in the pattern, this movement increases their difference by 1. Again, their relative order is unchanged and can be kept in a priority queue L-*PriorityQueue* (by their original difference) so that we can keep track of their maximum. Thus, given the distance at location i, in order to compute the distance at location $i + 1$, we only need to know the maximum difference of each type and compare to the difference of the new symbol (the new symbol and the one removed are easily handled).

To this end we keep track for each symbol j if it is currently to the left or to the right (this is stored in the array $location[\cdot]$), and the current position of the maximum in each type (stored in L-$PriorityQueue$ and R-$PriorityQueue$). In addition, we store for each symbol the point at which it moves from being at the right to being at the left (this is stored in the array $Trans$-$point[\cdot]$). Since P is simply the sequence $0, 1, \ldots, m - 1$, this $Trans$-$point[\cdot]$ can be easily computed. The important observation here is that when a symbol is added to L-$PriorityQueue$ (or R-$PriorityQueue$) we don't need to re-compute all differences (remember that their actual priority is not valid, only their relative order is valid) in order to find its place in the priority queue. We only re-compute the addition path of the new element. If the priority queues are kept as binary heaps, this takes at most $O(\log m)$ re-computations of differences each in time $O(1)$. If a priority queue is empty, we define its maximum function to return 0. In this way we are able to compute the new maximum for each location in $O(\log m)$ steps per location, for a total of $O(n \log m)$. A full description of the algorithm is provided in Fig. 1. Note that each symbol in the text participates in line 16 at most once, so the amortized cost of this line is $O(1)$. Also, note that the main part of the algorithm (lines 1–16) computes the distance correctly only for those locations which have bounded distance. However, by simple counting it is easy to eliminate (in $O(n)$ steps), all the locations of infinite distance. Thus, in line 17 we find the minimum among those which have bounded distance. We have proved the second part of Theorem 2.

3.2 Approximating the ℓ_∞ Distance

In this section we describe how to efficiently approximate the ℓ_∞ distance up to a factor of $1 + \varepsilon$ in time $O(\frac{1}{\varepsilon} n \log^2 m)$. First we show that it is enough to compute ℓ_p for $p \geq \log m/\varepsilon$ in order to approximate ℓ_∞ up to a factor of $1 + \varepsilon$. Then, we explain how to compute the ℓ_p distance in time $O(pn \log m)$ for even p. Choosing an even $p \geq \log m/\varepsilon$ gives Theorem 3.

Lemma 4. [Indyk et al.] [22] *For every* $p \geq \log m/\varepsilon$, $u, v \in \mathbb{R}^m$,

$$\|u - v\|_\infty \leq \|u - v\|_p \leq (1 + \varepsilon)\|u - v\|_\infty$$

By Lemma 4 we need only explain how to compute the ℓ_p distance for even p in time $O(pn \log m)$. For that we use the observation that the ℓ_2 algorithm of [17] can be generalized for any even p, using p convolutions instead of only one. This algorithm is based on the ℓ_2 pairing lemma (similar to Lemma 2) proved in [17]. Therefore, we need to show that the same pairing lemma is also correct for general p. This is given in Lemma 5. We have therefore explained the $O(pn \log m)$ algorithm.

Lemma 5. *Let* $x, y \in A^m$ *be two strings such that* $d_{\ell_p}(x, y) < \infty$. *Let* π_o *be the permutation that for all a and k, moves the k-th a in x to the location of the*

COMPUTING THE ℓ_∞ DISTANCE (all variables initialized to 0)
1   For $j = 0$ to $m - 1$ do
2   if $T[j] \leq j$ then $location[T[j]] \leftarrow Left$
3   else $location[T[j]] \leftarrow Right$
4   set $Trans\text{-}point[T[j]] \leftarrow j - T[j]$
5   For $j = 0$ to $m - 1$ do
6   add $T[j]$ to the list $Trans\text{-}symbols[Trans\text{-}point[T[j]]]$
7   $R\text{-}PriorityQueue \leftarrow |\{j \,|\, location[j] = Right\}|$
8   $L\text{-}PriorityQueue \leftarrow |\{j \,|\, location[j] = Left\}|$
9   $d[0] \leftarrow \max_{j=0}^{m-1} |j - T[j]|$
10 For $i = 1$ to $n - m$ do
11   set $t \leftarrow T[i + m]$
12   if $location[t] = Left$ then Add($L\text{-}PriorityQueue,t$)
13   else remove t from $Trans\text{-}symbols[Trans\text{-}point[t]]$
14   add t to the list $Trans\text{-}symbols[i + m - t]$
             and set $Trans\text{-}point[t] \leftarrow i + m - t$
15       $d[i] \leftarrow \max\{\max(L\text{-}PriorityQueue), \max(R\text{-}PriorityQueue)\}$
16   for each $t' \in Trans\text{-}symbols[i]$ do Add($L\text{-}PriorityQueue,location[t']$)
17 $d_{\min} \leftarrow \min\{d[i] \,|\, T^{(i)}$ is a permutation of $[0..m - 1]\}$
18 return $d_{\min}$

Fig. 1. Computing the ℓ_∞ Rearrangement Distance for $P = (0, 1, \ldots, m - 1)$

k-th a in y. Then,

$$d_{\ell_p}(x, y) = cost(\pi_o).$$

I.e. π_o is a permutation of the least cost.

Proof. Recall that π_o is characterized by having no *reversals*. Now we show that it has the least cost. Let τ be a permutation converting x to y of minimal cost that has the minimal number of reversals. If there are no reversals in τ, then there is nothing to prove, since it is exactly the permutation π_o. Otherwise, suppose τ reverses j and k ($j < k$). Let τ' be the permutation which is identical to τ, except that $\tau'(j) = \tau(k)$ and $\tau'(k) = \tau(j)$. Then, clearly τ' also converts x to y. We show that $cost(\tau') \leq cost(\tau)$. Consider two cases:

Case 1: $\tau(j) \geq k > \tau(k) \geq j$ or $\tau(k) \leq j < \tau(j) \leq k$. Consider the case $\tau(j) \geq k > \tau(k) \geq j$. We get:
$cost(\tau) - cost(\tau') =$
$= |\tau(j) - j|^p + |\tau(k) - k|^p - |\tau'(j) - j|^p - |\tau'(k) - k|^p$
$= |\tau(j) - j|^p + |\tau(k) - k|^p - |\tau(k) - j|^p - |\tau(j) - k|^p$
$= |(\tau(j) - k) + (k - \tau(k)) + (\tau(k) - j)|^p + |\tau(k) - k|^p - |\tau(k) - j|^p - |\tau(j) - k|^p$
$\geq |\tau(j) - k|^p + |k - \tau(k)|^p + |\tau(k) - j|^p + |\tau(k) - k|^p - |\tau(k) - j|^p - |\tau(j) - k|^p$
$= 2|\tau(k) - k|^p \geq 0.$
The argument for $\tau(k) \leq j < \tau(j) \leq k$ is symmetrical.

Case 2: $j < \tau(k) < \tau(j) < k$. Then,

$cost(\tau) - cost(\tau') =$
$= |\tau(j) - j|^p + |\tau(k) - k|^p - |\tau'(j) - j|^p - |\tau'(k) - k|^p$
$= |\tau(j) - j|^p + |\tau(k) - k|^p - |\tau(k) - j|^p - |\tau(j) - k|^p$
$= |(\tau(j) - \tau(k)) + (\tau(k) - j)|^p + |(k - \tau(j)) + (\tau(j) - \tau(k))|^p - |\tau(k) - j|^p - |\tau(j) - k|^p$
$\geq |\tau(j) - \tau(k)|^p + |\tau(k) - j|^p + |k - \tau(j)|^p + |\tau(j) - \tau(k)|^p - |\tau(k) - j|^p - |\tau(j) - k|^p$
$= 2|\tau(j) - \tau(k)|^p > 0.$

Thus, the cost of τ' is at most that of τ, and there is one less reversal in τ', in contradiction.

References

1. Amir, A., Lewenstein, M., Porat, E.: Faster algorithms for string matching with k mismatches. J. Algorithms 50(2), 257–275 (2004)
2. Galil, Z., Giancarlo, R.: Improved string matching with k mismatches. SIGACT News 17(4), 52–54 (1986)
3. Landau, G.M., Vishkin, U.: Efficient string matching with k mismatches. Theoretical Computer Science 43, 239–249 (1986)
4. Abrahamson, K.: Generalized string matching. SIAM J. Comp. 16(6), 1039–1051 (1987)
5. Karloff, H.: Fast algorithms for approximately counting mismatches. Information Processing Letters 48(2), 53–60 (1993)
6. Cole, R., Hariharan, R.: Approximate string matching: A faster simpler algorithm. In: Proc. 9th ACM-SIAM Symposium on Discrete Algorithms (SODA), pp. 463–472. ACM Press, New York (1998)
7. Levenshtein, V.I.: Binary codes capable of correcting, deletions, insertions and reversals. Soviet Phys. Dokl. 10, 707–710 (1966)
8. Cole, R., Gottlieb, L., Lewenstein, M.: Dictionary matching and indexing with errors and don't cares. In: STOC 2004: Proceedings of the thirty-sixth annual ACM symposium on Theory of computing, pp. 91–100. ACM Press, New York (2004)
9. Ferragina, P., Grossi, R.: Fast incremental text editing. In: Proc. 7th ACM-SIAM Symposium on Discrete Algorithms, pp. 531–540. ACM Press, New York (1995)
10. Gu, M., Farach, M., Beigel, R.: An efficient algorithm for dynamic text indexing. In: Proc. 5th Annual ACM-SIAM Symposium on Discrete Algorithms, pp. 697–704. ACM Press, New York (1994)
11. Sahinalp, S.C., Vishkin, U.: Efficient approximate and dynamic matching of patterns using a labeling paradigm. In: Proc. 37th FOCS, pp. 320–328 (1996)
12. Lowrance, R., Wagner, R.A.: An extension of the string-to-string correction problem. J. of the ACM, 177–183 (1975)
13. Berman, P., Hannenhalli, S.: Fast sorting by reversal. In: Hirschberg, D.S., Meyers, G. (eds.) CPM 1996. LNCS, vol. 1075, pp. 168–185. Springer, Heidelberg (1996)
14. Carpara, A.: Sorting by reversals is difficult. In: RECOMB 1997, pp. 75–83. ACM Press, New York (1997)
15. Bafna, V., Pevzner, P.: Sorting by transpositions. SIAM J. on Discrete Mathematics 11, 221–240 (1998)

16. Christie, D.A.: Sorting by block-interchanges. Information Processing Letters 60, 165–169 (1996)
17. Amir, A., Aumann, Y., Benson, G., Levy, A., Lipsky, O., Porat, E., Skiena, S., Vishne, U.: Pattern matching with address errors: Rearrangement distances. In: Proc. 17th SODA, pp. 1221–1229 (2006)
18. Amir, A., Aumann, Y., Kapah, O., Levy, A., Porat, E.: Approximate string matching with address bit errors Manuscript
19. Amir, A., Hartman, T., Kapah, O., Levy, A., Porat, E.: On the cost of interchange rearrangement in strings. In: Arge, L., Welzl, E. (eds.) ESA. LNCS, vol. 4698, Springer, Heidelberg (2007)
20. Zolotarev, V.: One-dimensional stable distributions. Translations of Mathematical Monographs 65 (1986)
21. Indyk, P.: Stable distributions, pseudorandom generators, embeddings and data stream computation. In: FOCS, pp. 189–197 (2000)
22. Indyk, P., Lewenstein, M., Lipsky, O., Porat, E.: Closest pair problems in very high dimensions. In: Díaz, J., Karhumäki, J., Lepistö, A., Sannella, D. (eds.) ICALP 2004. LNCS, vol. 3142, pp. 782–792. Springer, Heidelberg (2004)

Generalized LCS

Amihood Amir[1,*], Tzvika Hartman[2], Oren Kapah[2], B. Riva Shalom[2],
and Dekel Tsur[3]

[1] Department of Computer Science, Bar-Ilan University, Ramat-Gan 52900, Israel and
Department of Computer Science, Johns Hopkins University, Baltimore, MD 21218
amir@cs.biu.ac.il
Tel.: +972 3 531-8770
[2] Department of Computer Science, Bar-Ilan University, Ramat-Gan 52900, Israel
{hartmat,kapaho,gonenr1}@cs.biu.ac.il
[3] Department of Computer Science, Ben Gurion University, Be'er Sheva 84105, Israel
dekelts@cs.bgu.ac.il

Abstract. The Longest Common Subsequence (LCS) is a well studied
problem, having a wide range of implementations. Its motivation is in
comparing strings. It has long been of interest to devise a similar measure
for comparing higher dimensional objects, and more complex structures.
In this paper we give, what is to our knowledge, the first inherently
multi-dimensional definition of LCS. We discuss the Longest Common
Substructure of two matrices and the Longest Common Subtree problem
for multiple trees including a constrained version. Both problems cannot be solved by a natural extension of the original LCS solution. We
investigate the tractability of the above problems. For the first we prove
$\mathcal{NP}$-Completeness. For the latter $\mathcal{NP}$-hardness holds for two general
unordered trees and for k trees in the constrained LCS.

1 Introduction

The *Longest Common Subsequence* problem, whose first famous dynamic programming solution appeared in 1974 [19], is one of the classical problems in Computer Science. The widely known string version is given two strings of length n,
find the length of the maximal subsequence common to both strings. For example, for $A = abcddabef$ and $B = efbadeaab$, $LCS(A, B)$ is 4, where a possible
such subsequence is $adab$.

The LCS problem, has been well studied. For a survey, see [3]. The main
motivation for the problem is comparison of different strings. An immediate
example from computational biology is finding the commonality of two DNA
molecules. Most previous works deals with the one dimensional (string) version
of the problem. However, there has been increasing motivation for considering
generalizations of the LCS to higher dimensions (e.g. matrices) and different
data structures (e.g. trees). For example, the secondary and tertiary structure of
proteins and RNA play an important role in their functionality [5], thus it has

---

* Partly supported by NSF grant CCR-01-04494 and ISF grant 35/05.

N. Ziviani and R. Baeza-Yates (Eds.): SPIRE 2007, LNCS 4726, pp. 50–61, 2007.
© Springer-Verlag Berlin Heidelberg 2007

been an interesting challenge to devise an inherently multi-dimensional method of comparing multidimensional objects, as the LCS compares strings.

The first task we tackle in this paper is to give a natural definition generalizing the LCS. All generalizations until now are, essentially, linearizations [2]. Edit distance, a closely related problem, has also been generalized, and again the errors are within a single dimension [12], [1]. To our knowledge, our definition is the first inherently multi-dimensional generalization in the literature. It elegantly and naturally generalizes the string definition.

Unfortunately, it turns out that the LCS problem between two matrices is $\mathcal{NP}$-hard. LCS applied to trees has been previously defined for two trees, via the tree edit distance. We consider the problem for multiple trees, in the case of ordered and unordered trees. We also define a constrained version of the trees LCS (Con-LCS) applied to the above cases. Besides creating a generalized framework for the LCS problem applied to higher dimensions we aim at discovering the tractability of these problems, that is, when does the problem become unsolvable in polynomial time. The tractability results of all the above problems appear in the table of Figure 1, where LCStree refers to Largest Common Subtree and CON-LCStree stands for Constrained LCStree. Known previous results appear in the table with the appropriate citation.

	Matrix	Unordered LCStree	Unordered Con-LCStree	Ordered LCStree	Ordered Con-LCStree
2	NPC	NPC	P [?]	P [6,10,16]	P [21]
$k > 2$	NPC	NPC	NPC	$O(kn^{2k})$	$O(kn^{2k} + k^{k/2}n^{2k-1})$

Fig. 1. Tractability results

This paper is organized as follows: Section 2 pinpoints the essence of the string LCS problem, enabling us to generalize it to more complex structures and higher dimensions. The definition of two dimensional LCS as well as the $\mathcal{NP}$-hardness result are shown in Section 3. The generalization to trees is defined in Section 4. Section 5 proves $\mathcal{NP}$-hardness of unordered two trees and for k trees of the constrained LCS. Section 6 provides a dynamic programming algorithm for polynomially computing the LCS of two ordered trees and shows that the problem is solvable in $O(kn^{2k})$ for k trees, which is polynomial for a constant number of trees.

2 Preliminaries

The known solutions for LCS use dynamic programming algorithms, in which they compute at each step the $LCS(A[1,i], B[1,j])$ till they reach $LCS(A[1,n], B[1,n])$. The following observations are trivial, yet important to understand the main characteristic of the problems, which will assist us defining generalized LCS problems consistently to the original LCS 'spirit'.

Observation 1. *A string is a collection of objects (characters) with total precedence order between them, for every two distinct objects, one precedes another.*

Observation 2. *The LCS problem is ordered on a line. An LCS solution, matching $A[i]$ to $B[j]$ can match $A[i']$, where $i' > i$, only to a $B[j']$ where $j' > j$, and vice versa.*

Lemma 1. *The above characteristics of the LCS problem allows its optimal solution to consider at every step increasing prefixes of the input strings.*

Proof. The dynamic programming solution has a single possible direction of enlarging the substrings to which it computes their LCS, since all characters are ordered in precedence order. Therefore, computing $LCS(A[1, i], B[1, j])$ depends merely on the LCS of prefixes of A and B shorter in one or zero symbols.   □

Observation 3. *The LCS of strings A, B is the reverse of the LCS of A^r, B^r, where S^r is the reversed string of S.*

The above observations suggest a more combinatorial definition of the LCS problem, one that naturally generalizes to higher dimensions. Below is a combinatorial definition of the string LCS problem that supports all above observations.

Definition 1. *The* String Longest Common Subsequence *(LCS) Problem:*
Input: Two strings A, B of length n over alphabet Σ.
Output: The maximum domain size of a one-to-one function $f : \{1, ..., n\} \to$
         $\{1, ..., n\}$, where $A[i] = B[f(i)]$, for every i in the domain, and where,
         for $i, j \in Dom(f)$, $i < j$ iff $f(i) < f(j)$.

The advantage of this definition is that it abstracts the LCS into an order preserving matching. A similar order preserving matching that support the above observations is the natural generalization. However, when dealing with partially ordered structures, the dynamic programming method of computing the necessary calculations on prefixes of increasing size is meaningless, as a prefix can not be defined. A more general approach is used in our dynamic programming solutions of the tree LCS problems.

3   Two Dimensional LCS

Extending the Longest Common Subsequence problem to a two dimensional problem, the input should be two symbols matrices, in which we seek identical symbols, preserving their order in the matrix. This will not necessarily result in an array, but rather the symbols common to both matrices, that preserve their order in both matrices. For this reason we name the problem *2 Dimensional Longest Common Substructure (2D_LCS)*. As far as we know, no inherently two dimensional version of LCS was previously defined.

We define the problem in a way that Observations 2 and 3 are applicable in the two dimensional problem as well.

Definition 2. *The* 2 *Dimensional Longest Common Substructure* ($2D\_LCS$):
Input: Two matrices A, B of size $n \times n$ each, over alphabet Σ.
Output: The maximum domain size of a one-to-one function $f : \{1, ..., n\}^2 \rightarrow$
* $\{1, ..., n\}^2$, where $A[i, j] = B[f(i, j)]$, for every (i, j) in $Dom(f)$,*
*   and where, for $(i_k, j_k), (i'_k, j'_k) \in Dom(f)$ and for $f(i_k, j_k) = (ii_g, jj_g)$*
*   and $f(i'_k, j'_k) = (ii'_g, jj'_g)$, the following hold:*
1. $i_k < i'_k$ iff $ii_g < ii'_g$.
2. $i_k = i'_k$ iff $ii_g = ii'_g$.
3. $j_k < j'_k$ iff $jj_g < jj'_g$.
4. $j_k = j'_k$ iff $jj_g = jj'_g$.

An example for two matrices can be seen in Figure 2. The $2D\_LCS$ of these matrices is 4 and can be obtained by the boldface letters.

A	**C**	c
B	**D**	A
A	A	C

A	A	**C**
D	**B**	C
B	D	**D**

Matrix A   Matrix B

Fig. 2. An example of two matrices

The importance of the combinatorial definition is that it can be easily extended to higher dimensions, in which the matching symbols, required to be identical, preserve the order relation in space, meaning that for a d-dimensional structure other $2d$ constraints should be added, two for every axis. We get the following general definition:

Definition 3. *The d Dimensional Longest Common Substructure (dD_LCS):*
Input: Two arrays A, B of size n^d each, over alphabet Σ
Output: The maximum domain size of a one-to-one function $f : \{1, ..., n\}^d \rightarrow$
* $\{1, ..., n\}^d$, where $A[i_1, ..., i_d] = B[f(i_1, ..., i_d)]$, for every $(i_1, ..., i_d)$*
*   in $Dom(f)$, and where, for $(i_1, ..., i_d), (i'_1, ..., i'_d) \in Dom(f)$ and for*
* $f(i_1, ..., i_d) = (ii_1, ..., ii_d)$ $f(i_1, ..., i_d) = (ii_1, ..., ii_d)$ and $f(i'_1, ..., i'_d) =$*
* $(ii'_1, ..., ii'_d)$, the following hold for every $1 < j \leq d$:*
1. $i_j < i'_j$ iff $ii_j < ii'_j$.
2. $i_j = i'_j$ iff $ii_j = ii'_j$.

The requirement for order preservation between every two selected entries, though necessary for the LCS generalization, is very strong. As will be seen, this causes the high complexity solution to the $2D\_LCS$ problem.

3.1   2D_LCS Is $\mathcal{NP}$-Hard

We claim the problem is not polynomially computable. To this aim we define the $2D\_LCS$ decision version:

Definition 4. *The 2 Dimensional Common Substructure $(2D\_CS)$:*
Input: Two matrices A, B of size n^2 each, and a constant L.
Output: Does there exists a $2D\_LCS$ of A, B of size L.

Theorem 1. *The $2D\_CS$ problem is $\mathcal{NP}$-hard.*

We prove the hardness of the problem by a reduction from Clique.

Lemma 2. *Clique $\propto_m^p 2D\_CS$.*

Proof. Given a graph $G = (V, E)$ with n vertices and a constant K, We construct two matrices. Matrix A, of size $K \times K$, contains 1 in all entries except those on the main diagonal, where 2 is placed. Matrix B is the adjacency matrix of G with a slight change. B is of size $n \times n$, where $B[i, j] \in \{0, 1, 2\}$ is defined as:

$$A[i,j] = \begin{cases} 2\ i = j \\ 1\ otherwise \end{cases} \qquad B[i,j] = \begin{cases} 1\ (v_i, v_j) \in E \\ 2\ i = j \\ 0\ otherwise \end{cases}$$

Finally, take $L = K^2$.

Obviously the construction is done in polynomial time in the size of G, as L must be bounded by n. For an example of the construction see Fig. 3.

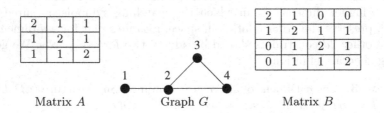

Fig. 3. Matrices A, B constructed for graph G and $K = 3$

Lemma 3. *G contains a clique of size $\geq K$ iff there exists a 2 dimensional common substructure of size $\geq L$ between A and B.*

Proof. ($\Rightarrow$) Suppose G contains a clique of size K. Let $v_{i_1}, ..., v_{i_K}$ be the nodes participating in the clique, listed in increasing indices order. Hence, $B[i_s, i_t] = 1$ for all $1 \leq s, t \leq K$, $s \neq t$. We get that matrix A, the whole of it can be matched, in the common substructure sense, to $B[i, j]$ entries, where $i, j \in \{i_1, .., i_K\}$ in the following order: The first row of A, will be matched to the $i_1, ..., i_K$ entries of row i_1 in B. There are K such entries, so the number of the matched entries is equal. Note that $B[i_1, i_1] = 2$ as it is located on the main diagonal, thus

$B[i_1, i_1] = A[1, 1]$. The rest of the entries $B[i_1, i_j]$, $i_j \in \{i_2, ..., i_K\}$ contain 1, just like the rest of $A[1, j]$, $1 < j \leq K$, so the symbols matched are identical. In the same way, row h of A is matched to the $i_1, ..., i_K$ entries of row i_h in B. Here again all selected symbols, in both matrices are 1 except the hth that is 2. All in all we get that there are $L = K^2$ identical symbols in the different matrices that can be matched.

We show that every pair of such symbols from A and its corresponding pair from B preserve the order relation in the plane. Let $A[i, j], A[i', j']$ be two symbols matched to $B[ii, jj], B[ii', jj']$ correspondingly. Due to the way the one-to-one matching function was chosen we have that $B[ii, jj] = B[i_i, i_j]$ and $B[ii', jj'] = B[i_{i'}, i_{j'}]$.

Consider all possible cases:
1. $i = i'$, so we have $ii = i_i = ii'$.
2. $i > i'$, implies $ii = i_i > i_{i'} = ii'$ since every $i_j < i_{j+1}$.
3. $j = j'$, again $jj = i_j = i_{j'} = jj'$.
4. $j > j'$, implies $jj = i_j > i_{j'} = jj'$ since every $i_j < i_{j+1}$.

($\Leftarrow$) The rest of the proof is similar to the previous case, and will appear in the journal version.   ∎

4 Largest Common Subtree (LCStree)

The problem of comparing trees occurs in many areas such as computer vision, compiler optimization, natural language processing and computational biology. In the latter field, for example, it is possible to represent the RNA secondary structure as a rooted ordered tree [15].

A comparison between two given trees T_1, T_2, can be defined in diverse ways, searching isomorphism between two trees, looking for homeomorphism, seeking a a maximum agreement subtree problem and more. Another possible comparison between objects is, of course, the edit distance. In a recent paper [6], Demaine et. al. improve Klein [10] and Shasha and Zhang [16] and give an $O(n^3)$ time algorithm for tree edit distance, for two n sized rooted *ordered* trees.

We define the problem in a way consistent with the String Longest Common Subsequence, that is all parts of the main structure are labelled and each of them can be deleted to give a common substructure. As a consequence, The Largest Common Subtree Problem should require the matching of inner nodes as well as the leaves, and should enable pruning nodes of the trees. Where pruning a node v implies all children of v become the children of the parent node of v. Our general definition easily adapts to trees.

Definition 5. *The* Largest Common Subforest Problem (LCStree):
Input: *Given two node-labelled trees $T_1 = (V_1, E_1), T_2 = (V_2, E_2)$, where the labels are from alphabet Σ.*
Output: The maximum domain size of a one-to-one function $f : V_1 \to V_2$, where the following hold:
1. Label of node v equals the label of node $f(v)$, for every $v \in Dom(f)$.
2. For $v, w \in Dom(f)$, v is an ancestor of w iff $f(v)$ is an ancestor of $f(w)$.

Observe that in case T_1, T_2 are degenerated trees, in the form of chains, the definition is reduced to the traditional strings LCS (Definition 1). This definition applies to unordered trees, for ordered trees, where order among siblings is significant, a third requirement should be added to the LCStree definition:

for $v, w \in Dom(f)$, v is to the right of w iff $f(v)$ is to the right of $f(w)$.
Where a node v is to the right of node w if v is a right sibling of w, or that an ancestor of v is a right sibling of an ancestor of w.

It is important to note that the *LCStree* is not necessarily a tree, but could be a forest. The above definition can be viewed as an extension of the Tree Inclusion problem [18,9,14] where we are given a pattern tree P and a text tree T both with labels on the nodes and we seek the smallest subtree of T that includes P. A tree is included in another tree if it can be obtained from the larger one by deleting nodes and in case of unordered trees, by also permuting siblings. The tree inclusion problem on unordered trees is $\mathcal{NP}$-hard [9]. For ordered trees it is polynomially solved [9], [14] in $O(|P| \cdot |T|)$.

It is a well known fact that the LCS problem can be looked at as a special case of the edit distance transforming one string to another by operations of substitution, deletion and insertion. Suppose the substitution operation is assigned a high cost, such that it will never be profitable to use it, the edit distance problem is then equivalent to finding the LCS of the strings. It can be easily seen that this notion can be applied to generalized LCS and Edit Distance problems, in case they are consistently defined. As the edit distance between two trees is a well studied problem [4], in the following sections we will use some edit distance algorithms for trees to solve equivalent LCStree questions.

4.1   Constrained-LCStree

An interesting version of the edit distance problem posed by Zhang [20,4] is the constrained edit distance, in which a natural constraint is added to the known tree edit distance, namely that disjoint subtrees must be mapped to disjoint subtrees. The constrained edit distance is motivated from classification tree comparison. We similarly define the constrained LCStree:

Definition 6. *The* Constrained Largest Common Subforest Problem(Con-LCStree):
Input: Given two node-labelled trees $T_1 = (V_1, E_1), T_2 = (V_2, E_2)$, where the labels
*   are from alphabet Σ.*
Output: The maximum domain size of a one-to-one function $f : V_1 \to V_2$, where
*   the following hold:*
*   1. Label of node v equals the label of node $f(v)$, for every $v \in Dom(f)$,*
*   2. For $v, w \in Dom(f)$, v is an ancestor of w iff $f(v)$ is an ancestor of $f(w)$.*
*   3. For $v, w \in Dom(f)$, v and w are siblings iff $f(v)$ and $f(w)$ are siblings.*

When two unordered trees T_1, T_2 are concerned, Zhang [21,4] suggested an algorithm solving the constrained trees edit distance consuming $O(|T_1||T_2| \cdot (deg(T_1) +$

$deg(T_2))log(deg(T_1)+deg(T_2)))$ time. Slight modifications of his algorithm make it applicable to Constrained-LCStree of two unordered trees.

Applying the Constrained Edit Distance to ordered trees was also done by Zhang [20]. All rooted subtrees were considered as relevant subproblems. Comparing the LCStree of two subtrees $T_1[t_1]$ with $T_2[t_2]$, t_1 and t_2 can be compared and then their children should be matched, in way that the order among the children would be adhered. To this aim Zhang suggested reducing the problem to string edit distance. In addition, t_1 can be matched to every child of t_2 and vice versa. For the Constrained-LCStree for two ordered trees, we can use his algorithm running in time $O(|T_1||T_2|)$. As a consequence, we discuss the problem of Constrained-LCStree merely when applied to multiple trees.

4.2 LCStree of 3 and More Trees (LCSktrees)

Having defined the LCStree problem for two trees, it is only natural to consider the LCStree problem of three and more trees. The definition automatically derives from the definition of the problem between two trees. We formally define here the most general problem, though for ordered trees or for constrained LCStree the modifications are obvious:

Definition 7. *The* Largest Common k Subforests Problem (LCSktree)*:*
Input: Given k node-labelled trees $T_i = (V_i, E_i)$, where the labels
are from alphabet Σ.
Output: The maximum size of the intersection of the domains of $k-1$ one-to-one
functions $f_i : V_1 \rightarrow V_{i+1}$, $1 \leq i < k$, where the following holds $\forall i\ 1 \leq i < k$:
1. Label of node v equals the label of node $f_i(v)$, for every $v \in Dom(f_i)$.
2. For $v, w \in Dom(f_i)$, v is an ancestor of w iff $f_i(v)$ is an ancestor of $f_i(w)$.

To our knowledge, this problem has not been considered hitherto even for ordered trees, where order among siblings is fixed. In the following subsections we prove $\mathcal{NP}$-Completeness even for the Constrained LCS of k unordered trees.

5   Unordered Trees

Our LCStree definition can be viewed as an extension of the Tree Inclusion problem [18,9,14] where we are given a pattern tree P and a text tree T both with labels on the nodes and we seek the smallest subtree of T that includes P. The tree inclusion problem on unordered trees is $\mathcal{NP}$-hard [9]. We thus claim the LCStree problem is not polynomially computable and prove its hardness by a reduction from the tree inclusion problem.

Lemma 4. *Tree Inclusion problem $\leq_T^p$ CStree.*

Given two trees T and P, let $T_1 = T$, $T_2 = P$ and $L = |P|$. It follows that T includes P iff there exist common subtrees of total size $\geq L$ between T_1 and T_2. ∎

Theorem 2. *The LCStree problem applied to two unordered trees is $\mathcal{NP}-hard$.*

5.1 Constrained LCStree for 3 and More Unordered Trees

Though the two trees version of the problem is polynomial, for three and more trees we claim as follows, the proof will appear in the full version of the paper.

Theorem 3. *The Con − LCSktree problem is a $\mathcal{NP}$ hard problem.*

6 Ordered Trees

The edit distance problem for ordered trees, as defined by Demain et. al. [6], relates to our ordered case definition as the string edit distance relates to the string LCS. Demain et. al. [6] suggest a dynamic programming consuming $O(n^3)$ time. Using their algorithm we get:

Theorem 4. *The LCStree problem for Ordered Trees is solved in $O(n^3)$.*

6.1 LCSktree for Ordered Trees

To our knowledge, there is no algorithm for tree edit distance of $k > 2$ ordered trees, therefore we suggest a new algorithm for the LCSktree for ordered trees, by expanding the Shasha and Zhang [16] algorithm, solving the tree edit distance problem. The algorithm now consumes $O(kn^{2k})$ time. The modification of the Shasha-Zhang algorithm to the LCStree case, will be described in the journal version.

Theorem 5. *The LCSktree problem for k Ordered Trees can be polynomially solved for a constant k trees.*

6.2 Constrained LCSKtree for Ordered Trees (Con-LCSktree)

In contrast to general unordered trees, where we proved the Con-LCSktree to be $\mathcal{NP}$ hard, the problem for k ordered trees is is polynomial for a constant number of trees. The instance of the problem is $T = \{T_1, ..., T_k\}$ labelled trees where order among siblings is significant. We give a solution for Con-LCStree for k ordered trees consuming $O(kn^k + k^{k/2}n^{k-1})$ time. For the case of $k = 2$ we get a $O(n^2)$ solution for the LCStree problem, just as was attained by the Zhang [21] for constrained edit distance between two ordered trees.

Here again the relevant subproblems are all rooted subtrees of the k trees, therefore the dynamic programming table we fill is of size $O(n^k)$.

Solving the Constrained LCStree problem for multiple trees can be done similarly to the case of two trees, where we have two cases: First if all roots of the current trees match, we need to match between their children. On the other hand, if we choose to discard a root we need to match one of its children with the other roots.

However this generalization is not straightforward, for the former case, we need a k-matching algorithm, while the 3Matching problem is $\mathcal{NP}$-Complete. However, we suggest to reduce it to a non crossing matching problem. As we deal with k strings we must consider a k-matching, matching of k dimensions, and in order to preserve the order between siblings, it must be a noncrossing matching.

Definition 8. *The* Maximum k Dimensional Weighted NonCrossing Matching Problem (kMWNM)*:*
Input: *A complete weighted k-partite graph $G_{n_1,...,n_k} = (V_1, ..., V_k, E)$.*
Output: The maximum weighted matching of $G_{n_1,...,n_k}$, where matching
        k vertices $x_1, ..., x_k$ implies, all vertices proceeding x_i can be matched
        only to those proceeding x_j, $j \neq i$ and all those succeeding x_i
        to vertices succeeding x_j merely.

For the kMWNM problem we will present an algorithm consuming $O(kn^k)$ in the journal version.

For the second case of the problem, we need to consider $0 < k' < k$ out of k trees, whose roots are matched with themselves. The LCSktree can be obtained by comparing a subtree from each of the $k - k'$ 'unselected' trees with the k' complete trees. As each of the $k - k'$ trees is represented by a single subtree, we have to consider the cartesian product of these subtrees.

Let the subtree of tree T_i rooted at t_i has n_i children. Suppose k' trees whose roots are compared were selected and renamed $t_{i_1}, ..., t_{i_{k'}}$ and the other $k - k'$ trees renamed $t_{j_1}, ..., t_{j_{k-k}}$, will contribute a child representing each $t_{j_l}^{x_l}$. We get following lemma:

Lemma 5. $Con - LCStree(t_1, \ldots, t_k) =$

$$
max \begin{cases}
Equal(t_1, ..., t_k) + kMWNM(\{t_1^1, ..., t_1^{n_1}\}, ... \{t_k^1, ..., t_k^{n_k}\}), \\
max_{i_1, x_l, 1 \le l \le n_l}\{Con - LCStree(t_{i_1}, t_{j_1}^{x_1}, ..., t_{j_{k-1}}^{x_{k-1}})\}, \\
max_{i_1, i_2, x_l, 1 \le l \le n_l}\{Con - LCStree(t_{i_1}, t_{i_2}, t_{j_1}^{x_1}, ..., t_{j_{k-2}}^{x_{k-2}})\}, \\
\quad \vdots \\
max_{i_f, 1 \le f \le k/2, x_l, 1 \le l \le n_l}\{Con - LCStree(l_{i_1}, ..., t_{i_{k/2}}, t_{j_1}^{x_1} ..., t_{j_{k-k/2}}^{x_{k-k/2}})\}
\end{cases}
$$

Theorem 6. *The Con-LCSktrees for ordered trees is polynomially solvable for a constant number of trees.*

Proof. The dynamic programming table will be of size n^k, where every entry $[t_1, ..., t_k]$ refers to the Con-LCSktree value of subtrees originating in the different k trees. Every axis i, stands for the rooted subtrees of T_i and is filled according to increasing size of the subtrees.

Filling entry $[t_1, ..., t_k]$ we must consider all options of selecting k' out of k trees which we match by their roots. For the case of $k' = k$, we use the $O(kn^k)$ solution for the noncrossing matching, where n is the number of vertices in each of the vertices lists of the matching. However, each of the matching computation is done on a k-partite graph induced by the direct children of the current subtree root. Hence, all matching computations consume: $O(\sum_{t_1=1}^n \cdots \sum_{t_k=1}^n k \cdot dc(t_1)dc(t_2) \cdots dc(t_k))$ which by the Observation 4 equals (kn^k).

Observation 4. *Suppose t is a node from T of size n, and $dc(t)$ is the number of the direct children of t, $\sum_{t=1}^n dc(t) = n - 1$.*

For $k' \neq k$ we need to choose the best subtree representative from each of the $k - k'$ 'unselected' trees, meaning $\prod_{i=k'+1}^k dc(t_i)$. Again, in the worst case, for $k' = 1$ we have a product of $k - 1$ direct children of the current roots, summing this product over all entries computed we get $O(n^{k-1})$. However, this does not terminate the cost as we have $k - 1$ options to select k', therefore $O(k^{k/2})$ options of selected trees, aligned by their roots. Consequently, we get $O(k^{k/2}n^{k-1})$ maximized values for a single entry. All in all, the algorithm requires, $O(kn^k + k^{k/2}n^{k-1})$ time.   ∎

7 Conclusions and Open Problems

The main contribution of the paper is generalizing the concept of the traditional Longest common Subsequence. In this paper we introduced two new problems derived from the traditional Longest Common Subsequence. We have proved the problem applied to matrices or to 3 general labelled trees or k for the constrained version, is $\mathcal{NP}$ hard, whereas concerning k ordered trees it can be polynomial for constant k. We have also generalized a noncrossing matching to k dimensions and used it as an accessory to the constrained LCSktree solution. Consequently to our new definitions, LCS questions regarding other non trivial structures may be considered and their tractability explored.

References

1. Amir, A., Landau, G.M.: Fast Parallel and Serial Multidimensional Aproximate Array Matching. Theor. Comput. Sci. 81(1), 97–115 (1991)
2. Baeza-Yates, R.: Similarity in Two-dimensional strings. In: Hsu, W.-L., Kao, M.-Y. (eds.) COCOON 1998. LNCS, vol. 1449, pp. 319–328. Springer, Heidelberg (1998)
3. Bergroth, L., Hakonen, H., Raita, T.: A survey of longest common subsequence algorithms. In: SPIRE 2000, pp. 39–48 (2000)
4. Bille, P.: A Survey on Tree Edit Distance and Related Problems. TCS 337(1-3), 217–239 (2005)
5. Branden, C., Tooze, J.: Introduction to Protein Structure. Garland Publishing, New York (1999)
6. Demaine, E., Mozes, S., Rossman, B., Weimann, O.: An $O(n^3)$-time algorithm for tree edit distance (submitted)
7. Farach, M., Przytycka, T.M., Thorup, M.: The Maximum Agreement Subtree Problem for Binary Trees. In: Proc. of the second ESA (1995)
8. Hirschberg, D.S.: space algorithm for Computing Maximal Common Subsequences. Commun. ACM 18(6), 341–343 (1975)
9. Kilpelinen, P., Mannila, H.: Ordered and unordered tree inclusion. SIAM Journal on Computing 24(2), 340–356 (1995)
10. Klein, P.N.: Computing the edit distance between unrooted ordered trees. In: ESA 1998, pp. 91–102 (1998)
11. Kilpelinen, P., Mannila, H.: Ordered and unordered tree inclusion. SIAM Journal on Computing 24(2), 340–356 (1995)
12. Krithivasan, K., Sitalakshmi, R.: Efficient two-dimensional pattern matching in the presence of errors. Informatio Sciences 43(3), 169–184 (1987)
13. Pinter, R.Y., Rokhlenko, O., Tsur, D., Ziv-Ukelson, M.: Approximate labelled subtree homeomorphism. In: Sahinalp, S.C., Muthukrishnan, S.M., Dogrusoz, U. (eds.) CPM 2004. LNCS, vol. 3109, pp. 59–73. Springer, Heidelberg (2004)

14. Richter, T.: A new algorithm for the ordered tree inclusion problem. In: Hein, J., Apostolico, A. (eds.) Combinatorial Pattern Matching. LNCS, vol. 1264, pp. 150–166. Springer, Heidelberg (1997)
15. Shapiro, B.A., Zhang, K.Z.: Comparing multiple RNA secondary structures using tree comparisons. Computer Applications in the Biosciences 6(4), 309–318 (1990)
16. Shasha, D., Zhang, K.: Simple Fast Algorithms for the Editing Distance Between Trees and Related Problems. SIAM J. Comput. 18(6), 1245–1262 (1989)
17. Takahashi, Y., Satoh, Y., Suzuki, H., Sasaki, S.: Recognition of largest common structural fragment among a variety of chemical structures. Analytical sciences 3, 23–28 (1987)
18. Valiente, G.: Constrained tree inclusion. J. Discrete Algorithms 3(2-4), 431–447 (2005)
19. Wagner, R.A., Fischer, M.J.: The string-to-string correction problem. J. ACM 21, 168–173 (1974)
20. Zhang, K.: Algorithm for the constrained editing problem between ordered labeled trees and related problems. Pattern Recognition 28, 463–478 (1995)
21. Zhang, K.: A Constrained Edit Distance Between Unordered Labeled Trees. Algorithmica 15(3), 205–222 (1996)

Exploiting Genre in Focused Crawling

Guilherme T. de Assis[1], Alberto H.F. Laender[1],
Marcos André Gonçalves[1], and Altigran S. da Silva[2]

[1] Computer Science Department, Federal University of Minas Gerais
31270-901 Belo Horizonte, MG, Brazil
{gtassis,laender,mgoncalv}@dcc.ufmg.br
[2] Computer Science Department, Federal University of Amazonas
69077-000 Manaus, AM, Brazil
alti@dcc.ufam.edu.br

Abstract. In this paper, we propose a novel approach to focused crawling that exploits genre and content-related information present in Web pages to guide the crawling process. The effectiveness, efficiency and scalability of this approach are demonstrated by a set of experiments involving the crawling of pages related to syllabi (genre) of computer science courses (content). The results of these experiments show that focused crawlers constructed according to our approach achieve levels of F1 superior to 92% (an average gain of 178% over traditional focused crawlers), requiring the analysis of no more than 60% of the visited pages in order to find 90% of the relevant pages (an average gain of 82% over traditional focused crawlers).

Keywords: Web crawling, Focused crawling, SVM classifiers.

1 Introduction

Focused crawlers or topical crawlers [2,7,9,10,13,15] are special purpose crawlers that serve to generate smaller and more restricted collections of Web pages. They have as their main goal to crawl pages that are, in the best possible way, relevant to a specific topic or user interest. Focused crawlers are important for a great variety of applications, such as digital libraries [14], competitive intelligence [11], and large Web directories [8], to name a few. Additionally, when compared with traditional crawlers used by general purpose search engines, they reduce the use of resources and scale, since they do not need to cover the entire Web.

The challenge of identifying specific and relevant sub-spaces of the Web, according to a theme, is usually carried out by means of appropriate heuristics, which direct the crawling process. Such a strategy involves determining how relevant a certain page is to a specific topic of interest. Most of the current strategies rely on text classifiers to determine such relevance [2,4,12,13], with the additional cost of having to train the classifiers. We take a different approach in this paper by explicitly considering genre (text style) and content-related aspects of a page, and by evaluating, according to these aspects, the relevance of this page to the information needs expressed by a user.

N. Ziviani and R. Baeza-Yates (Eds.): SPIRE 2007, LNCS 4726, pp. 62–73, 2007.
© Springer-Verlag Berlin Heidelberg 2007

As such, the work described in this paper has, as its main goal, to establish a framework to allow the construction of effective, efficient and scalable focused crawlers that take into consideration both the genre and the content of the desired pages. More specifically, we propose a focused crawling approach designed to situations in which the specific topic of interest can be expressed by two distinct sets of terms: the first expressing genre aspects of the desired pages, and the second one related to the subject or content of these pages. Examples where this happens include syllabi of specific courses, curricula vitae of professionals or job offers in a particular field, sale offers of specific products, software documentation, etc. Thus, distinctly from previous approaches in the literature [4,7,9,11,12,13], our approach exploits both the genre and the content of the Web pages to guide the crawling process.

In order to demonstrate the effectiveness, efficiency and scalability of the proposed approach, we conducted a set of experiments, involving topics related to the theme "education in the field of computing", more precisely syllabi of certain courses. As we shall see, the results of these experiments show that a focused crawler constructed according to our approach achieves F1 levels superior to 92%, which represents an average gain of 178% over traditional focused crawlers. In addition, our focused crawler required the analysis of no more than 60% of the visited pages in order to find 90% of the relevant pages, while traditional focused crawlers have to analyze at least 75% of the visited pages to achieve the same goal. Thus, the main contribution of this work is a novel approach to focused crawling that does not require a training phase and exploits genre and content-related aspects of the desired pages, which results in an effective, efficient and scalable strategy for focused crawling.

The rest of this paper is organized as follows. Section 2 addresses related work. Section 3 presents our proposed approach to focused crawling. Section 4 describes the set of experiments conducted to evaluate our approach and discusses the results obtained. Finally, Section 5 concludes and gives perspectives for future work.

2   Related Work

Since the first proposals for focused crawlers, such as FishSearch [3], a great variety of methods has been proposed for this purpose. According to Pant and Srinivasan [13], many crawling algorithms are variations of the Best-First crawler, in which a list of non-visited URLs is kept as a priority queue. Each non-visited URL has a score associated to it, which reflects the benefits of following this URL, determining, therefore, its priority to be visited. Variations of the Best-First crawler can be created changing the heuristics used to give scores to a URL in the priority queue. The Naive Best-First crawler [10] measures the relevance of a page to a specific topic by calculating the cosine between the vector of terms that represent that page and the vector of terms that represent the topic of interest. Then, it uses this result to estimate the benefit of following the URLs found in that page. FishSearch [3] uses a similar measure combined with the notion of depth limit, which prevents the exploration of paths leading to a sequence of irrelevant pages. SharkSearch [5] proposes a more sophisticated technique to classify the non-visited URLs, which is based on the anchor text, i.e., the text surrounding a link, and the inherited scores of ancestor pages to influence the scores of the non-visited URLs.

In [9], algorithmic aspects of focused crawlers are analyzed. For this purpose, a set of crawling algorithms considered as representatives in the literature was implemented, including: Breadth-First, PageRank and SharkSearch. Based on the evaluation of such crawling strategies, a new class of crawling algorithms, named InfoSpiders, was designed and implemented, presenting an improved performance. These algorithms include adaptive mechanisms, that is, mechanisms that promote the change of the crawler's behavior according to the context found during the crawling. It was concluded that a crawler that uses evolutionary algorithms reaches high scalability, due to the distribution of work through their concurrent agents, resulting in a better performance/cost ratio. Another distinct approach for focused crawling is proposed in [7]. This approach uses two probabilistic models, HMMs (Hidden Markov Models) and CRFs (Conditional Random Fields), to model the link structure and the content of documents leading to target pages by learning from user's topic-specific browsing. A focused crawler constructed according to this approach is treated as a random surfer, over an underlying Markov chain of hidden states, defined by the distance from the targets, from which the actual topics of a Web page are generated. Experimental results show that the proposed crawler often outperforms Best-First crawlers.

Another class of solutions for focused crawlers comes from Machine Learning and makes extensive use of classifiers. In this context, a classifier has the role of deciding whether a URL found during the page traversal is to be followed or not. According to [13], a traditional focused crawler guided by a classifier works as follows. Firstly, a priority queue is initialized with the URLs of the seed pages. Then, a non-visited URL is selected from the queue based on a score assigned by the classifier, and the corresponding page is fetched from the Web and abstractly represented in terms of its content (e.g., by a feature vector). At this moment, the abstract representation of the page is analyzed by a classifier and a score that measures the pertinence of the page to the desired topic is given. All URLs of the analyzed page are placed in the priority queue with the same score of this page. This process is repeated until there are no more URLs in the queue. Notice that, previously to the crawling process, the classifier must be trained with positive and negative examples of pages to be crawled.

The work described in [2] was the first one to use a classifier, in this case a Naive Bayes classifier, in order to guide a focused crawler. The basic idea behind such a crawler is to classify the crawled pages within the categories of a given taxonomy. The crawler, then, uses examples of URLs to create a Bayesian classifier, which is capable of determining the probability "$P(c—p)$" of a crawled page "p" to belong to a category "c" in the taxonomy. The work in [4] has proposed a variation of this crawler, naming it a context-focused crawler. Such a crawler uses a set of Naive Bayes classifiers that are trained to estimate the link distance between a crawled page and the relevant pages. In [12], the effects of various definitions of link contexts (i.e., terms that appear in the text around a hyperlink within a Web page) on the performance of SVM classifier-guided focused crawlers are investigated. The results show that a crawler that exploits words both in the immediate vicinity of a hyperlink as well as the entire parent page performs significantly better than a crawler that depends on just one of those clues. Moreover, a crawler that uses the tag tree hierarchy within Web pages provides effective coverage.

The work in [13] presents a systematic and comparative study involving experiments that explore many versions of the Naive Bayes, SVM and Neural Network classifying schemes. The experiments were performed in a collection of over 100 topics, therefore allowing statistically valid conclusions to be drawn. A crawling environment was designed and developed, which allowed new classifiers to be flexibly added. The results show that Naive Bayes is a weaker choice, when compared to SVM and Neural Network, to control a focused crawler. SVM seems to be the best choice among the three, since it performs similarly to a Neural Network, but requiring a lower training cost. Another framework to evaluate different focused crawling strategies is described in [15].

While previous work on focused crawling rely on a single concept space (i.e., the topic of the pages) for driving the crawling process, the use of other equally important concept spaces has been neglected in the literature, being the genre of the pages a notable example. The approach presented in this paper is, to best of our knowledge, a first attempt towards this direction.

3   Overview of the Proposed Approach to Focused Crawling

Our approach relies on the fact that some specific topics of interest can be represented by considering two separate aspects: its genre and content-related information. For instance, if a user wants to crawl Web pages that include syllabi of database courses, a focused crawler should analyze a specific Web page considering, separately, the terms present in that page that indicate that it corresponds to a syllabus (terms that characterize the course syllabus genre) and the terms that indicate that the page is related to the field of databases (terms that appear in a syllabus content of a database course). On the other hand, a traditional focused crawler guided by a classifier analyzes the content of a specific Web page, but does not consider separately the two kinds of information. Therefore, pages that include syllabi of other courses, as well any page referring to topics in the field of databases, could be selected by this crawler as being related to a database course (precision errors), whereas pages with genre and content poorly specified but including a syllabus of a database-related course would be classified as belonging to the "other" category (recall errors).

Thus, our focused crawling approach considers a set of heuristics to guide a crawler, in such way that it allows the separate analysis of the genre and the content of a Web page. In addition, it also considers the page URL string, since it may include terms related to both the genre and the content of the desired topic of interest. Our set of heuristics has been designed with two main objectives: improving the level of F1 of the crawling process and speeding up the crawling of relevant pages.

To crawl Web pages related to a specific topic of interest, we use the procedure FCrawl described in Fig. 1. This procedure takes as input sets of terms that represent the genre (*GenreTerms*) and the desired information content (*Content-Terms*), and outputs a set of relevant pages (*RelevantPages*). Eventually, another set of terms, representing the URL string (*URLTerms*) of a page related to this topic, can also be specified. Each URL string term is related to the page genre or to the desired content. This procedure also takes as input a set of URLs pointing to seed pages (*SeedPages*) and two additional parameters: (1) the similarity threshold (*SimilarityThreshold*), that indicates whether a specific visited page is relevant or not, and (2) the change threshold (*ChangeThreshold*), that indicates whether the score of a URL may be changed during the crawling process. A list of non-visited URLs is kept in a priority queue called *Frontier*. In addition, this procedure invokes the following pre-defined procedures and functions:

- InsertURL(*URL,score,URLtype*): inserts in *Frontier* a *URL* with a specific *score* and *URLtype* ("seed" or "non-seed");
- RemoveURL(*URL,URLtype*): removes from *Frontier* the URL with the highest score, returning such a *URL* and its *URLtype*;
- FetchParsePage(*URL,page,terms,links*): fetches and parses the page pointed by *URL*, returning the actual *page* and its sets of *terms* and *links*;
- CosineDistance(*page,terms*): returns the cosine distance between a *page* and a set of *terms*;
- Mean(*similarity1,similarity2,weight1,weight2*): returns the weighted mean between two similarity values according to weights assigned to them;
- ChangeScores(*URL,newscore*): changes to *newscore* the score of the URLs in *Frontier* that correspond to sibling pages of a given *URL*.

Procedure Fcrawl works as follows. Step 01 initializes *Frontier* with the URLs of the seed pages, setting the URL scores to 1. For each URL in *Frontier* (step 02), the corresponding page is visited (step 04) and its content analyzed (steps 05 to 09). The cosine similarity function is applied separately to each set of terms (steps 05, 06 and 08), generating a specific similarity score between the current page and the sets of terms that represent, respectively, the genre, the content and the URL string of the desired pages. Then, these scores are combined into a final single one (steps 07 and 09) and compared with a given threshold. If this final score is greater or equal to this threshold, the visited page is included in the set of relevant pages (step 10). As we shall see, determining the relevance of a page to a specific topic by considering separately pieces of evidence related to its genre and content is the main reason for the F1 levels we have achieved with our approach. Next, if the current page is a non-seed one, the final similarity score is compared with a second threshold to determine whether the scores of URLs in *Frontier* that correspond to children of the current page's parent may

```
Procedure FCrawl
Input:   GenreTerms, ContentTerms, URLTerms, SimilarityThreshold,
         ChangeThreshold, SeedPages;
Output:  RelevantPages;
Begin
    let GW, CW, GCW and UW be weights of importance for the genre, the
    content, the genre-content combination and the URL string respectively;
01  foreach (pageURL in SeedPages) do
        InsertURL(pageURL,1,"seed");
    end-foreach
02  while (there are non-visited URLs in Frontier) do
03    RemoveURL(CurrentURL,URLType);
04    FetchParsePage(CurrentURL,CurrentPage,PageTerms,PageLinks);
05    GenreSimilarity := CosineDistance(PageTerms,GenreTerms);
06    ContentSimilarity := CosineDistance(PageTerms,ContentTerms);
07    CombinedSimilarity := Mean(GenreSimilarity,ContentSimilarity,GW,CW);
08    URLSimilarity := CosineDistance(PageTerms,URLTerms);
09    FinalSimilarity := Mean(CombinedSimilarity,URLSimilarity,GCW,UW);
10    if (FinalSimilarity >= SimilarityThreshold) then
        RelevantPages := RelevantPages + CurrentPage;
      end-if
11    if (URLType = "non-seed" and FinalSimilarity >= ChangeThreshold) then
        ChangeScores(CurrentURL,FinalSimilarity);
      end-if
12    foreach (link in PageLinks) do
        InsertURL(link,0,"non-seed");
      end-foreach
    end-while
End
```

Fig. 1. The procedure FCrawl

be changed (step 11). Finally, the links previously extracted from the current page are inserted into *Frontier* (step 12) having their scores set to 0.

The strategy of dynamically changing the crawling priority of the non-visited pages is the reason for the performance of our crawling approach since it allows for relevant pages to be crawled as soon as possible. This occurs because in situations where a topic can be separately represented in terms of genre and content, we observed that the parent of a relevant page often contains other URLs that also point to relevant pages. On the other hand, a relevant page does not usually include a link to another relevant page. For example, considering that a crawler located a page that includes the syllabus of a database course, URLs of other relevant pages can be found in the page of the graduate program which that course is part of or in the page that includes the list of courses taught by a certain professor, whereas it is unlikely to find URLs of relevant pages in the actual page that includes the course syllabus. In fact, this strategy is a very common navigation pattern found on the Web [6].

4   Experimental Evaluation

4.1   Experimental Design and Setup

In order to conduct our experiments, we implemented a crawler based on the procedure FCrawl described in Fig. 1. In our experiments, this crawler was run in the context of the Brazilian Web with the purpose of crawling pages related to syllabi of specific computer science courses. Three disciplines with distinct characteristics were chosen as specific subjects: "databases", which is a discipline whose content is well-consolidated, "data structures", which is also a well-consolidated discipline but whose content may be dispersed into several courses, and "information retrieval", which is a discipline whose content is not yet well-consolidated due to its fast recent evolution.

Five seed pages of Brazilian higher education institutions were used. To represent the genre (course syllabi), we specified a set of 15 terms (or phrases) such as "syllabus", "course description", "instructor", "prerequisites", "tentative schedule", "topics" and "required text", no matter the discipline. For each chosen discipline, a specialist on the subject specified a set of approximately 20 terms to represent the content of its syllabus and another one with five terms to represent the URL string of good links to pages related to that discipline. Examples of terms used for representing the "databases" subject were "database", "data model", "entity-relationship", "relational model", "sql" and "normalization". It is important to notice that the specification of these sets of terms is a simple task when compared with the effort required to train a classifier. For instance, Fig. 2 illustrates a typical course syllabus page found on the Web. As we can see, the terms that characterize this page as such are easily recognized. Finally, for each discipline, an answer set containing the relevant pages among those visited by the crawler was also specified in order to allow us to assess the effectiveness of our crawler.

As known from conventional IR applications [1], in order to apply the cosine measure to evaluate a query against to a specific collection of documents, it is necessary to compute first, for each query term, an IDF (Inverse Document Frequency) value that indicates the importance of that term inside the collection. Thus, to be able to implement the function CosineDistance used by the procedure FCrawl, it would be necessary to compute IDF values for all terms specified by the specialists. However, our experiments have shown that calculating the IDF value for each term does not have a strong influence on the effectiveness of our approach, since it reaches similar F1 levels even when we consider the IDF value equal to 1 for all terms. This allows us to disregard the IDF value calculation step in our approach, which has obvious implications in terms of performance.

Thus, for this reason and due to the lack of space, in this paper we only report the results of the crawling processes that considered the IDF value equal to 1 for all terms. We will refer to these crawling processes as DS (topic "data structure course syllabi"), DB (topic "database course syllabi"), and IR (topic "information retrieval course syllabi").

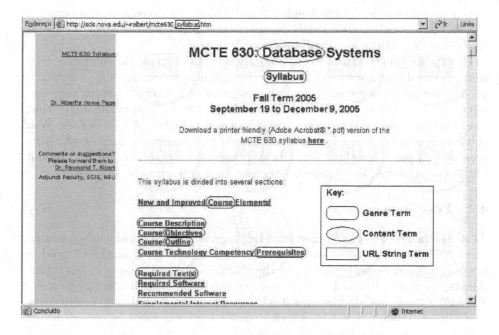

Fig. 2. A relevant page for the "databases" subject

Baselines. Classifiers are a natural option to guide a focused crawler. Therefore, in order to establish a baseline for comparing the results obtained with our approach, we developed a crawler guided by an SVM Radial Basis Function (RBF) classifier, one of the best classifiers for this kind of application according to [12], following the traditional focused crawling strategy described in Section 2. Thus, we considered as our baselines the following crawling processes executed with this crawler: RBF-DS (topic "data structure course syllabi"), RBF-DB (topic "database course syllabi"), and RBF-IR (topic "information retrieval course syllabi").

Each SVM RBF classifier was trained considering around 40 pages related to the desired topic and around 500,000 pages related to different topics, representing a sub-collection of the Brazilian Web. The F1 levels obtained by the crawling processes RBF-DS, RBF-DB and RBF-IR were, respectively, 46.48%, 45.83% and 23.26%.

Parameters Setting. According to steps 07 and 09 of the procedure FCrawl, our focused crawling approach performs arithmetic combinations among previously calculated similarity values. Each combination is performed, simply, by calculating the weighted average of the similarity scores involved. Thus, it was necessary to establish a "weight of importance" for each similarity value. We established such a weighting scheme experimentally, varying the values of the weights and analyzing the results obtained. With respect to step 07, the best results were reached considering the weight 5 to both genre (GW) and content (CW), since they are equally important. With respect to step 09, the best results were reached considering the weights 7 and 3 to the genre-content combination (*GCW*) and the URL string (*UW*) respectively, since for most pages the terms that form their URLs are usually not so relevant.

Finally, for practical reasons, in our experiments we made the following decisions: (1) we disregarded pages of certain Brazilian Web domains (e.g., ".gov.br") or that included some type of undesired content (e.g., ".exe"), which were not relevant to the context of the topics considered; (2) we limited the depth that the crawlers could reach within a site (maximum number of levels of the Web sub-graph of a site); (3) likewise, we limited the maximum number of URLs present in a page that should be followed; and (4) we standardized the content of the visited pages, by eliminating accents, capital letters and stop-words.

4.2 Effectiveness

As mentioned before, our focused crawling approach uses a set of guiding heuristics that has been designed with two main objectives: improving the F1 level of the crawling process and speeding up the crawling of relevant pages. In our experiments, we measured the F1 level after the execution of each crawling process. We also notice that, in the crawling processes we executed, our crawler visited almost 60,000 pages. Fig. 3 shows the F1 levels obtained by each crawling process, considering different threshold options. As we can see, for all topics, our focused crawler reached F1 levels superior to 92%, with different similarity thresholds. That is, even considering disciplines with distinct characteristics, the results achieved by our focused crawler are very good. Moreover, once the F1 value starts to diminish, it never surpasses its peak value again since the recall measure also starts to diminish, meaning that it is not necessary to vary the threshold anymore. The crawling process DS shows the worst results due to the fact that the content of the discipline "data structures" is dispersed in many distinct courses; thus, some syllabi have not been classified correctly because many content terms specified for this discipline did not appear in these syllabi.

We also notice that most crawling processes did not achieve 100% of F1. By manual inspection, we verified that this was mainly due to precision errors

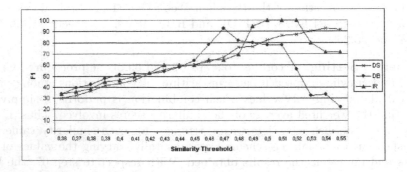

Fig. 3. F1 x Similarity Threshold (our approach)

because certain course syllabi included the content of other courses (e.g., a software engineering course included a database module). Moreover, some recall errors were due to ill designed syllabi (e.g., some database course syllabi did not show any term that could properly characterize the genre).

In addition, the results presented in Fig. 3 also show that our crawler clearly outperforms the crawler guided by the SVM RBF classifier, since it achieves, for all crawling processes, F1 levels that are significantly higher than the ones achieved by the crawler guided by the SVM RBF classifier (see Subsection 4.1).

4.3   Efficiency and Scalability

A key issue of any crawler is efficiency, i.e., its capability of crawling relevant pages as fast as possible. Fig. 4 shows the percentage of relevant pages retrieved in comparison with the percentage of visited pages during all crawling processes. For the crawling processes DS, DB and IR, we considered those with the similarity thresholds that achieved the best levels of F1. As we can see, our crawler is much more efficient than the crawler guided by the SVM RBF classifier since, for all crawling processes, it presents a better convergence to 100% of coverage of the relevant pages.

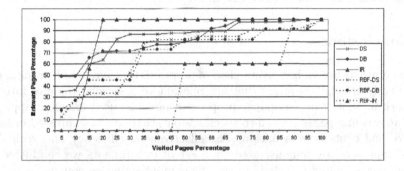

Fig. 4. Percentage of relevant pages x Percentage of visited pages

Considering the crawling processes executed according to our approach, a large percentage of relevant pages are retrieved with only 15% of the visited pages and, with 60% of the visited pages, at least 90% of all relevant pages are retrieved, which represents an average gain of 82% over the baseline crawling processes. This is a consequence of our strategy of dynamically changing the crawling priority of the non-visited pages, as discussed in Section 3. Once again, the crawling process DS shows the worst results due to the fact that the content of the discipline "data structures" is often scattered among pages from many distinct courses, which means that there would be more relevant pages to be crawled in other sites. On the other hand, the crawling process IR, that corresponds to a discipline not yet well-consolidated, shows the best results (only 20% of the visited pages required to find all relevant pages). This good performance might be due to the fact that there are few relevant pages on this subject to be crawled.

To evaluate the scalability of our approach, an analysis of the average amount of visited pages per relevant page retrieved was performed. The results are shown in Table 1, considering only the crawling processes that yielded the best results. As we can see, the baseline crawling processes present an average amount of visited pages per relevant page retrieved much higher than those presented by the crawling processes executed according to our approach.

Table 1. Amount of visited pages per relevant page retrieved

Crawling (our approach)	Average amount of visited pages	Crawling (baseline)	Average amount of visited pages
DS	710.55	RBF-DS	1675.85
DB	1105.40	RBF-DB	3984.00
IR	1199.00	RBF-IR	9178.00

All crawling processes were executed on the same computer. Our focused crawler took 3 to 4 hours to complete each crawling process, due to the simplicity of the set of heuristics it uses. On the other hand, the baseline crawling processes took 3 to 4 days to perform the same tasks; moreover, they required an additional training phase that took around 4 hours.

5   Conclusions

Focused crawlers are an important class of programs that have as their main goal to efficiently crawl Web pages that are relevant to a specific topic of interest. The work presented in this paper proposes a novel focused crawling approach aimed at crawling pages related to specific topics that can be expressed in terms of genre and content information. The effectiveness, efficiency and scalability of this approach are demonstrated by a set of experiments we conducted for crawling pages related to syllabi of specific computer science courses.

As shown by our experimental results, the major benefits of our focused crawling approach, compared with traditional ones based on classifiers, are: (1) improvement of the level of F1 in the crawling process, (2) more efficiency in the crawling process since only a small percentage of pages is required to be visited to crawl a large percentage of relevant pages, and (3) higher scalability since our approach adopts a simple set of heuristics to determine the relevance of a page as well as to guide the crawling, and does not require a training phase.

At the moment, we are conducting new experiments involving other domains for which the set of pages to be crawled can be expressed in terms of their genre and content (preliminary results with job and sale offers have also achieved high levels of F1). As future work, we intend to: (1) develop a strategy to facilitate the specification of the genre and content terms required to guide the crawling process, although the specification of these terms is not a complex task, and (2) devise a strategy to find the optimum similarity threshold to be used in a crawling process.

Acknowledgments. This research was partially funded by MCT/CNPq/CT-INFO projects GERINDO (grant number 552.087/02-5) and 5S-VQ (grant number 551.013/2005-2), by UOL (www.uol.com.br), through its UOL Bolsa Pesquisa program, process number 20060520105932, and by the authors' individual grants from CAPES and CNPq.

References

1. Baeza-Yates, R.A., Ribeiro-Neto, B.A.: Modern Information Retrieval. ACM Press/Addison-Wesley, New York (1999)
2. Chakrabarti, S., Berg, M., Dom, B.: Focused Crawling: A New Approach to Topic-Specific Web Resource Discovery. Journal of Computer Networks 31(11-16), 1623–1640 (1999)
3. De Bra, P.M.E., Post, R.D.J.: Information Retrieval in the World Wide Web: Making Client-Based Searching Feasible. Journal of Computer Networks and ISDN Systems 27(2), 183–192 (1994)
4. Diligenti, M., Coetzee, F., Lawrence, S., Giles, C.L., Gori, M.: Focused Crawling Using Context Graphs. In: Proc. 26th Int'l Conference on Very Large Data Bases, pp. 527–534 (2000)
5. Herscovici, M., Jacovi, M., Maarek, Y.S., Pelleg, D., Shtalhaim, M., Ur, S.: The Shark-Search Algorithm - An Application: Tailored Web Site Mapping. Journal of Computer Networks 30(1-7), 317–326 (1998)
6. Lage, J.P., Silva, A.S., Golgher, P.B., Laender, A.H.F.: Automatic Generation of Agents for Collecting Hidden Web Pages for Data Extraction. Data & Knowledge Engineering 49(2), 177–196 (2004)
7. Liu, H., Janssen, J.C.M., Milios, E.E.: Using HMM to Learn User Browsing Patterns for Focused Web Crawling. Data & Knowledge Engineering 59(2), 270–291 (2006)
8. McCallum, A., Nigam, K., Rennie, J., Seymore, K.: Automating the Construction of Internet Portals with Machine Learning. Journal of Information Retrieval 3(2), 127–163 (2000)
9. Menczer, F., Pant, G., Srinivasan, P.: Topical Web Crawlers: Evaluating Adaptive Algorithms. ACM Transactions on Internet Technology 4(4), 378–419 (2004)
10. Menczer, F., Pant, G., Srinivasan, P., Ruiz, M.E.: Evaluating Topic-driven Web Crawlers. In: Proc. 24th Annual Int'l ACM SIGIR Conference on Research and Development in Information Retrieval, pp. 241–249 (2001)
11. Pant, G., Menczer, F.: Topical Crawling for Business Intelligence. In: Koch, T., Sølvberg, I.T. (eds.) ECDL 2003. LNCS, vol. 2769, pp. 233–244. Springer, Heidelberg (2003)
12. Pant, G., Srinivasan, P.: Link Contexts in Classifier-Guided Topical Crawlers. IEEE Transactions on Knowledge and Data Engineering 18(1), 107–122 (2006)
13. Pant, G., Srinivasan, P.: Learning to Crawl: Comparing Classification Schemes. ACM Transactions on Information Systems 23(4), 430–462 (2005)
14. Pant, G., Tsioutsiouliklis, K., Johnson, J., Giles, C.L.: Panorama: Extending digital libraries with topical crawlers. In: Proc. 4th ACM/IEEE-CS Joint Conference on Digital Libraries, pp. 142–150 (2004)
15. Srinivasan, P., Menczer, F., Pant, G.: A General Evaluation Framework for Topical Crawlers. Journal of Information Retrieval 8(3), 417–447 (2005)

Admission Policies for Caches of Search Engine Results

Ricardo Baeza-Yates[1], Flavio Junqueira[1], Vassilis Plachouras[1],
and Hans Friedrich Witschel[2]

[1] Yahoo! Research, Barcelona, Spain
[2] University of Leipzig, Germany
{rby,fpj,vassilis}@yahoo-inc.com,
witschel@informatik.uni-leipzig.de

Abstract. This paper studies the impact of the tail of the query distribution on caches of Web search engines, and proposes a technique for achieving higher hit ratios compared to traditional heuristics such as LRU. The main problem we solve is the one of identifying infrequent queries, which cause a reduction on hit ratio because caching them often does not lead to hits. To mitigate this problem, we introduce a cache management policy that employs an admission policy to prevent infrequent queries from taking space of more frequent queries in the cache. The admission policy uses either stateless features, which depend only on the query, or stateful features based on usage information. The proposed management policy is more general than existing policies for caching of search engine results, and it is fully dynamic. The evaluation results on two different query logs show that our policy achieves higher hit ratios when compared to previously proposed cache management policies.

1 Introduction

Without search engines, finding new content on the Web is virtually impossible. Thus, a large number of users submit queries to search engines on a regular basis in search of content. As the number of users is large and the volume of data involved in processing a user query is high, it is necessary to design efficient mechanisms that enable engines to respond fast to as many queries as possible. An important mechanism of this kind is caching. In search engines, users submitting popular queries can benefit from a mechanism that stores the results for such queries in a cache memory, as the engine does not need to recompute results that are requested frequently enough. Caching query results then improves efficiency if the cached queries occur in the near future.

A cache comprises a memory space and an implementation of a cache management policy. As cache memories are limited in size, it is necessary to evict entries when the cache is full and there is a new entry to add. To evict entries, a cache has to implement an *eviction policy*. Such a policy ideally evicts entries that are unlikely to be a *hit*, *i.e.*, to be requested while in the cache. A simple and popular strategy is to evict the least recently used item from the cache [1]. This policy is known as LRU (Least Recently Used). In search engines, the entries in a cache can comprise, for example, query results and posting lists [2]. In this paper, we focus on query results. Henceforth, we say that a query is cached to denote that the results of the query are cached.

N. Ziviani and R. Baeza-Yates (Eds.): SPIRE 2007, LNCS 4726, pp. 74–85, 2007.

The problem with using only eviction policies in Web search engines is that the results for *all* queries are admitted to the cache, including those that will never appear again. Storing the results for these queries turns out to be fruitless since they cannot be cache hits. Until they are evicted from the cache, they "pollute" it in the sense that they use cache space, but they do not generate any cache hit. Since the frequency of queries follows a power law, there is a great percentage of queries that never appear again – at least not in a near future.

In this work, we propose the use of *admission policies* to prevent infrequent or even singleton queries from polluting the cache. Such policies are implemented in the form of an estimator that predicts whether a query is infrequent or whether it is frequent enough to be cached. The estimator can use either stateless features, which depend on each query, or stateful features, which are computed from usage information. We assume that there is a relation between the features and the future frequency of queries, even though we do not explicitly model it. To evaluate the efficiency of our policies, we use the hit ratio of the cache, instead of accuracy of frequency prediction.

Our cache management policy for search engine results comprises an eviction policy and an admission policy, and it divides the memory allocated for caching into two parts. The queries cached in the first part are the ones that the admission policy predicts as frequent or, more generally, likely to be a hit in the future. The remainder of the queries are cached in the second part of the memory. The experimental results from applying an admission policy, as described above, show that we obtain improvements over LRU and SDC, which is currently one of the best management policies for caching results in Web search engines.

Related work. The observation on filtering out infrequent queries has not been directly used in the design of caches for Web search engines so far. Markatos [3] investigated the effectiveness of caching for Web search engines. The reported results suggested that there are important efficiency benefits from using caches, due to the temporal locality in the query stream. Xie and O'Hallaron [4] also found that the distribution of query frequencies follows a Zipf distribution, very popular queries are issued by different users, and longer queries are less likely to be shared by many users.

On cache management policies, Lempel and Moran [5] proposed one that considers the probability distribution over all queries submitted by the users of a search engine. Fagni *et al.* [6] described a Static Dynamic Cache (SDC), where part of the cache is *read-only* or *static*, and it comprises a set of frequent queries from a past query log. The dynamic part is used for caching the queries that are not in the static part. This last work is the most closely related to the problem of polluting queries, because it actively protects the most frequent queries and always keeps them cached. The solution of SDC, however, only addresses the problem of polluting queries indirectly, while it introduces the constraint of the static cache.

In a different context, on caching of memory pages in operating systems, there has been work on splitting the memory available for caching according to the type of traffic. For example, the adaptive replacement cache (ARC) [7] uses two lists, one for recent references and one for frequent references. One the main features of ARC is that it adapts the lengths of the two lists. Compared to ARC, we instead try to separate very frequent references from infrequent ones, and we investigate different policies.

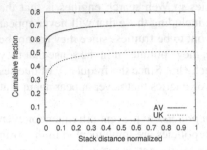

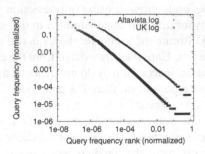

Fig. 1. Measure of temporal locality   **Fig. 2.** Query frequency distributions

Using multiple levels of caches for storing query results and postings of query terms [8] or even intersections of posting lists for query terms [9] result in improved hit ratio values (the hit ratio is the total number of hits over the total query volume). In this work we focus on caching of query results and we do not consider the problem of caching posting lists.

Finally, caching has been considered for Web applications that generate content dynamically [10,11]. One main difference from our work to the work of Sivasubramanian *et al.* and Olston *et al.* is their focus on applications that access one or more databases. One of the main challenges is to maintain database consistency as transactions change the state of the backend database. Search does not have a similar constraint because queries do not change the state of the data structures used to compute results. For federated databases, Malik *et al.* [12] have used admission policies for caching to minimize network communication overhead. Although they use the same concept of admission policy, the policies they use are different compared to the ones we propose.

2   Data Characterization

We use two query logs in our experiments. The first one corresponds to queries submitted to the Altavista Web search engine during a week in autumn 2001. The second query log corresponds to queries submitted to yahoo.co.uk during one year.

The queries in both logs have very similar length characteristics. The average length in characters and words of queries in the Altavista log are 17 and 2.6, respectively. The queries in the UK log consist on average of 17 characters and 2.5 words.

For caching, temporal locality is an essential property because if consecutive occurrences of the same query happen close together in time, then eviction policies can be highly efficient. To measure and compare temporal locality of the two query logs, we use a traditional technique that consists of computing the stack distance between two occurrences of the same query [13]. The stack abstraction works as follows. When processing a query stream, we process one query at a time. For each new query q, if q is not in the stack, then we push it onto the stack, and count it as an infinite stack distance. Otherwise, suppose that d is the depth of the query q in the stack, remove q from its current place in the stack and push it onto the top. The stack distance in this case is d.

Figure 1 shows the cumulative stack distance distribution for normalized distance values of both query logs. We use the total volume of each log to normalize the stack distance of the corresponding log. Note, however, that the total volume of the Altavista log is smaller compared to the total volume of the UK log. Thus, the same normalized value corresponds to a smaller absolute value in the Altavista log.

In this graph, there are two important observations. First, the highest probability is 0.7 for the Altavista log and 0.5 for the UK log. These probability values are not 1.0 because of singleton queries and compulsory misses. For such queries, the stack distance is infinite. Thus, in the Altavista log, we observe that a larger fraction of the query volume comprises repeated queries. Second, for the same distance value, the cumulative probability is higher for the Altavista log. If we pick $x = 0.1$ as an example, then the difference between the cumulative probabilities is roughly 0.2. Note that the difference between the two curves is roughly the same across the values in the x range. It happens because the difference for the maximum value of x is 0.2 and the increments are small for values of x larger than 0.1 due to the small fraction of large distance values.

As the stack distance between two consecutive occurrences of the same query has a higher probability of being short for the Altavista log, we conclude that the Altavista log presents significantly more temporal locality compared to the UK log. This is not surprising because of the higher volume of the UK log and its time span.

The actual frequency distributions of queries of the two logs, shown in Figure 2, confirm the conclusions above. From the figure, both distributions follow a power law distribution: the distribution for the Altavista log has slope -1.71 and the distribution for the UK log has slope -1.84. For the Altavista log, the singleton queries, which appear only once, correspond to 19% of the total query volume and to 62% of the unique queries. A cache that has an infinite amount of memory to store the results for all observed queries without any eviction, would achieve a hit ratio of 70.21%. For the UK query log, the singleton queries correspond to a higher percentage of the total volume of queries. More specifically, the set of singleton queries comprise 44% of the total volume and 88% of the unique queries. An infinite cache achieves 50.41% hit ratio. In the next section, we elaborate on the impact of the infrequent queries on caching.

3 Polluting Queries

A predominant characteristic of query streams is the presence of infrequent queries, in particular of queries that appear just once. For such queries, caching is often not effective because either there is a large number of other queries separating consecutive occurrences, or simply they never occur again. Moreover, using cache space for such a query might imply evicting the results of another more frequent query, thus increasing the number of cache misses. The basic idea of our approach consists of determining which queries are infrequent ones, and caching them in a separate part of the memory.

Table 1 illustrates the benefits of our approach. We compare two different caching policies: the optimal-admission policy and the least recently used (LRU) policy. The optimal-admission policy knows when a query will never appear again, and it does not cache such a query. On the other hand, if a query will appear again in the future, then the policy accepts it and it uses the LRU policy to determine which query to evict, assuming the cache is already full.

From the table, we observe that for the Altavista log the LRU policy achieves a high hit ratio already (between 59% and 65%), but the optimal-admission policy is still able to obtain even higher values (between 67% and 70%). For the UK log, however, using the optimal-admission policy enables an absolute increase in the hit ratio of over 10% in all three cases. This difference is mainly due to the amount of temporal locality in these logs, observed in the previous section. As the UK log presents less temporal locality, the policy controlling admission has a higher impact.

Table 1. Hit-ratio (%) comparison between the optimal-admission policy and LRU for different cache sizes in number of queries

Cache size	Optimal		LRU	
	AV	UK	AV	UK
50k	67.49	32.46	59.97	17.58
100k	69.23	36.36	62.24	21.08
250k	70.21	41.34	65.14	26.65

These results show that if we design heuristics that accurately determine which queries do not occur frequently, then we can improve significantly the hit ratio of caches for query results.

4   Admission Policies and Caching

This section describes a family of cache management policies (AC) that use an admission policy to separate infrequent queries from frequent ones. There are different ways of making a module implementing an admission policy interact with the cache. For example, we can have each query q sequentially evaluated: the admission policy first evaluates q, and, depending on the outcome, the cache processes the query in different ways. Alternatively, the admission policy evaluates the query and the cache verifies if the results are stored in parallel. The cache, however, has to wait for the outcome of the admission policy evaluation before applying its eviction policy. In this paper, we use the former, as the latter is mainly an optimization.

The proposed cache has two fully-dynamic parts. The first part is an admission-controlled cache. We call this part *controlled cache* (CC), because it only admits those queries that the admission policy classifies as future cache hits. All queries the admission policy rejects are admitted to the second part of the cache, which we call *uncontrolled cache* (UC). In our experiments in the next section, the uncontrolled cache implements a regular cache, more specifically LRU. We also use LRU for the controlled cache. Figure 3 shows how a query stream is processed with AC.

The rationale behind this management policy is similar to that of SDC [6], but it makes the first part of the cache more flexible and dynamic rather than static: the controlled part will contain those queries that are likely to be hits. Ideally, with a perfect

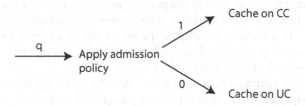

Fig. 3. Sequence of actions performed when a new query q arrives at a cache

admission policy, there would be no need for the uncontrolled cache. However, implementing such an admission control policy is in general difficult. More specifically, if this admission control policy is based, for example, on frequency, then its estimation on hits and misses is not perfect due to the temporal locality of some infrequent queries.

The uncontrolled cache can therefore handle queries that are infrequent, but appear in short bursts. Recall that the admission policy will reject queries that it determines to be infrequent. Infrequent (or unpopular) queries then may be asked again by the same user and within a short period of time. The uncontrolled cache handles these cases. Thus, AC guarantees that fewer infrequent queries enter the controlled cache, which is expected to handle temporal locality better.

To decide upon which queries to cache in the controlled part and the ones to cache in the uncontrolled part, we use an admission policy. Given a stream of queries Q_s, an admission policy is a function $f : Q_s \rightarrow \{0, 1\}$ that decides, for each query $q \in Q_s$ of the stream, whether it should be cached in the controlled part ($f(q) = 1$) or in the uncontrolled part ($f(q) = 0$). The generality of AC lies in the fact that various admission policies can be applied. For example, SDC is now a special case, which results from the following admission policy:

$$f(q) = \begin{cases} 1 \text{ if } q \text{ is among the } |CC| \text{ most frequent queries in a given training set} \\ 0 \text{ else} \end{cases} \tag{1}$$

With this admission policy, CC is static (no query ever needs to be evicted) and after some time, it will contain the same queries that would be in the static part of an SDC cache.

Now, other admission policies – that admit more queries – can be used to make CC more dynamic and hence more effective. To design a good admission policy, one needs to think of *features* that may be useful in distinguishing future cache hits from future misses. A feature in this context is some property of a query that the admission policy uses to determine in which part of the memory to cache it. Some options for such features will be discussed in the next sections.

4.1   Stateful Features

Stateful features are based on historical usage information of a search engine, and in general, they require extra memory space to hold statistics. Typically, these statistics are related to the frequency of query substrings (*e.g.*, words, *n-grams*) or the whole query. For example, a simple admission policy with a "stateful" feature admits all queries whose frequency in a training set of past queries is above a threshold k. It may be necessary to tune this threshold before applying the policy. In addition, the frequency statistics may need to be updated in regular intervals.

With respect to the extra amount of memory needed for the frequency-based feature, there are three important observations. First, this amount of memory is small compared to the total cache memory. For example, if the cache stores 20Kbytes for 100K queries, then it requires approximately 2GB of memory, while the amount of memory required for the state of the Altavista query log, corresponding to 1.4 million queries, is approximately 30MB, which is less than 2% of the space used for the cache. Second, we did not try to make the data structure holding frequency information efficient with respect to space. In fact, the 30MB value includes all the queries in the training set, although we do not need to keep all of them depending on the value of the frequency threshold. Alternatively, one can use a more space-efficient representation for queries, such as Bloom filters or other hash-based schemes. Hence, the 30MB value is an upper bound in the case of the Altavista query log. Even considering this additional amount of memory for LRU and SDC, preliminary experiments have shown that the resulting increase in hit ratio is negligible, as we have verified. Third, search engines often maintain statistics about the query stream for reasons such as improving the quality of the results returned. In this case, the information the feature requires may be readily available, and no extra amount of memory is necessary. For these reasons, we decided not to consider this extra amount of memory for LRU and SDC in our experiments.

4.2 Stateless Features

A stateless feature is a feature that can be readily computed from the query stream itself, without making use of collected information. The advantage of stateless features is that they neither require keeping track of statistics (and hence no update over time) nor memory space for storing such information.

Examples of potentially useful stateless features include the length of a query (in characters or words) or the number of non-alphanumerical characters in the query. The idea behind these features is that long queries or those containing many non-alphanumerical characters have lower probability of being popular. The corresponding admission policies require a threshold k. Polluting queries are then queries longer than k words or characters, or containing more than k non-alphanumerical characters.

5 Evaluation

In this section, we evaluate AC. We describe in detail the experimental setting, and then we present our results.

5.1 Experimental Setting

To evaluate the performance of AC, we conduct a number of experiments using the logs described in Section 2. More precisely, we use three different admission policies and compare them to an LRU baseline cache without admission policy and an SDC cache. In SDC, we have a hit when a query exists in either its static or its dynamic part. In the case of a miss, the query is introduced in the dynamic part. Here, the dynamic part of SDC implements LRU.

We divide both logs into a training set of queries M, which we use to train an admission policy, and a test set T. The training set consists of the same absolute number (namely 4.8 million) of queries in both cases. For our experiments, such a number of queries is sufficient to obtain reliable frequency estimates and a test set that is large enough for experiments with the Altavista log. Fagni *et al.* use the same two thirds/one third split for the Altavista log [6]. We do not consider pages of results in our experiments. That is, we consider two queries requesting different pages of results as the same query. The underlying assumption is that it is not expensive to cache all (or at least enough) results for each query in order to satisfy all these different requests.

The performance of a cache was measured by computing the hit ratio of a cache simulator on the test set T. Since SDC needs to be started with a warm cache, all caches start the evaluation phase warm. For SDC, we warm (cf. [6], section 5) its static part by populating it with the most frequent queries from M, and we warm its dynamic part by submitting the remaining queries to its dynamic part. For LRU and AC, we run the policy on the training set to warm their respective caches. In all cases, however, we count hits only for the test set T. We also report the hit ratio of the test set of an infinite cache warmed with the corresponding training set. Note that the hit ratio values for an infinite cache are different from the ones we report in Section 2 due to the split between test and training.

For the Altavista log, we use caches of size 50K and 100K and for the UK log of size 100K and 500K. For AC and SDC, the ratio of the controlled vs. uncontrolled (or static vs. dynamic) parts varies according to the parameters of the experiment. This is because the ratio that gives the highest hit ratio depends on the cache size and on properties of the query log.

All admission policies that we experiment with consist of just one feature and a corresponding threshold. Among these, there is one stateful feature, namely the frequency of the query in M (*PastF*), and two stateless features, namely the length of the query in characters (*LenC*) and words (*LenW*).

5.2 Results

Here we present the evaluation results for our proposed admission-controlled dynamic caches, using both the stateful and the stateless features. In each table, we present the results from the best split between static and dynamic cache for SDC, and controlled and uncontrolled cache for AC.

We start with the stateful feature *PastF*. The results are shown in Table 2 and Figure 4. When we use the past frequency to decide whether to admit queries, we obtain improved hit ratios over SDC. We can see that small caches require higher frequency thresholds because small caches can only effectively store queries that appear frequently enough so that they are not evicted. In the case of the AC cache with a capacity of 50K on the Altavista log, we can see that the best thresholds k_f are 7 and 8, where k_f is the frequency threshold.

Table 2. Hit ratios (%) for the Altavista and the UK query logs using AC, where a query is admitted in the controlled part of the cache if its frequency in the past is greater than k_f. The last row shows the percent of the LRU hit ratio relative to the hit ratio of an infinite cache.

	AV			AV			UK
Infinite	72.32		Infinite	72.32		Infinite	51.78
Sizes	50K		Sizes	100K		Sizes	100K 500K
LRU	59.49		LRU	61.88		LRU	21.03 30.96
SDC	62.25		SDC	64.49		SDC	29.61 35.91
AC $k_f = 6$	63.16		AC $k_f = 1$	65.04		AC $k_f = 0$	28.55 **37.45**
AC $k_f = 7$	63.19		AC $k_f = 2$	65.32		AC $k_f = 1$	29.94 34.62
AC $k_f = 8$	**63.19**		AC $k_f = 3$	**65.39**		AC $k_f = 2$	**30.28** 32.00
AC $k_f = 9$	63.16		AC $k_f = 4$	65.35		AC $k_f = 3$	29.32 30.28
AC $k_f = 10$	63.08		AC $k_f = 5$	65.22		AC $k_f = 4$	27.94 29.00
LRU/Infinite	82.26		LRU/Infinite	85.56		LRU/Infinite	40.61 59.79

Fig. 4. Hit ratios of the stateful feature for the Altavista and UK query logs

Table 3. Hit ratios (%) for the Altavista and the UK query logs using AC, where a query is not admitted to the controlled part of the cache if its length in characters (words) is greater than k_c (k_w). The last row shows the percent of the LRU hit ratio relative to the hit ratio of an infinite cache.

	AV		UK	
Infinite	72.32		51.78	
Sizes	50K	100K	100K	500K
LRU	59.49	61.88	21.03	30.96
SDC	**62.25**	**64.49**	**29.61**	**35.91**
AC $k_c = 10$	60.01	59.53	17.07	27.33
AC $k_c = 20$	58.05	62.36	22.85	32.35
AC $k_c = 30$	56.73	61.91	21.60	31.06
AC $k_c = 40$	56.39	61.68	21.19	30.53
AC $k_w = 2$	59.92	62.33	23.10	32.50
AC $k_w = 3$	59.55	61.96	21.94	31.47
AC $k_w = 4$	59.18	61.60	21.16	30.51
AC $k_w = 5$	59.01	61.43	20.81	30.02
LRU/Infinite	82.26	85.56	40.61	59.79

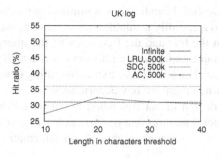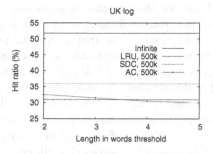

Fig. 5. Hit ratios of the stateless features *LenC* and *LenW* for the UK query log and an AC cache of 500K

Next we consider the stateless features. Table 3 presents the results when the AC cache uses the length in characters or words of the query, to predict whether a query is worth caching. In all cases, AC outperforms baseline LRU. The threshold that resulted in the best performance is $k_c = 20$, except for the case of the Altavista log with an AC cache of 50K. The results are similar when we use the length in words of queries, and the best threshold value is $k_w = 2$. Figure 5 shows the resulting hit ratios for the two features, *LenC* and *LenW*, for an AC cache of 500K used with the UK log.

Compared to LRU, AC achieves a higher performance because it is able to filter out some queries that pollute the cache. However, the stateless features are not as efficient as a predictor based on frequency, and consequently they do not outperform SDC. In the next section, we discuss further the advantages of using such features, as well as other possibilities for features that may improve performance.

6   On the Design of New Features

Table 1 shows that identifying perfectly if there will be another occurrence of a given query increases the hit ratio significantly. Thus, there is an opportunity for selecting features that can approximate an optimal admission policy. As the results of the previous section show, an admission policy using the stateful feature based on the past frequency of queries outperforms SDC. The stateless features *LenC* and *LenW* we have selected, however, were not sufficient to outperform SDC, although they still outperform LRU.

Although the stateful feature *PastF* performs well, there are two main issues with this feature. First, it requires the system to maintain the frequency of past queries, and consequently to use more memory space. Second, in very dynamic environments, the past may not correspond to the current query traffic, thus leading to poor performance. Thus, in settings with tight memory constraints or that rapidly change the distribution of incoming queries, stateless features are more appropriate. Designing stateless features that perform well proved not to be an easy task, however. We have presented only two stateless features, but we have experimented with more features which gave similar results. It is hence an open problem if there exists a feature (or combination of features) that can achieve a performance as good as the one of stateful features.

There are other stateless features that are potential candidates for improving the hit ratio, in particular features that benefit from data readily available in a search engine. Two such cases are the frequency of terms in the text and the frequency of the query in the text. The former needs extra memory, but that frequency information is usually already available in the index of a search engine. The latter needs extra computing time, but could be estimated quickly, in particular when there are few occurrences.

Another interesting open problem is the design of stateful features not based on frequency. It is not clear whether there are stateful features that perform at least as well as features based on frequency. For example, user sessions might contain information useful to determine queries that will be hits.

7 Conclusions

We have shown that a simple admission policy improves upon two well-known cache management policies: LRU and SDC. An important observation is that even a small improvement in hit ratio represents an important increase in hits for large query logs. For example, if we assume a fraction of 0.01 more hits out of 10 million queries, then we have 10 thousand more hits. Moreover, as we approach the hit ratio for an infinite cache, every small improvement is significant.

Our best admission policy uses the past frequency of queries to predict their frequency in the future. To compute the past frequencies, we considered an initial training period. Alternatively, one can use a sliding window scheme, and maintain frequency information on the current window instead of a previous training period. As for the benefits of our scheme with this feature, we obtained:

1. A relative improvement of 6% over LRU and 5% over SDC for the UK log with a cache holding results for 500k queries;
2. A relative improvement of 21% over LRU and 4% over SDC for the Altavista log with a cache holding results for 100k queries.

We have also experimented with an interesting set of policies that do not require maintaining usage information. These policies are interesting because they improved over LRU and they consume less memory resources. It is part of future work to determine if there exists such a policy based on stateless features that outperforms SDC.

There are plenty of open problems for future work. Among them we can mention:

- Combining different features to optimize the hit ratio using different machine learning techniques. This means optimizing the parameters and the weights of features;
- Using efficiently the space for different parts of the cache memory, as well as for information relevant to features;
- Modeling the behavior of such cache policies along with all other parts of a search engine to evaluate the trade-offs, in particular with respect to the infrastructure cost;
- Defining for each feature the function that establishes the relation between its threshold value and the cache size.

References

1. Denning, P.J.: Virtual memory. ACM Computing Surveys 2, 153–189 (1970)
2. Baeza-Yates, R., Gionis, A., Junqueira, F., Murdock, V., Plachouras, V., Silvestri, F.: The Impact of Caching on Search Engines. In: Proceedings of the 30th ACM SIGIR Conference, ACM Press, New York (2007)
3. Markatos, E.P.: On caching search engine query results. Computer Communications 24, 137–143 (2001)
4. Xie, Y., O'Hallaron, D.R.: Locality in search engine queries and its implications for caching. In: INFOCOM (2002)
5. Lempel, R., Moran, S.: Predictive caching and prefetching of query results in search engines. In: Proceedings of the 12th WWW Conference, pp. 19–28 (2003)
6. Fagni, T., Perego, R., Silvestri, F., Orlando, S.: Boosting the performance of Web search engines: Caching and prefetching query results by exploiting historical usage data. ACM Transactions on Information Systems 24, 51–78 (2006)
7. Megiddo, N., Modha, D.S.: Outperforming LRU with an adaptive replacement cache algorithm. IEEE Computer 37, 58–65 (2004)
8. Saraiva, P.C., de Moura, E.S., Ziviani, N., Meira, W., Fonseca, R., Riberio-Neto, B.: Rank-preserving two-level caching for scalable search engines. In: Proceedings of the 24th ACM SIGIR Conference, pp. 51–58. ACM Press, New York (2001)
9. Long, X., Suel, T.: Three-level caching for efficient query processing in large web search engines. In: Proceedings of the 14th WWW Conference, pp. 257–266 (2005)
10. Sivasubramanian, S., Pierre, G., van Steen, M., Alonso, G.: Analysis of caching and replication strategies for Web applications. IEEE Internet Computing 11, 60–66 (2007)
11. Olston, C., Manjhi, A., Garrod, C., Ailamaki, A., Maggs, B., Mowry, T.: A scalability service for dynamic Web applications. In: CIDR, Asilomar, California, USA, pp. 56–69 (2005)
12. Malik, T., Burns, R., Chaudhary, A.: Bypass Caching: Making Scientific Databases Good Network Citizens. In: ICDE, pp. 94–105 (2005)
13. Brehob, M., Enbody, R.: An analytical model of locality and caching. Technical Report MSU-CSE-99-31, Michigan State University (1999)

A Pocket Guide to Web History

Klaus Berberich, Srikanta Bedathur, and Gerhard Weikum

Max-Planck Institute for Informatics, Saarbrücken, Germany
{kberberi,bedathur,weikum}@mpi-inf.mpg.de

Abstract. Web archives like the Internet Archive preserve the evolutionary history of large portions of the Web. Access to them, however, is still via rather limited interfaces – a search functionality is often missing or ignores the time axis. *Time-travel search* alleviates this shortcoming by enriching keyword queries with a time-context of interest. In order to be effective, time-travel queries require *historical PageRank scores*. In this paper, we address this requirement and propose rank synopses as a novel structure to compactly represent and reconstruct historical PageRank scores. Rank synopses can reconstruct the PageRank score of a web page as of *any point during its lifetime*, even in the *absence* of a snapshot of the Web as of that time. We further devise a normalization scheme for PageRank scores to make them comparable across different graphs. Through a comprehensive evaluation over different datasets, we demonstrate the accuracy and space-economy of the proposed methods.

1 Introduction

The World Wide Web is increasingly becoming a key source of information pertaining not only to business and entertainment but also to a spectrum of sciences, culture, and politics. The *evolutionary history* of the Web, an even greater source of information, is preserved by web archives like the Internet Archive [1]. Access to these archival collections is nowadays mostly restricted to per-URL lookups; a comprehensive search functionality is often missing or ignores the time axis completely.

Time-travel search alleviates this shortcoming by extending the standard keyword querying with the inclusion of a time-context of interest. The keyword query is then evaluated over the state of the archive as of the specified time-context, i.e., only web page versions that existed during the specified time context are considered during query evaluation. Examples of such time-travel queries include:

 (i) *"Olympic Games" as of August 15th 2004* retrieves contemporary coverage about the 2004 Olympic Games in Athens.
(ii) *"Indiana Court Rules" as of May 18th 1999* finds historical statutes from the Indiana judiciary.

Recently, we addressed the problem of supporting time-travel queries with IR-style relevance measures [12]. In order to obtain high-quality results for these searches, it is necessary to combine IR-style relevance measures with authority measures like PageRank. Clearly, current PageRank scores cannot be used directly, since they do not appropriately reflect the web pages' authority as of the given time in the past. Instead, one would like to use *historical PageRank scores* as of the specified time.

N. Ziviani and R. Baeza-Yates (Eds.): SPIRE 2007, LNCS 4726, pp. 86–97, 2007.

One naïve solution to the problem would be to precompute and maintain full listings of PageRank scores for a large number of observation times. Based on the current conservative estimates on the size of the Web more than 1 Terabyte would be needed to store monthly rankings for a period of just 5 years. While this is still manageable in terms of storage, it is unclear i) how these rankings can be accessed efficiently at time-travel query-processing time, and ii) how to deal with time contexts for which no ranking was precomputed.

In this paper, we propose *rank synopses* as a novel structure to compactly represent and reconstruct historical PageRank scores. Rank synopses enable the accurate reconstruction of a web page's PageRank score as of any time (during its lifespan), even if no graph snapshot is available for that time, or, if PageRank scores are recomputed only infrequently. Rank synopses require 20%-50% the amount of storage required for retaining the full precomputed rankings and introduce only minimal disruptions in the reconstructed rankings.

Rank synopses are constructed from time series of PageRank scores. One obstacle when dealing with these time series is the incomparability of PageRank scores across different graph snapshots. Due to PageRank's probabilistic foundation and the fact that every node in the graph is guaranteed to be assigned a non-zero score, PageRank scores are sensitive even to non-local changes of the graph structure, such as the addition of a completely disconnected component. This leads to arbitrary fluctuations of PageRank scores across graph snapshots, deteriorating the performance of rank synopses. We overcome this by devising a computationally efficient score-normalization technique making PageRank scores comparable across different graph snapshots.

In summary, the primary contributions[1] of this paper are the following:

1. We present a novel *rank synopses* structure to compactly represent the evolution of PageRank scores and reconstruct historical PageRank scores with high accuracy.
2. We introduce a new normalization technique for PageRank that makes scores comparable across different graph snapshots.
3. Through experiments over a variety of datasets, we demonstrate the accuracy and space-economy of the presented techniques.

The remainder of this paper is organized as follows. Section 2 puts rank synopses in context with related work. The normalization technique for PageRank scores is described in Section 3. Section 4 introduces rank synopses, which are then evaluated in Section 5. Finally, in Section 6 we conclude the present work.

2 Related Work

This section gives a short overview of work that is related to the rank synopses presented in this paper. The *Web's dynamics* have attracted significant attention recently. Two large-scale studies [20,29] examined the evolution of its content and link structure. Ntoulas et al. [29] observed for the link structure that as much as 25% of new links are created every week, and that 80% of links are replaced over the period of one year. *Link*

---

[1] Specific aspects of this work have appeared as posters in CIKM'06 [10] and WWW'07 [11].

analysis has attracted a lot of attention [14,16,26] since the seminal PageRank [30] and HITS [24] methods were proposed. Several methods have been proposed [6,7,8,13,33] that bring together the Web's dynamics and link analysis by integrating temporal features of the web graph into link-analysis techniques. The *normalization of scores produced by link-analysis techniques* has only seldomly been addressed and has mostly been dealt with in an ad-hoc manner, e.g., by [21] who scale PageRank scores using the size of the underlying graph. One noticeable exception is [25] whose focus, however, is on centrality measures known from graph theory and not on PageRank and HITS.

The need for *web archives* has first been described by Kahle [22] who co-founded the Internet Archive – the biggest endeavor in this regard. Currently, web archives typically support only page-by-page access as implemented, for instance, by the Wayback Machine of the Internet Archive [1]. Comprehensive keyword-search functionality (as offered by today's web search-engines) is mostly missing for web archives. Only recently efforts to implement such functionality have been started with the open source NutchWAX [2] project being one notable effort.

Rank synopses are technically related to *time-series segmentation* techniques [23,32]. These methods are, however, aimed at accurately reconstructing individual time-series observations. Rank synopses, in contrast, aim at accurately reconstructing a ranking for a given time instance.

3   PageRank Score Normalization

PageRank is a well known and widely adopted link-based ranking technique. Given a directed graph $G(V, E)$ representing the link graph of the Web, the PageRank score $r(v)$ of a node v is defined as follows:

$$r(v) = (1 - \epsilon) \left(\sum_{(u,v) \in E} \frac{r(u)}{out(u)} \right) + \frac{\epsilon}{|V|} \qquad (1)$$

with $out(u)$ denoting the out-degree of node u and ϵ being the probability of making a random jump. As a consequence of its probabilistic foundation, PageRank scores are *not comparable across different graphs* as the following example demonstrates.

Consider the grey node in the two graphs shown in Figure 1. Intuitively, importance of neither the grey node nor the white nodes should decrease through the addition of

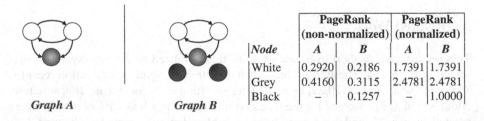

Node	PageRank (non-normalized) A	B	PageRank (normalized) A	B
White	0.2920	0.2186	1.7391	1.7391
Grey	0.4160	0.3115	2.4781	2.4781
Black	–	0.1257	–	1.0000

Graph A   *Graph B*

Fig. 1. PageRank Score Incomparability ($\epsilon = 0.15$)

the two black nodes, since none of these nodes are "affected" by the graph change. The non-normalized PageRank scores, however, as given in the corresponding table in Figure 1 convey a decrease in the importance of the grey node and the white nodes, thus contradicting intuition. These decreases are due to the random jump inherent to Page-Rank that guarantees the additional black nodes to have non-zero visiting probability. In contrast, using our normalization scheme, the PageRank scores of the grey and white nodes are retained across two snapshots of the graph.

Referring to Equation 1, we can see that the PageRank score of any node in the graph is lower bounded by $r_{low} = \frac{\epsilon}{|V|}$, which is the score assigned to a node without incoming edges. However, this definition does not account for dangling nodes (i.e., nodes without any outgoing edges) – which are shown to form a significant portion of the web graph crawled by search engines [19]. These nodes are treated by making a random jump whenever the random walk enters a dangling node. Under this model, with $D \subseteq V$ denoting the set of dangling nodes, the modified lower bound for PageRank scores is given by

$$r_{low} = \frac{1}{|V|}(\epsilon + (1 - \epsilon) \sum_{d \in D} r(d)) , \qquad (2)$$

which is again the score assigned to a node without incoming edges. We use this refined lower bound for normalizing PageRank scores – for a node v its normalized PageRank score is defined as

$$\hat{r}(v) = \frac{r(v)}{r_{low}} . \qquad (3)$$

In contrast to standard PageRank scores that correspond to *visiting probabilities* on the graph and thus depend on its size, the normalized PageRank scores convey *how much more likely a node is to be visited than a node having least possible importance*. Thus the normalization eliminates the dependence on the size of the graph. The computational cost associated with the proposed normalization is low: one pass over the dangling nodes is needed to sum up their scores (this can be done using one extra iteration in the power method that is commonly used to compute PageRank scores) and one additional pass over all nodes is required to normalize their scores. More details about the normalization technique, including a proof of its robustness, can be found in [11].

Note that the proposed normalization can be applied separately for any given graph. Thus, for instance, if PageRank scores obtained on two graphs are to be compared, the scores can be normalized on each graph separately, i.e., without knowing the other graph. This property is not common to all normalization schemes and centrality indices as pointed out in [25]. For our application where PageRank scores are computed and stored for snapshots of a large evolving Web graph, this forms a desired property from the efficiency perspective.

4   Rank Synopses

Having discussed the normalization of PageRank scores, we now present rank synopses as a technique to compactly represent and reconstruct historical PageRank scores. Let

$$G = \langle (V_0, E_0), \dots, (V_n, E_n) \rangle \qquad (4)$$

be an *evolving graph* represented as a sequence of *graph snapshots* $G_i(V_i, E_i)$. For the graph snapshot $G_i(V_i, E_i)$ taken at time t_i the set of vertices is denoted as V_i, and the corresponding set of edges is denoted as E_i. Note that no assumption is made about the frequency or regularity of observations. For each graph snapshot, one can compute PageRank scores for all nodes in the graph. Modern web search engines typically re-compute PageRank scores at periodic intervals to keep up with the evolution of the Web. For a *single page*, its PageRank scores across the graph snapshots can be regarded as a time series

$$\langle (t_0, r_0), \ldots, (t_n, r_n) \rangle .$$

This time series is represented by means of a *rank synopsis*

$$\Phi = \langle ([s_0, e_0], \Phi_0), \ldots, ([s_m, e_m], \Phi_m) \rangle .$$

The rank synopsis is a piecewise approximation of the time series. The elements $([s_i, e_i], \Phi_i)$ of Φ, which are referred to as *segments* in the remainder, contain a set of parameters Φ_i representing the function that is used to approximate the time series within the time interval $[s_i, e_i]$. The choice of parameters Φ_i depends only on the observations in the time interval $[s_i, e_i]$, and we do not require functions in temporally adjacent segments to be contiguous. The segments cover the whole time period spanned by the time series, i.e., $s_0 = t_0 \wedge s_m = t_n \wedge \forall_{1 \le i \le m} s_i \le e_i$. Moreover, time intervals of subsequent segments have overlapping right and left boundaries, i.e., $\forall_{1 \le i < m} e_i = s_{i+1}$.

The type of function used for the piecewise approximation is chosen a priori. This choice affects the scalability of the method as we detail below (e.g., depending on whether closed-form solutions of optimal parameters exist). For the scope of this work, we consider the two types of functions:

Linear. The linear function, represented as $\Phi_{linear}(t) = a_0 + a_1 t$ is frequently used for piecewise representations of time series [23]. The main advantages of using this representation is that linear segments can be fitted very efficiently and require only two parameters to be stored.

Logistic. Generalized logistic functions (for brevity referred to as logistic functions in the remainder) have the following form,

$$\Phi_{logistic}(t) = a_0 + \frac{a_1}{\left(1 + a_2\, e^{-a_3(t-a_4)}\right)^{(1/a_2)}} .$$

Logistic functions were previously used, although not as a piecewise approximation, in models of population growth, e.g., the so-called Verhulst model [27]. Recently, in the context of web dynamics, Cho et al. [18] proposed a model of page popularity evolution that is based on a logistic function. Since the PageRank score of a web page can be seen as an estimate of its popularity [18], we hypothesize that logistic functions provide a better approximation of the time series with significantly fewer segments (and therefore less storage) than the linear approximation, despite the larger number of parameters that need to be stored.

4.1 Rank Synopses Construction

When constructing a rank synopsis, we aim at finding a rank synopsis having a *small number of segments* while retaining a *guarantee on the approximation error* per observation. A tunable threshold θ defines an upper bound on the approximation error per segment, which is defined as the maximal relative error made on any observation covered by the segment, i.e.,

$$error(([s_i, e_i], \Phi_i)) = \max_{t_j \in [s_i, e_i]} |1 - \frac{\Phi_i(t_j)}{r_j}| \tag{5}$$

Rank synopses having a minimal number of segments while retaining this approximation guarantee are referred to as *optimal* in the remainder. Analogously, all other rank synopses are referred to as *approximate*.

We now describe how optimal and approximate rank synopses can be efficiently constructed. Rank synopses construction closely resembles time-series segmentation and histogram construction that are both well-studied problems. Therefore, for brevity, we only outline construction algorithms here and refer to relevant related work for details.

Optimal rank synopses can be constructed using dynamic programming as described for time-series segmentation [9,23,32]. Dynamic programming is applicable since every subsequence of segments

$$\langle ([s_i, e_i], \Phi_i), \ldots, ([s_j, e_j], \Phi_j) \rangle$$

must be optimal for observations $\langle (t_k, r_k), \ldots, (t_l, r_l) \rangle$ (with $t_k = s_i$ and $t_l = e_j$), which is an instance of the Bellman principle [9]. The time complexity of the algorithm is in $O(n^3 T_{fit})$ where T_{fit} denotes the time complexity for fitting parameters of the employed function type per segment. For linear segments T_{fit} is in $O(n)$ if closed-form optimal parameters are computed per segment and can be reduced to $O(1)$ by modest precomputations. For logistic segments, which need to be fit by an iterative method, T_{fit} is in $O(n)$ in practice. The space complexity of the algorithm is in $O(n^2)$.

For the construction of approximate rank synopses we adopt a simple sliding window algorithm [23] that we found to perform well in practice. The algorithm makes one pass over the observations, fixes the left boundary of a segment, and places the right boundary rightmost such that the guarantee on the approximation error is still retained. Requiring only one pass over the data, the algorithm has time complexity in $O(n T_{fit})$ and space complexity in $O(n)$. Thus, the approximate rank synopses produced by the sliding window algorithm retains our guarantee on the approximation error, but possibly requires a slightly larger number of segments.

4.2 Reconstructing Historical PageRank Scores

The evolution of a web page's PageRank score is compactly represented in a rank synopsis. To implement the desired time-travel ranking-functionality, given a time t, we need to reconstruct the web page's historical PageRank score (as of time t) from its rank synopsis. This reconstruction happens in two steps. In the first step, at most two segments are selected from the rank synopsis. If t happens to be the boundary of two adjacent segments, both these segments are selected. Otherwise, the one segment whose time interval $[s_i, e_i]$ contains t is selected. Let Φ_t denote the set of segments selected in the first step, then the PageRank score $\hat{r}_t$ as of time t is reconstructed in the second step as

$$\hat{r}_t = \sum_{([s_i, e_i], \Phi_i) \in \Phi_t} \frac{1}{|\Phi_t|} \Phi_i(t) ,$$

Historical scores are thus reconstructed from the functions that were fitted to approximate the observed scores. If two segments qualified in the first step, the mean of the two computed scores is used.

4.3 Rank Synopses Maintenance

The rank synopses algorithms presented above assume that all rankings across time are available at construction time. In practice, however, it is unrealistic to require that all precomputed rankings are kept for each round of rank synopses computation. Ideally, once rank synopses have been computed for a set of rankings, one requires that all rankings that have been already covered by rank synopses are either discarded or accessed infrequently. We now describe how to adapt the above algorithms to incrementally compute rank synopses whenever new observations $\langle(t_{n+1}, r_{n+1}), \ldots, (t_{n'}, r_{n'})\rangle$ become available. Apart from the new rankings, we require only the last segment $([s_m, e_m], \Phi_m)$ of the existing rank synopsis.

Given the last segment, the new observations are scanned in time order. For each new observation, an approximation is computed using the parameters Φ_m of the last segment to see if the approximation guarantee is retained. Let (t_i, r_i) be the first new observation for which the approximation guarantee is not retained, then the last segment of the existing rank synopsis is set as $([s_m, t_{i-1}], \Phi_m)$ – i.e., the segment is ended at the previous observation where it met the approximation guarantee. A new segment is opened from (t_{i-1}, r_{i-1}), and for the remaining new observations $\langle(t_i, r_i), \ldots, (t_{n'}, r_{n'})\rangle$ the algorithm as described above is applied to extend the existing rank synopsis.

4.4 Search Engine Integration

For rank synopses to be useful in practice, they must integrate smoothly into the query-processing pipeline of a search engine. Although today's search engines are not designed specifically for time-travel searching, we assume a similar query-processing strategy to be applicable. Brin and Page [17], in their original description of the Google search engine, describe that PageRank scores are combined in a final step with earlier computed IR scores reflecting query-specific relevance of web pages. Changing the standard lookup of PageRank scores into an reconstruction of historical PageRank scores as described in Section 4.2, rank synopses integrate seamlessly in this scheme of query-processing pipeline.

5 Experimental Evaluation

We demonstrate the accuracy and storage efficiency of the proposed rank synopses using the following three datasets:

- **DBLP:** Evolving citation graph derived from DBLP [3] restricted to the period between years 1989 through 2000 and not containing nodes that have neither incoming nor outgoing edges.
- **IA-UKGOV:** Evolving web graph derived from monthly snapshots of 11.gov.uk web sites over two years. This dataset was obtained from the European Archive [4] (available for online browsing at http://www.europarchive.org/ukgov.php).
- **Wikipedia:** Evolving graph extracted from the full revision history of the English Wikipedia [5] restricted to the period between years 2001 through 2005.

Table 1. Dataset Sizes and Characteristics

Dataset	# Snapshots	# Nodes	# Edges	Required Storage	Kendall's τ μ	σ
DBLP	(yearly) 12	21,531	109,048	1.46 MBytes	0.90	0.02
IA-UKGOV	(monthly) 24	560,496	4,913,060	30.65 MBytes	0.86	0.24
Wikipedia	(monthly) 60	1,618,650	58,845,136	303.69 MBytes	0.85	0.06

Table 1 summarizes the sizes and characteristics of datasets we have used. The reported number of nodes (edges) is the total number of nodes (edges) that ever existed in the graph – for DBLP, where nodes (edges) are only added but never removed, this corresponds to the numbers of nodes (edges) present in the final snapshot. PageRank scores were computed on each snapshot (using $\epsilon = 0.15$). The reported storage figures refer to the storage that is required to keep the precomputed rankings, assuming that one double-precision 8-byte value is needed per page per ranking. In order to quantify how dynamic the datasets are, we compared rankings computed on each pair of temporally adjacent snapshots. The mean μ and the standard deviation σ of the obtained Kendall's τ values are reported in the above table. As these figures show, DBLP is less dynamic than IA-UKGOV and Wikipedia that have lower mean Kendall's τ values. The relatively high σ value observed for IA-UKGOV is the result of few bulk changes that affect a large portion of the dataset.

A set of initial experiments revealed that optimal rank synopses are only marginally better than approximate rank synopses. Therefore, results obtained using optimal rank synopses are omitted here.

Implementation. All methods described in this paper were implemented using Java and the statistical software package R [31]. In particular, the actual rank synopses construction algorithms are implemented in R which provides rich function-fitting functionality. For logistic function fitting, R uses the restricted iterative Nelder-Mead optimization method [28] whose complexity is in $O(n)$ (although much slower than the $O(n)$ fitting of linear segments). All data was stored in an Oracle10g database, and experiments were conducted on a SUN V40z with 4 AMD Opteron 2.66GHz CPUs, 16GB of RAM, and a large disk-array.

Metrics. We use Kendall's τ to compare the distance between two rankings. We used the implementation provided by Boldi et al. [15] for computing the reported Kendall's τ

values. Kendall's τ values are in the range $[-1, 1]$, with 1 (-1) indicating a perfect agreement (disagreement) of the compared permutations.

5.1   Accuracy and Storage

We computed rank synopses for the following values of θ controlling the maximal approximation error per segment: 1%, 2.5%, 5%, 10%, 25%, and 50%. As a measure of accuracy we report the mean Kendall's τ between the original PageRank rankings and the ranking that we reconstruct from our rank synopses. The required storage is estimated assuming that linear and logistic segments require 2 and 5 double-precision 8-byte values, respectively. Storage costs for segment boundaries are ignored, since they do not depend on the type of function used.

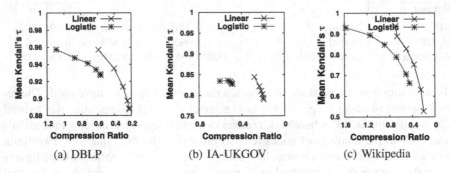

(a) DBLP   (b) IA-UKGOV   (c) Wikipedia

Fig. 2. Compression Ratio vs. Accuracy for Input Observations

In order to make results comparable across datasets, we plot compression ratios, i.e., ratios between the storage consumed by our rank synopses and the storage required by the original rankings (as reported in Table 1), against the accuracy achieved.

Figure 2 plots the resulting accuracies and compression ratios obtained for increasing values of θ. We observe that linear rank synopses consistently achieve higher compression than logistic rank synopses at the same level of accuracy. The gap between the methods widens with increasing accuracy. On DBLP (IA-UKGOV) we observe high accuracies above 0.88 (0.78). On the more dynamic Wikipedia, accuracies are strictly above 0.52. Interestingly, logistic synopses fail to achieve a compression for small values θ on DBLP (see Fig. 2(a)) and Wikipedia (see Fig. 2(c)).

5.2   Reconstructing Non-input Observations

In the preceding experiment, only rankings were reconstructed that had been used as an input when computing the rank synopses. The next experiment analyzes the accuracy achieved by rank synopses when reconstructing a ranking that was *not used* as input. To this end, we computed rank synopses using only every second precomputed ranking, and compared reconstructed and original rankings for the left-out time points.

Figure 3 plots the resulting accuracies and compression ratios. It can be observed that accuracies are slightly lower than the accuracies reported in the previous experiment. On DBLP (see Fig. 3(a)) and Wikipedia (see Fig. 3(c)) we observe accuracies

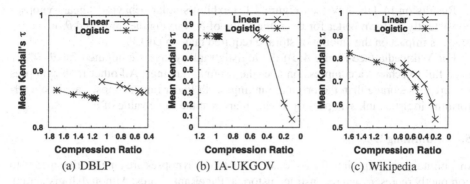

Fig. 3. Compression Ratio vs. Accuracy for Non-Input Observations

consistently above 0.85 (0.52). Logistic rank synopses achieve accuracies above 0.78 on IA-UKGOV (see Figure 3(b)), linear rank synopses, however, exhibit a sharp drop for large values of θ.

5.3 Scalability

Finally, in a third experiment, we examined how the storage required by rank synopses scales with the number of rankings that is used as an input. Figure 4 shows the growth in storage in MBytes against the number of input rankings, for different values of θ for both logistic and linear synopses.

For DBLP, linear synopses require consistently less storage than the original rankings, and grow modestly – if we increase the number of rankings from 2 to 12, the storage for linear synopses only doubles in contrast to an increase by more than 7 times observed for the original rankings. Not surprisingly, logistic synopses show good storage behavior only for *larger threshold values* (e.g., 5% and 25%) and *larger numbers of observations*, as they require more storage per individual segment. For both logistic and linear rank synopses, varying the threshold θ significantly affects the amount of required storage.

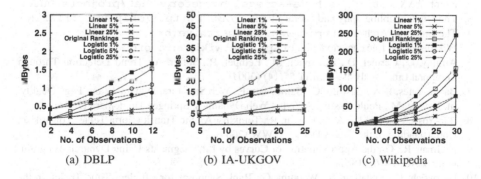

Fig. 4. Scalability of Rank Synopses

Results on IA-UKGOV (see Figure 4(b)) exhibit similar behavior: linear synopses consistently perform better for the same value of θ. Varying the threshold θ, however, has less impact on the amount of storage required than on DBLP.

For Wikipedia (see Figure 4(c)) the logistic rank synopses computed for $\theta = 1\%$ even fail to achieve a compression over the original rankings. All other rank synopses require less storage than the original rankings with linear rank synopses again outperforming logistic rank synopses by a wide margin at the same value of θ.

5.4 Summing Up

In summary, our experiments demonstrate that rank synopses are an effective means to compactly represent and reconstruct historical PageRank scores. Although logistic rank synopses provide slightly better accuracy values overall than linear rank synopses, they do so at a significantly higher storage and construction costs. Our initial hypothesis that motivated the use of logistic segments could thus not be consistently verified.

Currently we are building a storage efficient and accurate time-travel search engine for Web archives. As part of these efforts, we intend to focus on efficiently integrating content-level relevance measures with rank synopses techniques presented in this paper.

6   Conclusions

In this work we have presented rank synopses as an efficient and accurate technique to compactly represent and reconstruct historical PageRank scores. We demonstrated the incomparability of PageRank scores across graphs and presented a new normalization technique to overcome it. The presented techniques were experimentally evaluated using three datasets of different sizes and characteristics. One promising direction of future research is the extension of the presented techniques for the management of other time-varying statistics (e.g., idf-scores) that are used in time-travel search.

References

1. Internet Archive, http://www.archive.org
2. Nutch WAX, http://archive-access.sourceforge.net/projects/nutch
3. The Digital Bibliography and Library Project (DBLP), http://dblp.uni-trier.de
4. The European Archive, http://www.europarchive.org
5. Wikipedia, the free encyclopedia, http://en.wikipedia.org
6. Amitay, E., Carmel, D., Hersovici, M., Lempel, R., Soffer, A.: Trend Detection Through Temporal Link Analysis. JASIST 55(14) (2004)
7. Baeza-Yates, R.A., Castillo, C., Saint-Jean, F.: Web Structure, Dynamics and Page Quality. In: Levene, M., Poulovassilis, A. (eds.) Web Dynamics, Springer, Heidelberg (2004)
8. Bar-Yossef, Z., Broder, A.Z., Kumar, R., Tomkins, A.: Sic Transit Gloria Telae: Towards an Understanding of the Web's Decay. WWW (2004)
9. Bellman, R.: On the Approximation of Curves by Line Segments Using Dynamic Programming. CACM 4(6) (1961)
10. Berberich, K., Bedathur, S., Weikum, G.: Rank Synopses for Efficient Time Travel on the Web Graph. CIKM (2006)

11. Berberich, K., Bedathur, S., Vazirgiannis, M., Weikum, G.: Comparing Apples and Oranges: Normalized PageRank for Evolving Graphs WWW (2007)
12. Berberich, K., Bedathur, S., Neumann, T., Weikum, G.: A Time Machine for Text Search. SIGIR (2007)
13. Berberich, K., Vazirgiannis, M., Weikum, G.: Time-aware Authority Ranking. Internet Mathematics, 2(3) (2005)
14. Bianchini, M., Gori, M., Scarselli, F.: Inside PageRank. ACM TOIT, 5(1) (2005)
15. Boldi, P., Santini, M., Vigna, S.: Do your worst to make the best: Paradoxical Effects in PageRank incremental computations. Internet Mathematics, 2(3) (2005)
16. Borodin, A., Roberts, G.O., Rosenthal, J.S., Tsaparas, P.: Link Analysis Ranking: Algorithms, Theory, and Experiments. ACM TOIT, 5(1) (2005)
17. Brin, S., Page, L.: The Anatomy of a Large-Scale Hypertextual Web Search Engine. Computer Networks and ISDN Systems, 30(1–7) (1998)
18. Cho, J., Roy, S., Adams, R.E.: Page Quality: in Search of an Unbiased Web Ranking. SIGMOD (2005)
19. Eiron, N., McCurley, K.S., Tomlin, J.A.: Ranking the Web Frontier. WWW (2004)
20. Fetterly, D., Manasse, M., Najork, M., Wiener, J.L.: A Large-Scale Study of the Evolution of Web Pages. Software: Practice and Experience, 34(2) (2004)
21. Gyöngyi, Z., Garcia-Molina, H.: Link Spam Alliances. VLDB (2005)
22. Kahle, B.: Preserving the Internet. Scientific American, 276(3) (1997)
23. Keogh, E.J., Chu, S., Hart, D., Pazzani, M.J.: An Online Algorithm for Segmenting Time Series. ICDM (2001)
24. Kleinberg, J.M.: Authoritative Sources in a Hyperlinked Environment. JACM, 46(5) (1999)
25. Koschützki, D., Lehmann, K.A., Tenfelde-Podehl, D., Zlotowski, O.: Advanced Centrality Concepts. In: Brandes, U., Erlebach, T. (eds.) Network Analysis. LNCS, vol. 3418, Springer, Heidelberg (2005)
26. Langville, A.N., Meyer, C.: Deeper Inside PageRank. Internet Mathematics, 1(3) (2004)
27. Meyer, P.S., Yung, J.W., Ausubel, J.J.: A Primer on Logistic Growth and Substitution. Technological Forecasting and Social Change, 61(3) (1999)
28. Nelder, J.A., Mead, R.: A Simplex Algorithm for Function Minimization. Computer Journal, 7 (1965)
29. Ntoulas, A., Cho, J., Olston, C.: What's New on the Web?: The Evolution of the Web from a Search Engine Perspective. WWW (2004)
30. Page, L., Brin, S., Motwani, R., Winograd, T.: The PageRank Citation Ranking: Bringing Order to the Web. Tech. rep. Stanford Digital Library Technologies Project (1998)
31. R Development Core Team. R: A Language and Environment for Statistical Computing. R Foundation for Statistical Computing, Vienna, Austria (2005)
32. Terzi, E., Tsaparas, P.: Efficient Algorithms for Sequence Segmentation. SIAM-DM (2006)
33. Yu, P.S., Li, X., Liu, B.: On the Temporal Dimension of Search. WWW (2004)

Jump-Matching with Errors

Ayelet Butman[1], Noa Lewenstein[2], Benny Porat[3], and Ely Porat[3,4]

[1] Holon Academic Institute of Technology
[2] Netanya Academic College
[3] Bar Ilan University
[4] Google Inc

Abstract. Two equal-length integer-value strings jump-match if each of their corresponding (locationwise) elements differ by the same value d. In Jump matching one seeks all text substrings which jump-match the pattern. Strings approximate jump-match if all elements differ by the same value asides from at most k, where k is predefined. In approximate jump-matching one seeks the text substrings which approximate jump-match with the pattern.

We present innovative, efficient deterministic and randomized algorithms to solve the approximate jump-matching problem.

1 Introduction

Numerical string pattern matching has lately garnered a lot of research-interest. Several metrics have been considered, most notably the L_1 metric [Ata01, ALPU05, CCI05, ILLP04], Euclidean distance, the L_∞ metric [ILLP04, LP05] and in general the L_p metric. Other numerical string comparisons are less-than matching [AF95], transposition invariant matching [LU00, MNU05], and self-normalised distance [CC07].

In this paper we consider *jump-matching*, a.k.a. transposition matching.

Definition 1. *Let $S = s_1, \cdots, s_m$ and $R = r_1, \cdots, r_m$ be two numerical strings $\in \mathbb{N}^m$. We say that S and R jump-match if there exists $d \in \mathbb{Z}$ (the jump) s.t. $s_j = r_j + d$, $\forall 1 \leq j \leq m$.*

The *jump-matching* problem is defined as follows.

Input: Pattern $P = p_1, ..., p_m \in \mathbb{N}^m$ and text $T = t_1, ..., t_n \in \mathbb{N}^n$.
Output: All locations i of T, where P *jump-matches* $T_i = t_i, \cdots, t_{i+m-1}$.

Jump-matching rises in various applications, such as music information retrieval and image recognition. For example, in the latter, we seek a specific object in an image. The way to solve this problem is by scanning the image top-down and for each location in the image deciding if there is an appearance of the object. Given a specific object to look for, it can appear in the new image in different lighting conditions. Different lighting conditions can be interpreted as addition/substraction of a fixed number in the gray level.

N. Ziviani and R. Baeza-Yates (Eds.): SPIRE 2007, LNCS 4726, pp. 98–106, 2007.

Fig. 1. Elephants with different gray-levels (with some mismatches)

Moreover, we will have noise in these images, where noise refers to mismatches relating to the original image. The figures appearing in Figure 1 are taken from the caltech dataset (http://www.vision.caltech.edu/html-files/archive.html) and demonstrate (a) an image of elephant, and (b) the same image with constant pixel difference of 20 gray-levels. Some pixels have additional noise.

The exact version of jump-matching was implicitly solved in [LU00] and we mention there solution below. However, we are more interested in the approximate version. For the approximate version the results in [MNU05] consider various distance measures and propose solutions for them. However, we will be interested in a formulation, which we believe is natural, that was not researched there (in the pattern-matching sense). We propose two algorithms to solve these problems, one deterministic and one randomized.

2 Jump-Matching

Let $S = s_1, s_2, \cdots, s_n \in \mathbb{N}^n$. We define $D(S)$, the *difference string*, to be $s_2 - s_1, s_3 - s_2, ..., s_n - s_{n-1}$.

The difference string was introduced in [LU00] and the following is a simple extension of their work.

Lemma 1. *Let $P \in \mathbb{N}^m$ be a pattern and $T \in \mathbb{N}^n$ be a text. P jump-matches T_i iff $D(P)$ appears at location $i - 1$ of $D(T)$.*

Proof. Assume P jump-matches T at location i. By definition, there is a $d \in \mathbb{Z}$, such that $p_j - t_{i+j-1}$ for each $1 \le j \le m$. Consider $D(P)$ and $D(T_i)$, and one position q, $1 \le q \le m$ of $D(P)$ and $D(T_i)$. At this position $D(P)_q = p_{q+1} - p_q$ and $D(T_i)_q = t_{i+q} - t_{i+q-1}$. However, we know that $d = p_{q+1} - t_{i+q}$ and $d = p_q - t_{i+q-1}$ and hence, $p_{q+1} - t_{i+q} = p_q - t_{i+q-1}$ which implies that $p_{q+1} - p_q = t_{i+q} - t_{i+q-1}$. Therefore, $D(P)_q = D(T_i)_q$ and since the same holds at all positions, $D(P) = D(T_i)$. Lastly, note that $D(T_i)$ is a substring of $D(T)$ at location $i - 1$.

Conversely, if P does not jump-match T_i then there must be a location q for which there is a difference d, where location $q + 1$ has difference $\neq d$, or vice versa. By similar arguments $D(P)_q \neq D(T_i)_q$ and, hence, $D(P) \neq D(T_i)$ and, therefore, $D(P)$ does not match at location $i - 1$ of T. $\square$

It follows from the Lemma that one can solve jump-matching in $O(n + m)$ time by applying $D()$ to both P and T and using an arbitrary linear-time pattern matching algorithm.

3   Approximate Jump-Matching

While jump-matching is easy to solve efficiently, applications such as object recognition render it essential to consider jump-matching with errors. This requires clarification. Consider two strings $S = s_1, \cdots, s_m \in \mathbb{N}^m$ and $R = r_1, \cdots, r_m \in \mathbb{N}^m$ that jump-match. These jump-match with a single unique jump d for every position. Now consider S and R that do not jump-match. That means that for every possible $d \in \mathbb{Z}$ there exists at least one position j such that $s_j \neq r_j + d$. We call such positions j, d-errors. For each $d \in \mathbb{Z}$ we may have a different number of d-errors. Let $d_0 \in \mathbb{Z}$ be the jump d which obtains the minimum number, say k, of d-errors. We say that S and R *jump-match with k errors* and we refer to the d_0-errors simply as errors.

The *Approximate Jump-Matching problem* is defined as follows:

Input: Pattern $P = p_1, ..., p_m \in \mathbb{N}^m$ and text $T = t_1, ..., t_n \in \mathbb{N}^n$ and integer k.
Output: All locations i of T, where P and $T_i = t_i, \cdots, t_{i+m-1}$ *jump-match* with *at most k errors*.

The naive way of solving the Approximate Jump-Matching problem is to evaluate, for each location i of T, all differences $d_j = t_{i+j-1} - p_j$ over positions $1 \leq j \leq m$ and to choose d which appears at most positions. We then verify if d appears in at least $m - k$ positions. This takes $O(nm \log m)$ time. Our goal is to derive better algorithms.

Unfortunately, it does not seem that there is a claim that we can make similar to Lemma 1 for the jump-matching problem. However, we can claim the following:

Lemma 2. *Let $P \in \mathbb{N}^m$ be a pattern and $T \in \mathbb{N}^n$ be a text and $k \in \mathbb{N}$ s.t. $k < m$. If P jump-matches T_i with at most k errors then $D(P)$ appears with at most $2k$ mismatches at location $i - 1$ of $D(T)$.*

Proof. We omit the proof here and point out that it follows arguments similar to those in the proof of Lemma 1. $\square$

Example 1.

T:	2	6	14	23	8	5	74	3	97	42
P:			16	25	5	7	76	5		**d=2**
$D(T)$:	4	8	9	−15	−3	69	−71	94	−55
$D(P)$:			9	−20	−2	69	−71

Let S and $R \in \mathbb{N}^m$. We define a partition on $1, \cdots, m$, based on S and R, as follows. An interval $[i, j]$ is said to be a *block* if there exists d such that for all $i \leq l \leq j$, $s_l - r_l = d$, but $s_{i-1} - r_{i-1} \neq d$ and $s_{j+1} - r_{j+1} \neq d$. A *block-partition* is a partition of $1, \cdots, m$ into blocks. The following observation can be shown to be true relatively straightforwardly.

Observation 1. *Assume that the block-partition of S and $R \in \mathbb{N}^m$ has x blocks. Then*

1. *There are exactly $x - 1$ mismatches between $D(S)$ and $D(R)$,*
2. *Contiguous blocks have different jumps,*
3. *Non-contiguous blocks may have the same jumps.*

This leads to a useful fact.

Corollary 1. *Let S and $R \in \mathbb{N}^m$. If their block partition contains $2k + 2$, or more, blocks then S and R do not jump-match with k errors.*

Proof. Say, by contradiction that S and R jump-match with jump d. By Observation 1, contiguous blocks have different jumps. Since there are at least $2k + 2$ blocks, at least $k + 1$ blocks have a non-d jump and hence at least $k + 1$ errors, a contradiction. □

Note that while Lemma 2 states that there are at most $2k$ mismatches for k errors, it follows from Observation 1 that if there are only 2 blocks there will be only be 1 mismatch while there may be many errors.

3.1 Deterministic Algorithm

We base our algorithm on the block partition. Our method utilizes the "kangaroo" approach [GG86, LV85]. The problem solved is that of pattern matching with k mismatches. The kangaroo approach uses a suffix tree and compares strings by using LCA (longest common ancestor) information. Specifically, they check at each text location i whether P matches T_i with k mismatches as follows. The idea is to find the first mismatch between P and T_i by finding the longest common prefix (LCP) of P and T_i. The LCP is found by an LCA query in the suffix tree between nodes representing P and T_i (both appear in the suffix tree). Their LCA corresponds to their LCP. The LCP ends either at the end of the string or at the first error. This comparison, which can be done in $O(1)$ time, is called a *kangaroo hop*. If we meet an error, say at position q, we skip over it and do another kangaroo hop from the suffix $q + 1$ of P and suffix $i + q - 1$ of T, (both are represented in the suffix tree). We keep on kangaroo-hopping until we discover $k + 1$ errors or hit the end of the string. When we use this method we say that we are *kangarooing*.

Using the kangaroo method we can find a partition of x blocks in time of $O(|x|)$. In our case we will never do more than $2k + 2$ kangaroo hops because Corollary 1 renders this useless, as there cannot be a jump match with at most k errors at this location.

The high-level description of the text searching algorithm appears below.

Algorithm 1. Approximate Jump-Matching (Deterministic Algorithm)

```
1  for every text location i do
2      Establish the block-partition and number of blocks, x (Kangaroo method);
3      If x > 2k + 1 then quit – no match occurrence possible at i;
4      for j = 1 to x do
5          compute dⱼ, the jump for block j, Bⱼ;
6          lⱼ ← |Bⱼ|, the number of elements in block Bⱼ;
7          save the pair ⟨dⱼ, lⱼ⟩;
8      sort the ⟨dⱼ, lⱼ⟩ pairs, according to their dⱼ value;
9      sum the lⱼ's for each d over all pairs ⟨dⱼ, lⱼ⟩ for which d = dⱼ;
10     let d be the jump with maximal sum, l;
11     if l ≥ m − k then
12         announce "there is a k-error match with jump d";
13     else
14         announce "there is no match";
```

Running time: The running time of our algorithm is as follows. In order to use the kangaroo method there is initial preprocessing of a suffix tree for $D(T)$ and $D(P)$, along with LCA information, which takes $O(n + m)$ time. In the algorithmic process, each location is analyzed separately. The kangaroo method (line 2) costs $O(k)$ time to identify the blocks (if there are more than $2k+1$ blocks we stop "kangarooing" since we will quit in line 3 anyway). In the 4-7 loop d_j can be computed from evaluating one position in the block and l_j is immediately available from the kangaroo method. Hence, the loop takes $O(|x|) = O(k)$ time. Step 8 takes $O(k \log k)$ time, step 9 and 10 takes $O(k)$ time and the rest $O(1)$. Hence, the overall running time for Algorithm 1 is $O(nk \log k)$.

Correctness: Since we compute each text location separately, it is sufficient to be sure that the computation is correct at that text location. We first establish the block partition. Line 3 uses Corollary 1 to bound the block partition size we need to consider. From there on the algorithm simply computes brute force the dominating jump.

4   Randomized Algorithm

To get better results we generalize the idea of the difference string that we considered in the previous section. The difference string was defined for differences between each position and the previous position. The key observation in jump-matching was that two contiguous positions (in S and R) have the same jump iff the difference between their values is the same in S and R. However, there is no reason to consider only contiguous positions. In fact one may consider arbitrary permutations π on the position set. Formally,

Definition 2. *Let $S = s_1, ..., s_m \in \mathbb{N}^m$ and $\pi = (\pi_1, ..., \pi_m)$ be a permutation of the vector $(1, ..., m)$. We say that the* difference string *according to π is $D_\pi(S) = s_{\pi_1} - s_1, s_{\pi_2} - s_2, \cdots, s_{\pi_m} - s_m$.*

We now show that similar properties hold for the difference string under π.

Lemma 3. *Let $S, R \in \mathbb{N}^m$ and let π be a permutation. Let $1 \leq q \leq m$ be a position. Let $d = s_q - r_q$ and let $d' = s_{\pi(q)} - r_{\pi(q)}$. Then $d \neq d'$ iff $D_\pi(S)_q \neq D_\pi(R)_q$.*

Proof. By definition, $D_\pi(S)_q = s_{\pi_q} - s_q$ and $D_\pi(R)_q = r_{\pi_q} - r_q$. Since $d = s_q - r_q$ and $d' = s_{\pi(q)} - r_{\pi(q)}$ it follows that $d = d'$ iff $s_{\pi_q} - r_{\pi_q} = s_q - r_q$ which happens iff $s_{\pi_q} - s_q = r_{\pi_q} - r_q$ which is equivalent to $D_\pi(S)_q = D_\pi(R)_q$. □

We can deduce the following.

Corollary 2. *Let $S, R \in \mathbb{N}^m$. If S jump-matches R with l errors then $Ham(D_\pi(S), D_\pi(R)) \leq 2l$.*

Consider P and T_i that jump-match with l errors. If $l \leq k$ then by Corollary 2 we are guaranteed that $Ham(D_\pi(P), D_\pi(T_i)) \leq 2k$ for any permutation π. On the other hand, if $l > k$ then perhaps we could come up with a permutation π for which $Ham(D_\pi(P), D_\pi(T_i)) > 2k$. This would enable us to use a k-mismatches (or better said, $2k$-mismatches) algorithm on the D_π's to check whether $l \leq k$ or not, which would solve the approximate jump-matching problem. Of course, it may be that it is hard to find such π that forces $Ham(D_\pi(P), D_\pi(T_i)) > 2k$, as we mentioned with the block partitions. Moreover, even if for permutation π we have $Ham(D_\pi(P), D_\pi(T_i)) > 2k$ it may be that for a different location j, $Ham(D_\pi(P), D_\pi(T_j)) \leq 2k$ even if P and T_j jump-match with more than k errors. This leads us to the concept of families of permutations. To formalize this we give some necessary definitions. To simplify, we say that S and R do not k-jump-match if S and R jump-match with l errors where $l > k$.

Definition 3. *Let $S, R \in \mathbb{N}^m$ s.t. S and R do not k-jump-match. Let $\pi = (\pi_1, ..., \pi_m)$ be a permutation of the vector $(1, ..., m)$. We say that π is k-tight for S and R if $Ham(D_\pi(S), D_\pi(R)) > 2k$.*

Definition 4. *Let $\mathbb{F}_m^k$ be a family of permutations on $\{1, \cdots, m\}$. We say that $\mathbb{F}_m^k$ is ρ-k-tight if:*

$$\rho \leq \min_{S \text{ and } R \in \mathbb{N}^m \text{ do not } k-jump-match} \frac{|\{\pi \in \mathbb{F}_m^k \mid \pi \text{ is } k - tight \text{ for } S \text{ and } R\}|}{|\mathbb{F}_m^k|}$$

Finding such permutation families will be instrumental for our algorithm. Specifically, notice that the definition of ρ is over all S and R (that do not k-jump-match). Hence, if such a family exists with, say, a constant ρ then at every text location there exist at least a constant fraction of the permutations which are k-tight for P and T_i. Hence, if we choose a random permutation from the family then at any given text location it is likely to hit a k-tight permutation for P and T_i. In fact, without proof we claim that the family of all permutations will be a ρ-k-tight family for any constant ρ. However, the family must also be efficiently computable. One such family is the following, which we will show (1) to be ρ-k-tight for a constant ρ and (2) will be efficiently computable.

Let π_i be the shift permutation, $(i+1, i+2, \cdots, m, 1, \cdots, i)$. Note that $D(S)$ is exactly $D_{\pi_1}(S)$ besides for the last location. Let $\mathbb{F}_{\text{shift}} = \{\pi_i \mid 1 \le i \le m-1\}$ be the family of all shift permutations. We can claim the following:

Theorem 1. *If $m > 6k^2$ then $\mathbb{F}_{\text{shift}}$ is a $\frac{1}{(4+\epsilon)} - k$-tight family, for any ϵ.*

Proof. Let $S, R \in \mathbb{N}^m$ such that S and R do not k-jump-match. We will show that $\frac{1}{(4+\epsilon)} \le \frac{|\{\pi \in \mathbb{F}_{\text{shift}} \mid \pi \text{ is } k-tight \text{ for } S \text{ and } R\}|}{|\mathbb{F}_{\text{shift}}|}$. Say S and R jump-match with l errors. Obviously $l > k$. We will consider three cases, $k < l \le 2k$, $2k < l \le \frac{m}{4}$ and $l > \frac{m}{4}$.

Case $k < l \le 2k$: Let d be the dominant jump in the jump-match with l errors of S and R. That is, there are exactly $m - l \ge m - 2k$ positions where $s_j = r_j + d$. We say that a position q is a d'-*position* if $d' = r_q - s_q$. If $d' \ne d$ we say that the position is a *non-d-position*. For permutation π we say that position q is mapped to $q' = \pi(q)$ (by π).

For each $\pi_i \in \mathbb{F}_{\text{shift}}$, q is mapped to a unique q'. Let q be a non-d-position. q will be mapped to a different non-d-position q' in exactly $l - 1$ permutations. The same is true for each non-d-location q. Hence, in all but, at most, $(l-1)l$ permutations of $\mathbb{F}_{\text{shift}}$ every non-d-position is mapped to a d-position. Hence, in, at least, $m - (l-1)l$ permutations all non-d-positions map to d-positions and the position mapped to the non-d-positions are also d-positions. Hence, by Lemma 3 for all these $m - (l-1)l$ permutations π_i, $D_{\pi_i}(S)$ and $D_{\pi_i}(R)$ mismatch with $2l > 2k$ mismatches. Since, $l \le 2k$, $(l-1)l < 4k^2$. Since $m \ge 6k^2$ the ratio of k-tight permutations for S and R out of all $\mathbb{F}_{\text{shift}}$ permutations is at least $\frac{6k^2 - 4k^2}{6k^2} = \frac{1}{3} > \frac{1}{4+\epsilon}$.

Case $\frac{m}{4} < l$: Let d be the dominant jump in the jump-match with l errors of S and R. Since $l > \frac{m}{4}$ the number of d-positions is $< \frac{3}{4}m$. As d is dominant the number of d'-positions, for any d', is $< \frac{3}{4}m$. Hence, for any position, say a d'-position, there are at least $\frac{m}{4}$ positions which are non-d'-positions. So, for every position q, say a d'-position, there are at least $\frac{m}{4}$ mappings from q to a non-d'-position $q' = \pi_i(q)$ (for some i), for an overall $\frac{m^2}{4}$ mappings. Hence, by Lemma 3 there are $\frac{m^2}{4}$ mismatches between $D_{\pi_i}(S)$ and $D_{\pi_i}(R)$ over all permutations $\pi_i \in \mathbb{F}_{\text{shift}}$. Since for each permutation which is not k-tight, by definition, $Ham(D_{\pi_i}(S), D_{\pi_i}(R)) \le 2k$ it follows that there are at least $\frac{m^2}{4} - m2k$ mappings in k-tight permutations. However, $m > 6k^2$ so $\sqrt{\frac{m}{6}} > k$ and hence at least $\frac{m^2}{4} - 2m\sqrt{\frac{m}{6}} > \frac{m^2}{4} - m\sqrt{m} > \frac{m^2}{(4+\epsilon)}$ for any ϵ and large enough m. Since, each k-tight permutation can have at most m mappings, there are at least $\frac{\frac{m^2}{(4+\epsilon)}}{m} = \frac{m}{(4+\epsilon)}$ k-tight permutations.

Case $2k < l \le \frac{m}{4}$: Since $l \le \frac{m}{4}$ it must be that the number of positions with dominant d is at least $\frac{3}{4}m$. Now, choose (an arbitrary) $2k$ positions from the l error positions and consider the set of mappings M between d-positions and these $2k$ positions. There are overall, at least, $\frac{3}{4}m * 2k * 2 = 3mk$ such mappings. By Lemma 3 each mapping contributes a mismatch between $D_{\pi_i}(S)$ and $D_{\pi_i}(R)$

for some i. However, for any π_i which is not k-tight by definition, $D_{\pi_i}(S)$ and $D_{\pi_i}(R)$ can have at most $2k$ mismatches for at most $2mk$ over all permutations which are not k-tight. Hence, there are, at least, $3mk - 2mk = mk$ mappings from permutations which are k-tight. Yet, since there are only $2k$ chosen non-d-positions, there can be at most $4k$ mappings for each permutation. Hence, there are at least $\frac{mk}{4k} = \frac{m}{4}$ permutation which are k-tight.   □

The idea is to use the family of permutations F_{shift} to chose a random permutation $\pi \in F_{\text{shift}}$ and to evaluate every m-length substring T_i with P. For the analysis, we will be able to use Theorem 1 to obtain a sufficient probability of it being a k-tight permutation. However, the challenge is to compute the number of mismatches between the pattern's difference string and the difference string for each T_i. In the deterministic algorithm the difference string of each T_i was a substring of the difference string of T. Hence, we could compute the difference string for T and it was sufficient. Here the difference string is not even defined for T. Moreover, for π_j and a given text location it is defined differently according to the location of the pattern. Nevertheless, there is a nice feature of the F_{shift} family that will allow us to compute as desired. Consider permutation π_j. Split the patterns into two pattern $P^1 = p_1, p_2, \cdots, p_{m-j}$ and $P^2 = p_{m-j+1}, \cdots, p_m$. Note that in the difference string $D_{\pi_j}(P)$ the first $m - j$ positions are difference between j-distance positions, whereas the last j are distance $m - j$ positions. Hence, by splitting P into P^1 and P^2 we can form appropriate T^1 and T^2 in which to compute the number of errors. We can then use any algorithms for $2k$ mismatches that returns the number of errors at each location. The, currently, fastest such algorithm is that of Amir et.al. [ALP04] that runs in time $O(min(n\sqrt{k \log k}, \frac{nk^3}{m} \log m))$. We denote this algorithm as $ALP(k)$. The algorithm follows.

Algorithm 2. Approximate Jump-Matching (Randomized Algorithm)

```
1   for j=1 to 8log m do
2       Let F_shift = {π_i | 1 ≤ i ≤ m − 1} be the family of all shift permutations;
3       Choose a random 1 ≤ i ≤ m − 1 (with shift permutation π_i);
4       Set P¹ = p₁, p₂, ···, p_{m-i} and P² = p_{m-i+1}, ···, p_m;
5       Set D¹(P¹) = p_{i+1} − p₁, ···, p_m − p_{m-i} and D¹(T) = t_{i+1} − t₁, ···, t_n − t_{n-i};
6       Run ALP(k) with D¹(P¹) and D¹(T); store output in Q^{1,j}[1, ···, n − m + 1];
7       Set D²(P²) = p₁ − p_{m-i+1}, ···, p_i − p_m and
            D²(T) = t₁ − t_{m-i+1}, ···, t_{n-m+i} − t_n;
8       Run ALP(k) with D²(P²) and D²(T); store output in Q^{2,j}[1, ···, n − m + 1];
9   for i = 1 to n-m+1 do
10      NoMatch = False;
11      for j=1 to 8log m do
12          Q[i] = Q^{1,j}[i] + Q^{2,j}[i];
13          if Q[i] > 2k then
14              announce "there is no k-error jump-match at location l";
15              NoMatch = True;
16      if NoMatch = False then
17          announce "there is a k-error jump-match at location l";
```

Theorem 2. *The output of algorithm 2 is correct with probability* $1 - \frac{1}{n}$.

Proof. Let i be a text location such that P and T_i jump-match with l errors, where $l > k$. In each round we choose a π_j and can bound the error-probability as follows: $Prob((Ham(D_{\pi_j}(P), D_{\pi_j}(T_i)) \leq 2k) < 1 - \frac{1}{(4+\epsilon)} < \frac{4}{5})$. Since $(\frac{4}{5})^4 < \frac{1}{2}$ it follows that for $8 \log n$ random choices of permutations $Prob((Ham(D_{\pi_j}(P), D_{\pi_j}(T_i)) \leq 2k) < (\frac{4}{5})^{8 \log n} < (\frac{1}{2})^{2 \log n} = \frac{1}{n^2}$. Hence, over all locations j which jump-match with more than k errors we have probability $< \frac{1}{n}$ of not receiving $Ham(D_{\pi_j}(P), D_{\pi_j}(T_i)) > 2k$ in one of the rounds.   □

Time Analysis: The running time is $O(n\sqrt{k \log k} \log n)$ by $O(\log n)$ applications of the ALP algorithm [ALP04].

References

[AF95]   Amir, A., Farach, M.: Efficient 2-dimensional approximate matching of half-rectangular figures. Information and Computation 118(1), 1–11 (1995)

[ALP04]   Amir, A., Lewenstein, M., Porat, E.: Faster algorithms for string matching with k mismatches. J. of Algorithms, 257–275 (2004)

[ALPU05] Amir, A., Lipsky, O., Porat, E., Umanski, J.: Approximate matching in the l_1 metric. In: Apostolico, A., Crochemore, M., Park, K. (eds.) CPM 2005. LNCS, vol. 3537, pp. 91–103. Springer, Heidelberg (2005)

[Ata01]   Atallah, M.J.: Faster image template matching in the sum of the absolute value of differences measure. IEEE Transactions on Image Processing 10(4), 659–663 (2001)

[CC07]   Clifford, P., Clifford, R.: Self-normalised distance with don't cares. In: CPM 2007. LNCS, vol. 4580, pp. 63–70. Springer, Heidelberg (2007)

[CCI05]   Clifford, P., Clifford, R., Iliopoulos, C.S.: Faster algorithms for delta, gamma-matching and related problems. In: Apostolico, A., Crochemore, M., Park, K. (eds.) CPM 2005. LNCS, vol. 3537, pp. 68–78. Springer, Heidelberg (2005)

[GG86]   Galil, Z., Giancarlo, R.: Improved string matching with k mismatches. SIGACT News 17(4), 52–54 (1986)

[ILLP04] Indyk, P., Lewenstein, M., Lipsky, O., Porat, E.: Closest pair problems in very high dimensions. In: Díaz, J., Karhumäki, J., Lepistö, A., Sannella, D. (eds.) ICALP 2004. LNCS, vol. 3142, pp. 782–792. Springer, Heidelberg (2004)

[LP05]   Lipsky, O., Porat, E.: Approximate matching in the l_∞ metric. In: Consens, M.P., Navarro, G. (eds.) SPIRE 2005. LNCS, vol. 3772, pp. 331–334. Springer, Heidelberg (2005)

[LU00]   Lemström, K., Ukkonen, E.: Including interval encoding into edit distance based music comparison and retrieval. pp. 53–60 (2000)

[LV85]   Landau, G.M., Vishkin, U.: Efficient string matching in the presence of errors. In: Proc. 26th IEEE FOCS, p. 126 (1985)

[MNU05]   Mäkinen, V., Navarro, G., Ukkonen, E.: Transposition invariant string matching. J. Algorithms 56(2), 124–153 (2005)

Estimating Number of Citations Using Author Reputation

Carlos Castillo, Debora Donato, and Aristides Gionis

Yahoo! Research Barcelona
C/Ocata 1, 08003 Barcelona
Catalunya, Spain

Abstract. We study the problem of predicting the popularity of items in a dynamic environment in which authors post continuously new items and provide feedback on existing items. This problem can be applied to predict popularity of blog posts, rank photographs in a photo-sharing system, or predict the citations of a scientific article using author information and monitoring the items of interest for a short period of time after their creation. As a case study, we show how to estimate the number of citations for an academic paper using information about past articles written by the same author(s) of the paper. If we use only the citation information over a short period of time, we obtain a predicted value that has a correlation of $r = 0.57$ with the actual value. This is our baseline prediction. Our best-performing system can improve that prediction by adding features extracted from the past publishing history of its authors, increasing the correlation between the actual and the predicted values to $r = 0.81$.

1 Introduction

Editors in publishing houses (as well as producers for record labels and other industries) face often the following problem: given a work, or a promise of a work, what is a good method to predict if this work is going to be successful? Answering this question can be very useful in order to decide, for instance, whether to buy the rights over the work, or to pay in advance to the authors. The editor's prediction on the success of the work can, in principle, depend on the past publishing history or credentials of the author, and on the estimated quality of the item that is being examined. Of course, the estimation can be quite inaccurate, as the actual success of an item depends on many elements, including complex interactions among its audience plus external factors that cannot be determined in advance.

We are interested in the problem of estimating the *success* of a given item, understood as the impact of the item in its community. In the case of books, for instance, success can be measured in terms of book sales. In the case of scholarly articles, success is typically measured as a function of the number of citations an article receives over time.

In this work, we deal with the citation prediction task in the context of a large set of academic articles. Our main questions are:

N. Ziviani and R. Baeza-Yates (Eds.): SPIRE 2007, LNCS 4726, pp. 107–117, 2007.
© Springer-Verlag Berlin Heidelberg 2007

- Can we characterize the evolution of the citations of a paper over time?
- Can we predict the number of citations of a paper, given information about its authors?
- Can we improve such a prediction, if we know the number of citations a paper has received over a short timespan?

The method we describe in this paper receives as input an article and the past publication history of the authors of that article. The output is an estimation of how many citations the article will accumulate over its first few years. Such prediction can be further improved over time as the system receives information about how many citations the article received over the first few months after its publication.

The next section relates our work with previous papers on this problem. In Section 3 we describe the dataset we are using, in Section 4 we describe the features we extract for the prediction task, and in Section 5 we discuss the experimental results we obtained. The last section outlines our main conclusions and describes future work.

2   Related Work

The 2003 KDD Cup [7] included a citation prediction task resembling the one we undertake on this paper. The citation prediction task included estimating the change in the number of citations of papers between two different periods of time. Participants received data about the citation graphs and the contents of a set of about 30,000 papers from the e-print arXiv.[1] The training data covered a 3-months period (February to April 2003) and the testing data was the next 3-months period (May to July 2003). In contrast, in the current work we do not use content attributes from the papers and the time period covered by the prediction task is in the order of years, not months.

The problem of predicting the ranking of scientists was studied recently by Feitelson and Yovel [5]. They show that a multiplicative process gives a good approximation of the number of citations that authors in a certain position in the ranking receive. In their paper, Feitelson and Yovel want to estimate the rank of each author in the list ordered by citations, not the citation counts. The main idea is that authors will move up in the rank until the rate of change of their citations is equal to the rate of change of the citations of the authors in similar positions in the ranking. In contrast, in our work we focus on the number of citations of particular papers (not authors), and mostly in the absolute number of citations, not on the ranking.

Popescul and Ungar [14] use machine learning to try to predict specific citations among papers (e.g., if paper p_1 is going to cite paper p_2 or not), using features such as the authors, citation information, and the venues where papers appear. A related problem, predicting co-authorship relationships between authors, is studied by Liben-Nowell and Kleinberg [11]. In our paper, we estimate aggregate counts in the citation network, not specific links.

---

[1] http://arxiv.org/

The Web has provided an enormous amount of data about dynamic networks, including general Web pages, blogs, Wikis, and other information systems. In general the dynamics of blogs [1,9,12], Wikipedia [3], authorship networks [11] and other networks [10] have attracted considerable attention in the last years. In the specific case of blogs, Fujimura et al. [6] observe that the number of in-links to individual blog entries is quite small in general, and a certain amount time from the moment of the creation of each blog is needed to acquire those in-links. To overcome these problems and be able to rank blog postings, they propose the *EigenRumor* algorithm, in which a variation of the hubs and authorities algorithms [8] is applied to authors and blog postings. In the case of Web links, Baeza et al. [2] observed that PageRank is biased towards older pages (since they have had more time to accumulate citations. Cho et al. [4] propose to reduce this bias by considering a different quality metric: a weighted sum of the derivative of PageRank over time and the actual PageRank value.

Recent results on predicting popularity in social networks point out that when users influence each other, the ranking of items may be less predictable than when they judge independently. Salganik et al. [15] experiment with an artificial cultural market showing that, while high-quality items rarely have low rankings and low-quality items rarely have high rankings, basically any ranking outcome is possible.

3   Dataset

CiteSeer[2] is a dataset of academic papers, with many facilities for searching and exploring the dataset. It has been in operation since 1997 and currently indexes over 750,000 bibliographic records. CiteSeer data is available through an Open Archives Initiative (OAI) interface that allows users to download records from the dataset. The set of records we used covers 581,866 papers published from 1995 to 2003, including both years.

Given that we are interested in creating a model for the authors, we kept only papers for which at least 1 of the authors had 3 papers or more in the dataset. In this way, we obtained 519,542 papers, which is about 89% of the original set. The large temporal window available allow us to use part of the data available to build the reputation of each author. In particular, we select 1,500 papers written during 4 months in 1999 (i.e., right in the center of the temporal window) and use all the past data (papers, authors and citations) to extract the features related to the authors of each paper at the moment of its publication. We use the remaining 4.5 years in the future to monitor the popularity growth and to test our predictions.

Next we looked at the number of citations that papers received. We focused on two particular moments in time: first, 6 months after the publication of each paper. On average, papers in our dataset had 2 citations at that time. Second, we looked at 30 months (4.5 years) after publication. On average, papers in our

---

[2] http://citeseer.ist.psu.edu/

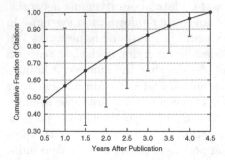
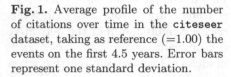

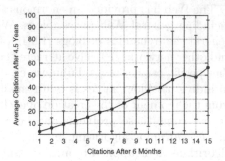

Fig. 1. Average profile of the number of citations over time in the `citeseer` dataset, taking as reference (=1.00) the events on the first 4.5 years. Error bars represent one standard deviation.

Fig. 2. Citations at the end of the first year versus average citations at the end of 5 years. Error bars represent one standard deviation. Std/avg ranges from 1.4 on the left to 0.7 on the right.

dataset had 5.9 citations at that time. Fig. 1 summarizes the *fraction* of citations that papers receive between 6 months and 30 months. Overall, papers seem to accumulate citations steadily over their first few years. On average, a paper in our dataset receives roughly half of its citations on the first 6 months, and 57% of them on the first year. However, the variance on these statistics is very large. The error bars in the figure represent one standard deviation, and show that different papers accumulate citations following very different profiles.

The number of citations after 6 months and the number of citations after 30 months are correlated, and the correlation coefficient is about 0.57. This means that on average a paper that receives many citations shortly after it is published, will receive many citations over the following years. In Fig. 2, we plot the number of citations after 6 months and the average number of citations after 30 months, including error bars. We can see that the correlation is not strong enough to make an accurate prediction. For instance, for papers with 10 citations in the first 6 months, one standard deviation means somewhere between 10 to 60 citations over 4.5 years, which is a rather large interval. Not surprisingly, the correlation improves as more data points are included, as we show later in Table 1. However, the goal of our task is to be able to estimate the number of citations of a paper shortly after it is published.

One remark about quality and citations is in order. The number of citations is not a perfect metric for the quality of a paper. The reasons behind the impact of a paper can vary, and it is not always true that quality is the key factor for the number of citations of a paper. Moreover, it is difficult to define an objective measure of quality itself. It is evident that surveys, methodological papers, or just papers addressing "hot topics", or in fields shared by large communities, are more likely to be read and cited than other papers, all other things being equal.

4   Features

Notation. We consider a graph $G = (V_a \cup V_p, E_a \cup E_c)$ that summarizes all the authorship and citation information available. The vertices of this graph are composed by the set V_a, which represents the authors, and the set V_p, which represents the papers. The set of edges include $E_a \subseteq V_a \times V_p$, that captures the authoring relationship, so that $(a, p) \in E_a$ if author a has co-authored paper p. A paper can have more than one author and we denote by $k_p \triangleq |\{a|(a,p) \in E_a\}|$ the number of authors of a paper. The edges in graph G also include $E_c \subseteq V_p \times V_p$, that captures the citation relationship, so that $(p_1, p_2) \in E_c$ if paper p_1 cites paper p_2.

In any bibliography dataset, authors have a double role: from one side, they deliver original content and produce new items, from the other side, they provide an implicit evaluation of other authors. The two types of edges in E_a and E_c capture the authoring and citation relationships respectively. These are denoted in the framework of Fujimura and Tanimoto [6] *information provisioning* and *information evaluation*, respectively.

Each paper $p \in V_p$ has also a timestamp associated to its creation: time(p). It is assumed that all citations go from a more recent paper to an older paper. In the graph G, we define the number of citations of a paper at time t, $C_t(p)$ as:

$$C_t(p) = |\{p'|(p',p) \in E_c \wedge \text{time}(p') < t\}|$$

that is, $C_t(p)$ is the number of papers citing p that were published in the first t units of time after p. We then extend this notation by defining the number of citations of an author a before time t as:

$$C_t(a) = |\{p'|(p',p) \in E_c \wedge (a,p) \in E_a \wedge \text{time}(p') < t\}|.$$

We are interested in determining whether the number of citations of a paper after a long time period, can be approximated by a function of some *a priori* features related to the authors of the paper, and/or to the number of citations of the paper after a shorter time period. More specifically, we use three different types of features: (1) *a priori* author-based features, (2) *a priori* link-based features, and (3) *a posteriori* features.

A priori **author-based features** try to capture how well previous papers from the same authors have performed in the past. At time t, the past publication history of a given author a can be expressed in terms of:

(*i*) Total number of citations received $C_t(a)$: the global number of citations received by the author i from all the papers published before time t.
(*ii*) Total number of papers (co)authored $M_t(a)$: the total number of papers published by the author a before time t

$$M_t(a) = |\{p|(a,p) \in E_a \wedge \text{time}(p) < t\}|.$$

(*iii*) Total number of coauthors $A_t(a)$: for papers published before time t

$$A_t(a) = |\{a'|(a',p) \in E_a \wedge (a,p) \in E_a \wedge \text{time}(p) < t \wedge a' \neq a\}|$$

Given that one paper can have multiple authors, we aggregate the values that capture the history of each of those authors. For aggregating the values of individual authors we use max, sum, and avg operators — a detailed description of the aggregation process is given in Appendix A. In total we obtain 12 *a priori* author-based features.

A priori **link-based features** try to capture the intuition that good authors are probably aware of the best previous articles written in a certain field, and hence they tend to cite the most relevant of them. Mutual reinforcement characterizes the relation between citing and cited papers; and this relationship also translates to an implied relationship among citing and cited authors. The intuition is that authors cited by good authors should have a higher probability to be cited, and also that good authors usually cite significant papers. This type of implicit endorsement provided by links is the basis of link-based ranking algorithms like PageRank [13] and HITS [8].

If two papers p_1 and p_2 written by different authors a_1 and a_2 respectively cite each other; that is, $(a_1, p_1) \in E_a$, $(a_2, p_2) \in E_a$ and $(p_1, p_2) \in E_c$; we can infer an implicit relationship between authors a_1 and a_2. In the *EigenRumor* algorithm introduced by Fujimura and Tanimoto [6], the relationships implied by both provisioning and evaluation of information are used to address the problem of correctly ranking items produced by sources that have been proved to be authoritative, even if the items themselves have not still collected a high number of inlinks.

The exact implementation of EigenRumor we use is explained in the Appendix. Basically, we compute 7 EigenRumor-based features related to the hub and authority score of the authors of each paper p.

A posteriori **features** simply monitor the evolution of the number of citations of a paper at the end of a few time intervals that are much shorter than the target time for the observation. We consider the number of citations that each paper receives in the first 6 months and in the first 12 months after its publication.

5   Experimental Results

Our goal is to produce an approximation of the number of citations of a paper p at time T, $\hat{C}_T(p) \approx C_T(p)$ using all the information available for items created before time $T' < T$. A first metric for the quality of such approximation is the correlation coefficient between the variables $\hat{C}_T(p)$ and $C_T(p)$.

The correlation coefficient weights papers equally independent on their number of citations. This may be a disadvantage as there are some applications (such as search engines) in which it is more important to produce an accurate prediction for highly-cited papers than for the rest. For this reason, we also consider the following metric: we say that a paper is *successful* if it is among the top 10% of the papers published at the same time t (remember that t represents actually a certain period of time, e.g. one month). Next we evaluate $\hat{C}_T(p)$ by measuring how accurately it can predict the "success" of a paper. Given a classification algorithm $\mathcal{C}$, we consider its confusion matrix:

		Prediction	
		Unsucceful	Successful
True Label	Unsuccessful	a	b
	Successful	c	d

where a represents the number of unsuccessful papers that were correctly classified, b represents the number of successful papers that were falsely classified as unsuccessful, c represents the successful papers that were falsely classified as unsuccessful, and d represents the number of successful papers that were correctly classified. For the evaluation, we use the F-measure, defined as $\frac{2 \times precision \times recall}{precision + recall}$, where $recall$ is $\frac{d}{c+d}$ and $precision$ is $\frac{d}{b+d}$.

We used the freely-available machine-learning package Weka [16]. In particular we used the Weka implementation of linear regression (for prediction of the number of citations) and C4.5 decision trees (for the prediction of success). In the case of decision trees, we applied an asymmetrical cost matrix to increase the recall, by making a misclassification of successful as unsuccessful 1.5 times more costly than a misclassification of unsuccessful as successful.

Tables 1 and 2 show the experimental results obtained over the 1,500 papers extracted as described in Section 3. The r and F values reported in this section are the average after 10-fold cross validation (in which 9/10 of the data are used to learn a model which is then tested in the remaining 1/10 of the data).

From Table 2 we can observe that by using *a priori* author information we obtains a clear improvement in the prediction, given that the correlation coefficient r of the predicted value goes from 0.57 to 0.81 in the prediction that uses the first 6 months of citations.

In the task of predicting success using 6 months of *a posteriori* data, the F-Measure increases significantly, from 0.15 to 0.55. The value F is a bit hard to interpret, but in this case, an F value of 0.55 reflects that about 57% of the top-10% papers are detected, with about 5% false positives. Remember that we are predicting the impact of a paper after 30 months using 6 months of data.

Table 1. Experimental results, using only *a posteriori* citation information. r is the correlation of the predicted value with the number of citations after 4.5 years. F is the F-Measure in the task of predicting "success" (defined as being in top 10% in citations).

A posteriori citations	Predicting Citations r	Predicting Success F
6 months	0.57	0.15
1.0 year	0.76	0.54
1.5 years	0.87	0.63
2.0 years	0.92	0.71
2.5 years	0.95	0.76
3.0 years	0.97	0.86
3.5 years	0.99	0.91
4.0 years	0.99	0.95

Table 2. Experimental results, using *a priori* and *a posteriori* features

A priori features	A posteriori features			
	First 6 months		First 12 months	
	r	F	r	F
None	0.57	0.15	0.76	0.54
Author-based	0.78	0.47	0.84	0.54
Hubs/Auth	0.69	0.39	0.80	0.54
Host	0.62	0.46	0.77	0.57
EigenRumor	0.74	**0.55**	0.83	**0.64**
ALL	**0.81**	**0.55**	**0.86**	0.62

6 Conclusions and Future Work

Our main conclusion is that, in the context of academic papers, information about the authors of a paper may help in predicting the number of citations it will receive in the future, even if we do not take into account other factors such as, for instance, the venue where the paper was published.

In the course of our experiments, we observed that *a priori* information about authors degrades quickly. When the features describing the reputation of an author are calculated at a certain time, and re-used without taking into account the last papers the author has published, the predictions tend to be much less accurate.

Whether is it possible to obtain similar results to other communities is left as future work. We have attempted the same prediction task over data from Flickr.[3] In Flickr, each "author" is a photographer, each "paper" is a photography, and "citations" are votes and/or comments a photographer places over somebody else's photography. This allows us to define information provisioning and information evaluation matrices similar to the ones we use for calculating the attributes we use in this prediction task.

We have used the same algorithms described in this paper, but so far we have not been able to improve the quality of a baseline prediction (using only *a posteriori* information) using *a priori* attributes. Of course, in Flickr the data is much more noisy and sparse, posting a photo is easier than publishing a paper, and in general the dynamics may be different from the dataset we have studied on this article. However, one objective of our research is to devise algorithms for predicting the popularity of items in web communities, and we turn our future work there.

Acknowledgments. We thank Paul Chirita for helping us frame this problem at an early stage. We also thank Carles Rius from DAMA-UPC, who provided us a pre-processed version of CiteSeer data.

---

[3] http://www.flickr.com/

References

1. Adar, E., Zhang, L., Adamic, L.A., Lukose, R.M.: Implicit structure and the dynamics of blogspace. In: WWE 2004, New York, USA (May 2004)
2. Baeza-Yates, R., Saint-Jean, F., Castillo, C.: Web structure, dynamics and page quality. In: Laender, A.H.F., Oliveira, A.L. (eds.) SPIRE 2002. LNCS, vol. 2476, Springer, Heidelberg (2002)
3. Buriol, L., Castillo, C., Donato, D., Leonardi, S., Millozzi, S.: Temporal evolution of the wikigraph. In: WI 2006, Hong Kong, pp. 45–51. IEEE CS Press, Los Alamitos (December 2006)
4. Cho, J., Roy, S., Adams, R.E.: Page quality: in search of an unbiased web ranking. In: SIGMOD 2005, pp. 551–562. ACM Press, New York (2005)
5. Feitelson, D.G., Yovel, U.: Predictive ranking of computer scientists using citeseer data. Journal of Documentation 60(1), 44–61 (2004)
6. Fujimura, K., Tanimoto, N.: The eigenrumor algorithm for calculating contributions in cyberspace communities. In: Falcone, R., Barber, S., Sabater-Mir, J., Singh, M.P. (eds.) Trusting Agents for Trusting Electronic Societies. LNCS (LNAI), vol. 3577, pp. 59–74. Springer, Heidelberg (2005)
7. Gehrke, J., Ginsparg, P., Kleinberg, J.: Overview of the 2003 kdd cup. SIGKDD Explor. Newsl. 5(2), 149–151 (2003)
8. Kleinberg, J.M.: Authoritative sources in a hyperlinked environment. Journal of the ACM 46(5), 604–632 (1999)
9. Kumar, R., Novak, J., Raghavan, P., Tomkins, A.: Structure and evolution of blogspace. Commun. ACM 47(12), 35–39 (2004)
10. Leskovec, J., Kleinberg, J., Faloutsos, C.: Graphs over time: densification laws, shrinking diameters and possible explanations. In: KDD 2005, pp. 177–187. ACM Press, New York (2005)
11. Liben-Nowell, D., Kleinberg, J.: The link prediction problem for social networks. In: CIKM 2003, pp. 556–559. ACM Press, New York (2003)
12. Mei, Q., Liu, C., Su, H., Zhai, C.: A probabilistic approach to spatiotemporal theme pattern mining on weblogs. In: WWW 2006, pp. 533–542. ACM Press, New York (2006)
13. Page, L., Brin, S., Motwani, R., Winograd, T.: The PageRank citation ranking: bringing order to the Web. Technical report, Stanford Digital Library Technologies Project (1998)
14. Popescul, A., Ungar, L.H.: Statistical relational learning for link prediction. In: IJCAI 2003 (2003)
15. Salganik, M.J., Dodds, P.S., Watts, D.J.: Experimental study of inequality and unpredictability in an artificial cultural market. Science 311(5762), 854–856 (2006)
16. Witten, I.H., Frank, E.: Data Mining: Practical Machine Learning Tools and Techniques with Java Implementations. Morgan Kaufmann, San Francisco (1999)

Appendix A

For repeatability of our experiments, we include here the specific list of features we used for the prediction task.

A priori **author-based features.** Remembering that k_p is the total number of the authors of the paper p, let $t = time(p)$, $C_t(a)$, $M_t(a)$, $A_t(a)$ the global number of citations, papers and coauthors of the author a at time t.

We computed for each paper p the following features at time t :

– Features based on the number of citations $C_t(a)$
  1. Sum of all citations collected by all the authors: $\sum_a^{k_p} C_t(a)$
  2. Average citations per author: $\frac{\sum_a^{k_p} C_t(a)}{k_p}$
  3. Maximum number of citations: $max_a\, C_t(a)$
  4. Sum of all citations collected by all the authors per paper: $\sum_a^{k_p} \frac{C_t(a)}{M_t(a)}$
  5. Average citations per author per paper: $\frac{\sum_a^{k_p} \frac{C_t(a)}{M_t(a)}}{k_p}$
  6. Maximum number of citations per paper: $max_a \frac{C_t(a)}{M_t(a)}$
– Features based on the number of papers $M_t(a)$
  7. Sum of all papers published by all the authors: $\sum_a^{k_p} M_t(a)$
  8. Average number of papers per author: $\frac{\sum_a^{k_p} M_t(a)}{k_p}$
  9. Maximum number of papers: $max_a\, M_t(a)$
– Features based on the number of coauthors $A_t(a)$
  10. Sum of all coauthors of each authors: $\sum_a^{k_p} A_t(a)$
  11. Average number of coauthors per author: $\frac{\sum_a^{k_p} A_t(a)}{k_p}$
  12. Maximum number of coauthors: $max_a\, A_t(a)$

A *priori* link-based features. Following [6], consider:

– the *provisioning matrix* $P_{a,t}$ is the matrix induced by the authoring relationship $E_{a,t}$, defined as:

$$E_{a,t} \subseteq V_a \times V_p \triangleq \{(a, p*) \in E_a \wedge \text{time}(p*) < t\}$$

where the $i-th$ row corresponds to the papers written by the author i.
– the *evaluation matrix* $P_{e,t}$ is the matrix induced by the evaluation relationship $E_{e,t}$, defined as:

$$E_{e,t} \subseteq V_a \times V_p \triangleq \{(e, \hat{p}) : (e, p*) \in E_{a,t} \wedge (p*, \hat{p}) \in E_c \wedge \text{time}(p*) < t\}$$

where the $i-th$ row corresponds to the papers cited by the author i

Naming $R(p)$ the score of the paper p, $A_{t'}(a)$ and $H_{t'}(a)$ respectively the authority the hub of the author a, the EigenRumor algorithm states that

– $R = P_{a,t}^T A_t$ that captures the intuition that good papers are likely written by good authors
– $R = P_{e,t}^T H_t$ that captures the intuition that good papers are likely cited by good authors
– $A_t = P_{a,t} R$ good authors usually write good papers
– $H_t = P_{e,t} R$ good authors usually cite good papers

Combining the previous equations we obtain the following formula for the score vector

$$R = \alpha P_{a,t}^T A_t + (1 - \alpha) P_{e,t}^T H_t$$

Aggregating the authority and hub scores for all the coauthors of each paper, we obtain 7 features:

- Relevance of the paper
  1. EigenRumor $R(p)$;
- Authority features
  2. Sum of the authority scores of all the coauthors: $\sum_a^{k_p} A_t(a)$
  3. Average authority per author: $\frac{\sum_a^{k_p} A_t(a)}{k_p}$
  4. Maximum authority: $max_a A_t(a)$
- Hub features
  5. Sum of the hub scores of all the coauthors: $\sum_a^{k_p} H_t(a)$
  6. Average hub per author: $\frac{\sum_a^{k_p} H_t(a)}{k_p}$
  7. Maximum hub: $max_a H_t(a)$

A Fast and Compact Web Graph Representation*

Francisco Claude and Gonzalo Navarro

Department of Computer Science, Universidad de Chile
{fclaude,gnavarro}@dcc.uchile.cl

Abstract. Compressed graphs representation has become an attractive research topic because of its applications in the manipulation of huge Web graphs in main memory. By far the best current result is the technique by Boldi and Vigna, which takes advantage of several particular properties of Web graphs. In this paper we show that the same properties can be exploited with a different and elegant technique, built on Re-Pair compression, which achieves about the same space but much faster navigation of the graph. Moreover, the technique has the potential of adapting well to secondary memory. In addition, we introduce an approximate Re-Pair version that works efficiently with limited main memory.

1 Introduction

A compressed data structure, besides answering the queries supported by its classical (uncompressed) counterpart, uses little space for its representation. Nowadays this kind of structures is receiving much attention because of two reasons: (1) the enormous amounts of information digitally available, (2) the ever-growing speed gaps in the memory hierarchy. As an example of the former, the graph of the static indexable Web was estimated in 2005 to contain more than 11.5 billion nodes [12] and more than 150 billion links. A plain adjacency list representation would need around 600 GB. As an example of (2), access time to main memory is about one million times faster than to disk. Similar phenomena arise at other levels of memory hierarchy. Although memory sizes have been growing fast, new applications have appeared with data management requirements that exceeded the capacity of the faster memories. Because of this scenario, it is attractive to design and use compressed data structures, even if they are several times slower than their classical counterpart. They will run much faster anyway if they fit in a faster memory.

In this scenario, compressed data structures for graphs have suddenly gained interest in recent years, because a graph is a natural model of the Web structure. Several algorithms used by the main search engines to rank pages, discover communities, and so on, are run over those Web graphs. Needless to say, relevant Web graphs are huge and maintaining them in main memory is a challenge, especially if we wish to access them in compressed form, say for navigation purposes.

---

* Partially funded by a grant from Yahoo! Research Latin America.

N. Ziviani and R. Baeza-Yates (Eds.): SPIRE 2007, LNCS 4726, pp. 118–129, 2007.

As far as we know, the best results in practice to compress Web graphs such that they can be navigated in compressed form are those of Boldi and Vigna [6]. They exploit several well-known regularities of Web graphs, such as their skewed in- and out-degree distributions, repetitiveness in the sets of outgoing links, and locality in the references. For this sake they resort to several ad-hoc mechanisms such as node reordering, differential encoding, compact interval representations and references to similar adjacency lists.

In this paper we present a new way to take advantage of the regularities that appear in Web graphs. Instead of different ad-hoc techniques, we use a uniform and elegant technique called Re-Pair [19] to compress the adjacency lists. As the original linear-time Re-Pair compression requires much main memory, we develop an approximate version that adapts to the available space and can smoothly work on secondary memory thanks to its sequential access pattern. This method can be of independent interest. Our experimental results over different Web crawls show that our method achieves space comparable to that of Boldi and Vigna, yet our navigation is several times faster.

2 Related Work

Let us consider graphs $G = (V, E)$, where V is the set of vertices and E is the set of edges. We call $n = |V|$ and $e = |E|$ in this paper. Standard graph representations such as the incidence matrix and the adjacency list require $n(n-1)/2$ and $2e \log n$ bits, respectively, for undirected graphs. For directed graphs the numbers are n^2 and $e \log n$, respectively[1]. We call the *neighbors* of a node $v \in V$ those $u \in V$ such that $(v, u) \in E$.

The oldest work on graph compression focuses on undirected unlabeled graphs. The first result we know of [30] shows that planar graphs can be compressed into $O(n)$ bits. The constant factor was later improved [17], and finally a technique yielding the optimal constant factor was devised [14]. Results on planar graphs can be generalized to graphs with constant *genus* [20]. More generally, a graph with genus g can be compressed into $O(g + n)$ bits [10]. The same holds for a graph with g *pages*. A page is a subgraph whose nodes can be written in a linear layout so that its edges do not cross. Edges of a page hence form a nested structure that can be represented as a balanced sequence of parentheses.

Some classes of planar graphs have also received special attention, for example trees, triangulated meshes, triconnected planar graphs, and others [15,17,13,28]. For dense graphs, it is shown that little can be done to improve the space required by the adjacency matrix [23].

The above techniques consider just the compression of the graph, not its access in compressed form. The first compressed data structure for graphs we know of [16] requires $O(gn)$ bits of space for a g-page graph. The neighbors of a node can be retrieved in $O(\log n)$ time each (plus an extra $O(g)$ complexity for the whole query). The main idea is again to represent the nested edges using parentheses, and the operations are supported using succinct data structures

---

[1] In this paper logarithms are in base 2.

that permit navigating a sequence of balanced parentheses. The retrieval was later improved to constant time by using improved parentheses representations [22], and also the constant term of the space complexity was improved [9]. The representation also permits finding the degree (number of neighbors) of a node, as well as testing whether two nodes are connected or not, in $O(g)$ time.

All those techniques based on number of pages are unlikely to scale well to more general graphs, in particular to Web graphs. A more powerful concept that applies to this type of graph is that of graph *separators*. Although the separator concept has been used a few times [10,14,8] (yet not supporting access to the compressed graph), the most striking results are achieved in recent work [5,4]. Their idea is to find graph components that can be disconnected from the rest by removing a small number of edges. Then, the nodes within each component can be renumbered to achieve smaller node identifiers, and only a few external edges must be represented.

They [4] apply the separator technique to design a compressed data structure that gives constant access time per delivered neighbor. They carefully implement their techniques and experiment on several graphs. In particular, on a graph of 1 million (1M) nodes and 5M edges from the Google programming contest[2], their data structures require 13–16 bits per edge (bpe), and work faster than a plain uncompressed representation using arrays for the adjacency lists. It is not clear how these results would scale to larger graphs, as much of their improvement relies on smart caching, and this effect should vanish with real Web graphs.

There is also some work specifically aimed at compression of Web graphs [7,1,29,6]. In this graph the (labeled) nodes are Web pages and the (directed) edges are the hyperlinks. Several properties of Web graphs have been identified and exploited to achieve compression:

Skewed distribution: The indegrees and outdegrees of the nodes distribute according to a power law, that is, the probability that a page has i links is $1/i^\theta$ for some parameter $\theta > 0$. Several experiments give rather consistent values of $\theta = 2.1$ for incoming and $\theta = 2.72$ for outgoing links [2,7].

Locality of reference: Most of the links from a site point within the site. This motivates in [3] the use of lexicographical URL order to list the pages, so that outgoing links go to nodes whose position is close to that of the current node. Gap encoding techniques are then used to encode the differences among consecutive target node positions.

Similarity of adjacency lists: Nodes close in URL lexicographical order share many outgoing links [18,6]. This permits compressing them by a reference to the similar list plus a list of edits. Moreover, this translates into source nodes pointing to a given target node forming long intervals of consecutive numbers, which again permits easy compression.

In [29] they partition the adjacency lists considering popularity of the nodes, and use different coding methods for each partition. A more hierarchical view of the nodes is exploited in [26]. In [1,27] they take explicit advantage of the similarity

---

[2] www.google.com/programming-contest, not anymore available.

property. A page with similar outgoing links is identified with some heuristic, and then the current page is expressed as a reference to the similar page plus some edit information to express the deletions and insertions to obtain the current page from the referenced one. Finally, probably the best current result is from [6], who build on previous work [1,27] and further engineer the compression to exploit the properties above.

Experimental figures are not always easy to compare, but they give a reasonable idea of the practical performance. Over a graph with 115M nodes and 1.47 billion (1.47G) edges from the Internet Archive, [29] require 13.92 bpe (plus around 50 bits per node, bpn). In [27], over a graph of 61M nodes and 1G edges, they achieve 5.07 bpe for the graph. In [1] they achieve 8.3 bpe (no information on bpn) over TREC-8 Web track graphs (WT2g set), yet they cannot access the graph in compressed form. In [7] they require 80 bits per node plus 27.2 bpe (and can answer reverse neighbor queries as well).

By far the best figures are from [6]. For example, they achieve space close to 3 bpe to compress a graph of 118M nodes and 1G link from WebBase[3]. This space, however, is not sufficient to access the graph in compressed form. An experiment including the extra information required for navigation is carried out on a graph of 18.5M nodes and 292M links, where they need 6.7 bpe to achieve access times below the microsecond. Those access times are of the same order of magnitude of other representations [29,26,27]. For example, the latter reports times around 300 nanoseconds per delivered edge.

A recent proposal [24] advocates regarding the adjacency list representation as a text sequence and use compressed text indexing techniques [25], so that neighbors can be obtained via text decompression and reverse neighbors via text searching. The concept and the results are interesting but not yet sufficiently competitive with those of [6].

3   Re-Pair and Our Approximate Version

Re-Pair [19] is a phrase-based compressor that permits fast and local decompression. It consists of repeatedly finding the most frequent pair of symbols in a sequence of integers and replacing it with a new symbol, until no more replacements are convenient. This technique was recently used in [11] for compressing suffix arrays. More precisely, Re-Pair over a sequence T works as follows:

1. It identifies the most frequent pair ab in T
2. It adds the rule $s \rightarrow ab$ to a dictionary R, where s is a new symbol that does not appear in T.
3. It replaces every occurrence of ab in T by s.[4]
4. It iterates until every pair in T appears once.

Let us call C the resulting text (i.e., T after all the replacements). It is easy to expand any symbol c from C in time linear on the expanded data (that is,

---

[3] www-diglib.stanford.edu/~testbed/doc2/WebBase/
[4] As far as possible, e.g. one cannot replace both occurrences of aa in aaa.

optimal): We expand c using rule $c \rightarrow c'c''$ in R, and continue recursively with c' and c'', until we obtain the original symbols of T. In [11] they propose a new way of compressing the dictionary R which further reduces the space. This compression makes it worthy to replace pairs that appear just twice in the text.

Despite its quadratic appearance, Re-Pair can be implemented in linear time [19]. However, this requires several data structures to track the pairs that must be replaced. This is usually problematic when applying it to large sequences, see for example [31]. Indeed, it was also a problem when using it over suffix arrays [11], where a couple of techniques were proposed to run with restricted memory. A first one was an $O(n \log n)$ time exact method (meaning that it obtained the same result as the algorithm as described), which was in practice extremely slow. A second one was an approximate algorithm (which does not always choose the most frequent pair to replace), which was much faster and lost some compression, yet it applies only to suffix arrays.

We present now an alternative approximate method that (1) works on any sequence, (2) uses as little memory as desired on top of T, (3) given an extra memory to work, can trade accurateness for speed, (4) is able to work smoothly on secondary memory due to its sequential access pattern.

We describe the method assuming we have $M > |T|$ main memory available. If this is not the case, we can anyway run the algorithm by maintaining T on disk since, as explained, we will access it sequentially (performing several passes).

We place T inside the bigger array of size M, and use the remaining space as a (closed) hash table H of size $|H| = M - |T|$. Table H stores unique pairs of symbols ab occurring in T, and a counter of the number of occurrences it has in T. The key ab is represented as a single integer by its position in T (any occurrence works). We traverse $T = t_1 t_2 \ldots$ sequentially and insert all the pairs $t_i t_{i+1}$ into H. If, at some point, the table surpasses a load factor $0 < \alpha < 1$, we do not insert new pairs anymore, yet we keep traversing of T to increase the counters of already inserted pairs.

After the traversal is completed, we scan H and retain the k most frequent pairs from it, for some parameter $k \geq 1$. A heap of k integers is sufficient for this purpose. Those pairs will be simultaneously replaced in a second pass over T. For this sake we must consider that some replacements may invalidate others, for example we cannot replace both ab and bc in abc. Some pairs can have so many occurrences invalidated that they are not worthy of replacement anymore (especially at the end, when even the most frequent pairs occur a few times).

The replacement proceeds as follows. We empty H and insert only the k pairs to be replaced. This time we associate to each pair a field pos, the position of its first occurrence in T. This value is $null$ if we have not yet seen any occurrence in this second pass, and $proceed$ if we have already started replacing it. We now scan T and use H to identify pairs that must be replaced. If pair ab is in H and its pos value is $proceed$, we just replace ab by sz, where s is the new symbol for pair ab and z is an invalid entry. If, instead, pair ab already has a first position recorded in pos, and we read this position in T and it still contains ab (after possible replacements that occurred after we saw that position), then we make both replacements and set the *pos* value to *proceed*. Otherwise, we set the *pos* value of pair ab to the current occurrence we are processing. This method ensures that we create no new symbols s that will appear just once in T.

After this replacement, we compact T by deleting all the z entries, and restart the process. As now T is smaller, we can have a larger hash table of size $|H| = M - |T|$. The traversal of T, regarded as a circular array, will now start at the point where we stopped inserting pairs in H in the previous stage, to favor a uniform distribution of the replacements.

Assume the exact method carries out r replacements. This approximate method can carry out r replacements (achieving hopefully similar compression) in time $O(\lceil r/k \rceil (n + h \log k))$ average time, where $h = |H| = O(n)$. Thus we can trade time for accurateness by tuning k. This analysis, however, is approximate, as some replacements could be invalidated by others and thus we cannot guarantee that we carry out k of them per round. Yet, the analysis is useful to explain the space/time tradeoff involved in the choice of k.

Note that even $k = 1$ does not guarantee that the algorithm works exactly as Re-Pair, as we might not have space to store all the different pairs in H. In this respect, it is interesting that we become more accurate (thanks to a larger H) for the later stages of the algorithm, as by that time the frequency distribution is flatter and more precision is required to identify the best pairs to replace.

As explained, the process works well on disk too. This time T is on disk and table H occupies almost all the main memory, $|H| \approx M$. In H we do not store the position of pair ab but instead ab explicitly, to avoid random accesses to T. The other possible random access is the one where we check *pos*, the first occurrence of a pair ab, when replacing T. Yet, we note that there are at most k positions in T needing random access at any time, so a buffer of k disk pages totally avoids the cost of those accesses. That is, we maintain in main memory the disk blocks containing the position *pos* of each pair to be replaced. When *pos* changes we can discard the disk block from main memory and retain the new one instead (which is the block we are processing). It is possible, however, that we have to write back those pages to disk, but this occurs only when we replace the first occurrence of a pair ab, that is, when *pos* changes from a position to the value *proceed*. This occurs at most k times per stage.

Thus the worst-case I/O cost of this algorithm is $O(\lceil r/k \rceil (n/B + k)) = O(\lceil r/k \rceil \, n/B + r + k)$, where B is the disk block size (again, this is is an approximation).

4   A Compressed Graph Representation

Let $G = (V, E)$ be the graph we wish to compress and navigate. Let $V = \{v_1, v_2, \ldots, v_n\}$ be the set of nodes in arbitrary order, and $adj(v_i) = \{v_{i,1}, v_{i,2}, \ldots v_{i,a_i}\}$ the set of neighbors of node v_i. Finally, let $\overline{v_i}$ be an alternative identifier for node v_i. We represent G by the following sequence:

$$T = T(G) = \overline{v_1} \; v_{1,1} \; v_{1,2} \ldots v_{1,a_1} \; \overline{v_2} \; v_{2,1} \; v_{2,2} \ldots v_{2,a_2} \; \ldots \; \overline{v_n} \; v_{n,1} \; v_{n,2} \ldots v_{1,a_n}$$

so that $v_{i,j} < v_{i,j+1}$ for any $1 \leq i \leq n$, $1 \leq j < a_i$. This is essentially the concatenation of all the adjacency lists with separators that indicate the node each list belongs to.

Applying Re-Pair to this representation $T(G)$ has several advantages:

- Re-Pair permits fast local decompression, as it is a matter of extracting successive symbols from C (the compressed T) and expanding them using the dictionary of rules R.
- This works also very well if $T(G)$ must be anyway stored in secondary memory because the accesses to C are local and sequential, and moreover we access fewer disk blocks because it is a compressed version of T. This requires, however, that R fits in main memory, which can be forced at compression time, at the expense of losing some compression ratio.
- As the symbols $\overline{v_i}$ are unique in T, they will not be replaced by Re-Pair. This guarantees that the beginning of the adjacency list of each v_i will start at a new symbol in C, so that we can decompress it in optimal time $O(|adj(v_j)|)$ without decompressing unnecessary symbols.
- If there are similar adjacency lists, Re-Pair will spot repeated pairs, therefore capturing them into shorter sequences in C. Actually, assume $adj(v_i) = adj(v_j)$. Then Re-Pair will end up creating a new symbol s which, through several rules, will expand to $adj(v_i) = adj(v_j)$. In C, the text around those nodes will read $\overline{v_i}s\overline{v_{i+1}} \ldots \overline{v_j}s\overline{v_{j+1}}$. Even if those symbols do not appear elsewhere in $T(G)$, the compression method for R developed in [11] will represent R using $|adj(v_i)|$ numbers plus $1 + |adj(v_i)|$ bits. Therefore, in practice we are paying almost the same as if we referenced one adjacency list from the other. Thus we achieve, with a uniform technique, the result achieved in [6] by explicit techniques such as looking for similar lists in an interval of nearby nodes.
- Even when the adjacency lists are not identical, Re-Pair can take partial advantage of their similarity. For example, if we have $abcde$ and $abde$, Re-Pair can transform them to scs' and ss', respectively. Again, we obtain automatically what in [6] is done by explicitly encoding the differences using bitmaps and other tools.
- The locality property is not exploited by Re-Pair, unless its translates into similar adjacency lists. This, however, makes our technique independent of the numbering. In [6] it is essential to be able of renumbering the nodes according to site locality. Despite this is indeed a clever numbering for other reasons, it is possible that renumbering is forbidden if the technique is used inside another application. However, we show next a way to exploit locality.

The representation $T(G)$ we have described is useful for reasoning about the compression performance, but it does not give an efficient method to know where a list $adj(v_i)$ begins. For this sake, after compressing $T(G)$ with Re-Pair, we remove all the symbols $\overline{v_i}$ from the compressed sequence C (as explained, those symbols must remain unaltered in C). Using exactly the same space we have gained with this removal, we create a table that, for each node v_i, stores a pointer to the beginning of the representation of $adj(v_i)$ in C. With it, we can obtain $adj(v_i)$ in optimal time for any v_i.

If we are allowed to renumber the nodes, we can exploit the locality property in a subtle way. We let the nodes be ordered and numbered by their URL, and every adjacency list encoded using differential encoding. The first value is absolute and the rest represents the difference to the previous value. For example the list 4 5 8 9 11 12 13 is encoded as 4 1 3 1 2 1 1.

Differential encoding is usually a previous step to represent small numbers with fewer bits. We do not want to do this as it hampers decoding speed. Our main idea to exploit differential encoding is that, if every node tends to have local links, there will be many small differences we could exploit with Re-Pair, say pairs like $(1,1)$, $(1,2)$, $(2,1)$, etc. We also present results for this variant and show that the compression is improved at the expense of some extra decompression.

5   Experimental Results

The experiments were run on a Pentium IV 3.0 GHz with 4GB of RAM using Gentoo GNU/Linux with kernel 2.6.13 and g++ with -O9 and -DNDEBUG options. The compression ratios r we show are the compressed file size as a percentage of the uncompressed file size.

We first study the performance of our approximate technique described in Section 3, as compared to the original technique. Table 1 shows the results for different M (amount of main memory for construction) and k parameters, over a 400 MB suffix array built from a 100 MB XML file (see [11]). The same file with the method proposed in [11] achieves $r = 20.08\%$ after 7.22 hours. The approximate version that works only over suffix arrays achieves 21.3% in 3 minutes.

As it can be seen, our approximate method obtains compression ratios reasonably close to the original one, while being practical in time and RAM for construction. In the following we only use the approximate method, as our graphs are much larger and the exact method does not run.

We now study our graph compression proposal (Section 4). Fig. 1 shows the results for four Web crawls, all downloaded from http://law.dsi.unimi.it/. UK is a graph with 18,520,486 nodes and 298,113,762 edges, EU has 862,664 nodes and 19,235,140 edges, Arabic has 22,744,080 nodes and 639,999,458 edges, and Indochina has 7,414,866 nodes and 194,109,311 edges.

Table 1. Compression ratios and times for different memory usage for construction M, and parameter k

k	M (MB)	r %	time(min)	k	M (MB)	r %	time(min)
10,000,000	1126	23.41%	22	10,000,000	891	27.40%	21
5,000,000	930	23.25%	13	5,000,000	763	27.86%	12
1,000,000	840	22.68%	13	1,000,000	611	29.42%	14
500,000	821	22.44%	18	500,000	592	28.30%	21
100,000	805	22.03%	59	100,000	576	30.49%	67

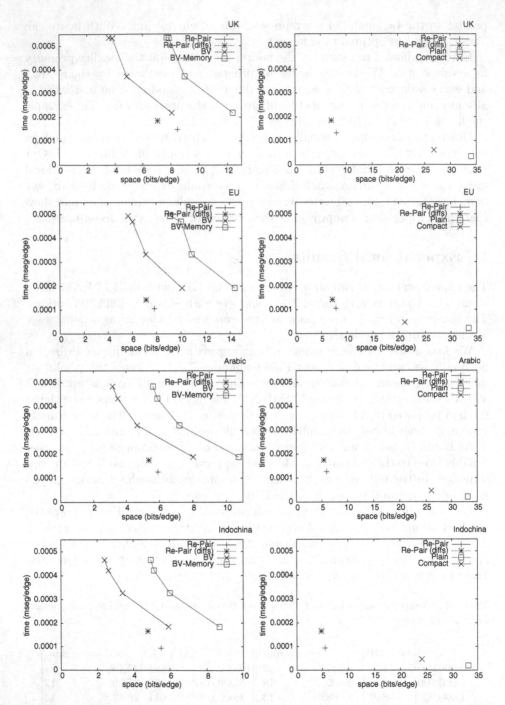

Fig. 1. Space and time to find neighbors for different graph representations, over different Web crawls. BV-Memory represents the minimum heap space needed by the process to run.

We show on the left side the behavior of our Re-Pair-based method with and without differential encoding, compared to Boldi & Vigna's implementation [6] run on our machine with different space/time tradeoffs. The space is measured in bits per edge (bpe), where the total space cost is divided by the number of edges. The implementation of Boldi & Vigna gives a bpe measure that is consistent with the sizes of the generated files. However, their process (in Java) needs more memory to run. This could suggest that they actually need to build more structures that are not stored on the file, but this is difficult to quantify because of other space overheads that come from Java itself and from the WebGraph framework their code is inside. To account for this, we draw a second line that shows the minimum amount of RAM needed for their process to run. In all cases, however, the times we show are obtained with the garbage collector disabled and sufficient RAM to let the process achieve maximum speed. Although our own code is in C++, we found that the Java compiler achieves very competitive results (in unrelated tests over a similar code).

On the right side we compare our method with two fast uncompressed representations: a plain one using 32-bit integers to represent the adjacency lists, and a compact representation using $\lceil \log_2 n \rceil$ bits for every link and $\lceil \log_2 m \rceil$ for every node (to point inside the adjacency list).

The results show that our method is a very competitive alternative to Boldi & Vigna's technique, which is currently the best by a wide margin for Web graphs. In all cases, our method gives a comparable amount of space. Moreover, using the same amount of space, our method is always faster (usually twice as fast, even considering their best line). In addition, one of our versions does not impose any particular node numbering.

Compared to an uncompressed graph representation, our method is also a very interesting alternative. It is 3–5 times smaller than the compact version and 2–3 times slower than it; and it is 4–6 times smaller than the the plain version and 3–6 times slower. In particular, a graph like Arabic needs 2.4 GB of RAM with a plain representation, whereas our compressed version requires only 420 MB of RAM. This can be easily manipulated in a normal 1 GB machine, whereas the plain version would have to resort to disk.

6   Conclusions

We have presented a graph compression method that takes advantage of similarity between adjacency lists by using Re-Pair [19], a phrase-based compressor. The results over different Web crawls show that our method achieves compression ratios similar to the best current schemes [6], while being significantly faster to navigate the compressed graph. Our scheme adapts well to secondary memory, where it can take fewer accesses to disk than its uncompressed counterpart for navigation. In passing, we developed an efficient approximate version of Re-Pair, which also works well on secondary memory. Our work opens several interesting lines for future work:

1. More thorough experiments, considering also the case where the construction and/or the navigation must access secondary memory.
2. Thorough exploration of the performance of our approximate Re-Pair method by itself. Apart from studying it in more general scenarios, we are considering tuning it in different ways. For example we could use a varying k across the compression stages.
3. Further study of the compression format itself and tradeoffs. For example it is possible to compress sequence C with a zero-order compressor (the zero-order entropy of our C sequences tells that its size could be reduced to 61%-77%), although expansion of symbols in C will be slower. Another tradeoff is obtained by replacing the vector of pointers from each v_j to its list in C by a bitmap of $|C|$ bits that mark the beginning of the lists. A $select(j)$ operation (which gives the position of the j-th bit set [21]) over this bitmap would give the position of $adj(v_j)$ in C. This is slower than a direct pointer but will usually save space. For example we would save 1.0 bits/edge in the UK crawl (estimated without an implementation).
4. Combine the current representation with the ideas advocated in [24], so as to have a kind of self-index which, with some overhead over the current representation, would be able of finding reverse neighbors and answer other queries such as indegree and outdegree of a node, presence of a specific link, and so on[5].

References

1. Adler, M., Mitzenmacher, M.: Towards compressing Web graphs. In: Proc. IEEE DCC, pp. 203–212. IEEE Computer Society Press, Los Alamitos (2001)
2. Aiello, W., Chung, F., Lu, L.: A random graph model for massive graphs. In: Proc. ACM STOC, pp. 171–180. ACM Press, New York (2000)
3. Bharat, K., Broder, A., Henzinger, M., Kumar, P., Venkatasubramanian, S.: The Connectivity Server: Fast access to linkage information on the web. In: Proc. WWW, pp. 469–477 (1998)
4. Blandford, D.: Compact data structures with fast queries. PhD thesis, School of Computer Science, Carnegie Mellon University, Also as TR CMU-CS-05-196 (2006)
5. Blandford, D., Blelloch, G., Kash, I.: Compact representations of separable graphs. In: Proc. SODA, pp. 579–588 (2003)
6. Boldi, P., Vigna, S.: The webgraph framework I: compression techniques. In: Proc. WWW, pp. 595–602 (2004)
7. Broder, A., Kumar, R., Maghoul, F., Raghavan, P., Rajagopalan, S., Stata, R., Tomkins, A., Wiener, J.: Graph structure in the web. J. Computer Networks 33(1–6), 309–320 (2000) Also in Proc. WWW9
8. Chakrabarti, D., Papadimitriou, S., Modha, D., Faloutsos, C.: Fully automatic cross-associations. In: Proc. ACM SIGKDD, ACM Press, New York (2004)

---

[5] This line of work is in cooperation with P. Ferragina and R. Venturini, University of Pisa.

9. Chuang, R., Garg, A., He, X., Kao, M.-Y., Lu, H.-I.: Compact encodings of planar graphs with canonical orderings and multiple parentheses. In: Larsen, K.G., Skyum, S., Winskel, G. (eds.) ICALP 1998. LNCS, vol. 1443, pp. 118–129. Springer, Heidelberg (1998)
10. Deo, N., Litow, B.: A structural approach to graph compression. In: Brim, L., Gruska, J., Zlatuška, J. (eds.) MFCS 1998. LNCS, vol. 1450, pp. 91–101. Springer, Heidelberg (1998)
11. González, R., Navarro, G.: Compressed text indexes with fast locate. In: Ma, B., Zhang, K. (eds.) CPM 2007. LNCS, vol. 4580, pp. 216–227. Springer, Heidelberg (2007)
12. Gulli, A., Signorini, A.: The indexable web is more than 11.5 billion pages. In: Proc. WWW (2005)
13. He, X., Kao, M.-Y., Lu, H.-I.: Linear-time succinct encodings of planar graphs via canonical orderings. J. Discrete Mathematics 12(3), 317–325 (1999)
14. He, X., Kao, M.-Y., Lu, H.-I.: A fast general methodology for information-theoretically optimal encodings of graphs. SIAM J. Comput. 30, 838–846 (2000)
15. Itai, A., Rodeh, M.: Representation of graphs. Acta Informatica 17, 215–219 (1982)
16. Jacobson, G.: Space-efficient static trees and graphs. In: Proc. FOCS, pp. 549–554 (1989)
17. Keeler, K., Westbook, J.: Short encodings of planar graphs and maps. Discrete Applied Mathematics 58, 239–252 (1995)
18. Kumar, R., Raghavan, P., Rajagopalan, S., Tomkins, A.: Extracting large scale knowledge bases from the Web. In: Proc. VLDB (1999)
19. Larsson, J., Moffat, A.: Off-line dictionary-based compression. Proc. IEEE 88(11), 1722–1732 (2000)
20. Lu, H.-I.: Linear-time compression of bounded-genus graphs into information-theoretically optimal number of bits. In: Proc. SODA, pp. 223–224 (2002)
21. Munro, I.: Tables. In: Chandru, V., Vinay, V. (eds.) Foundations of Software Technology and Theoretical Computer Science. LNCS, vol. 1180, pp. 37–42. Springer, Heidelberg (1996)
22. Munro, I., Raman, V.: Succinct representation of balanced parentheses, static trees and planar graphs. In: Proc. FOCS, pp. 118–126 (1997)
23. Naor, M.: Succinct representation of general unlabeled graphs. Discrete Applied Mathematics 28, 303–307 (1990)
24. Navarro, G.: Compressing web graphs like texts. Technical Report TR/DCC-2007-2, Dept. of Computer Science, University of Chile (2007)
25. Navarro, G., Mäkinen, V.: Compressed full-text indexes. ACM Computing Surveys 39(1) article 2 (2007)
26. Raghavan, S., Garcia-Molina, H.: Representing Web graphs. In: Proc. ICDE (2003)
27. Randall, K., Stata, R., Wickremesinghe, R., Wiener, J.: The LINK database: Fast access to graphs of the Web. Technical Report 175, Compaq Systems Research Center, Palo Alto, CA (2001)
28. Rossignac, J.: Edgebreaker: Connectivity compression for triangle meshes. IEEE Transactions on Visualization 5(1), 47–61 (1999)
29. Suel, T., Yuan, J.: Compressing the graph structure of the Web. In: Proc. IEEE DCC, pp. 213–222. IEEE Computer Society Press, Los Alamitos (2001)
30. Turán, G.: Succinct representations of graphs. Discrete Applied Mathematics 8, 289–294 (1984)
31. Wan, R.: Browsing and Searching Compressed Documents. PhD thesis, Dept. of Computer Science and Software Engineering, University of Melbourne (2003)

A Filtering Algorithm for k-Mismatch with Don't Cares

Raphaël Clifford[1] and Ely Porat[2]

[1] University of Bristol, Dept. of Computer Science, Bristol, BS8 1UB, UK
clifford@cs.bris.ac.uk
[2] Bar-Ilan University, Dept. of Computer Science, 52900 Ramat-Gan, Israel
porately@cs.biu.ac.il

Abstract. We present a filtering based algorithm for the k-mismatch pattern matching problem with don't cares. Given a text t of length n and a pattern p of length m with don't care symbols in either p or t and a bound k, our algorithm finds all the places that the pattern matches the text with at most k mismatches. The algorithm is deterministic and runs in $O(nm^{1/3}k^{1/3}\log^{2/3}m)$ time. The location of the mismatches at each alignment is also given at no extra cost.

1 Introduction

We consider approximate string matching under the widely used Hamming distance. In particular our interest is in a bounded version of this problem which we call *k-mismatch with don't cares*. Given a text t of length n and a pattern p of length m with don't care symbols and a bound k, our algorithms find all the places that the pattern matches the text with at most k mismatches. If the distance is greater than k, the algorithm need only report that fact and not give the actual Hamming distance.

The problem of finding all the occurrences of a given pattern of length m in a text t of length n is a classic one in computer science whose linear time solutions were first presented in the 1970s [4, 12]. The problem of determining the time complexity of exact matching with optional single character *don't care* symbols has also been well studied. Fischer and Paterson [9] presented the first solution based on fast Fourier transforms (FFT) with an $O(n\log m\log|\Sigma|)$ time algorithm in 1974[1], where Σ is the alphabet that the symbols are chosen from. Subsequently, the major challenge has been to remove this dependency on the alphabet size. Indyk [10] gave a randomised $O(n\log n)$ time algorithm which was followed by a simpler and slightly faster $O(n\log m)$ time randomised solution by Kalai [11]. In 2002, the first deterministic $O(n\log m)$ time solution was given [6] which was then further simplified in [5].

---

[1] Throughout this paper we assume the RAM model when giving the time complexity of the FFT. This is in order to be consistent with the large body of previous work on pattern matching with FFTs.

N. Ziviani and R. Baeza-Yates (Eds.): SPIRE 2007, LNCS 4726, pp. 130–136, 2007.

The key observation given by [5] but implicit in previous work is that for numeric strings, if there are no don't care symbols then for each location $1 \leq i \leq n - m + 1$ we can calculate

$$\sum_{j=1}^{m}(p_j - t_{i+j-1})^2 = \sum_{j=1}^{m}(p_j^2 - 2p_j t_{i+j-1} + t_{i+j-1}^2) \tag{1}$$

in $O(n \log m)$ time using FFTs. Wherever there is an exact match this sum will be exactly 0. If p and t are not numeric, then an arbitrary one-to-one mapping can be chosen from the alphabet to the set of positive integers $\mathbb{N}$. In the case of matching with don't cares, each don't care symbol in p or t is replaced by a 0 and the sum is modified to be

$$\sum_{j=1}^{m} p_j' t_{i+j-1}' (p_j - t_{i+j-1})^2$$

where $p_j' = 0$ ($t_i' = 0$) if p_j (t_i) is a don't care symbol and 1 otherwise. This sum equals 0 if and only if there is an exact match with don't cares and can also be computed in $O(n \log m)$ time using FFTs.

Approximate matching is a key technique that is both widely studied and used in practice. The motivation might be to compensate for errors in a document, or for example to look for similar images in a library or in the case of bioinformatics to look for functional similarities between genes or proteins. However, there may still be values that are unknown in all these cases and any measure of approximation will ideally take this into account. As a more detailed example, a rectangular image segment may contain a facial image and the objective is to identify the face in a larger scene. However, background pixels around the faces may be considered to be irrelevant for facial recognition and these should not affect the search algorithm. Alternatively, a consensus sequence derived from multiple alignment in computational biology (see e.g. [8]) may contain unknown values and the aim is to perform approximate matching rapidly on a large DNA or protein database. Due to the asymmetry between query and database or pattern and text it is often the case that uncertainty lies in either the pattern or the text but not both. It is this model that is considered here.

2   Related Work and Previous Results

Much progress has been made in finding fast algorithms for the k-mismatch problem *without* don't cares over the last 20 years. $O(n\sqrt{m \log m})$ time solutions to the k-mismatch problem based on repeated applications of the FFT were given independently by both Abrahamson and Kosaraju in 1987 [1, 13]. Their algorithms are in fact independent of the bound k and report the Hamming distance at every position irrespective of its value. In 1985 Landau and Vishkin gave a beautiful $O(nk)$ algorithm that is not FFT based which uses constant time LCA operations on the suffix tree of p and t [14]. This was subsequently improved in [3] to $O(n\sqrt{k}\log k)$ time by a method based on filtering and FFTs again. Approximations within a multiplicative factor of $(1 + \epsilon)$ to the Hamming distance can also be found in $O(n/\epsilon^2 \log m)$ time [10]. A variant of the edit-distance problem (see e.g. [15]) called the k-difference problem with don't cares was considered in [2].

To the authors' knowledge, no non-naive algorithms have been given to date for the k-mismatch problem with don't cares. However, the $O(n\sqrt{m}\log m)$ divide and conquer algorithms of Kosaraju and Abrahamson [1, 13] can be easily extended to handle don't cares in both the pattern and text with little extra work. This is because the algorithm counts matches and not mismatches. First we count the number of non-don't care matches at each position i in $O(n\sqrt{m}\log m)$ time. Then we need only subtract this number from the maximum possible number of non-don't care matches in order to count the mismatches. To do this we create a new pattern string p' so that $p'_j = 1$ if p_j is not a don't care and $p_j = 0$ otherwise. A new text string t' is also made in the same way. $p' \otimes t'$ now gives us the maximum number of non-don't care matches possible at each position. This single cross-correlation calculation takes $O(n \log m)$ time. Therefore the overall running time remains $O(n\sqrt{m}\log m)$.

3   Problem Definition and Preliminaries

Let Σ be a set of characters which we term the *alphabet*, and let ϕ be the don't care symbol. Let $t = t_1 t_2 \ldots t_n \in \Sigma^n$ be the text and $p = p_1 p_2 \ldots p_m \in \Sigma^m$ the pattern. Either the pattern or the text may also include ϕ in their alphabet but not both. The terms *symbol* and *character* are used interchangeably throughout. Similarly, we will sometimes refer to a *location* in a string and synonymously at other times the *position*.

- Define $HD(i)$ to be the Hamming distance between p and $t[i, \ldots, i+m-1]$ and define the don't care symbol to match any symbol in the alphabet.
- Define $HD_k(i) = \begin{cases} HD(i) \text{ if } HD(i) \leq k \\ \bot \quad\quad\quad\quad \text{otherwise} \end{cases}$
- We say that at position i in t, p is a k-mismatch if $HD_k(i) \neq \bot$.
- We say that a non don't care symbol is *solid* and that a match between two non don't care symbols is a *solid match*.

Our algorithms make extensive use of the fast Fourier transform (FFT). An important property of the FFT is that in the RAM model, the cross-correlation,

$$(t \otimes p)[i] \overset{\text{def}}{=} \sum_{j=1}^{m} p_j t_{i+j-1}, \quad 0 \leq i \leq n - m + 1,$$

can be calculated accurately and efficiently in $O(n \log n)$ time (see e.g. [7], Chapter 32). By a standard trick of splitting the text into overlapping substrings of length $2m$, the running time can be further reduced to $O(n \log m)$.

4   Filtering for the k-Mismatch Problem

Our algorithm extends the approach taken in [3] and so we first give some technical background on how filtering can be used to speedup the k-mismatch problem *without* don't cares and the complications that don't cares bring.

1. Choose $\sqrt{k}$ characters that occur at least $\sqrt{2k}$ times in the pattern (assume for the moment that this is possible).
2. Construct a new filter subpattern by considering only the $\sqrt{k} * \sqrt{2k} = 2k$ positions in the pattern where one of these characters occurs.
3. Make a single pass over the text with the filter counting at each turn the total number of positions in the filter that could be aligned with the text. This takes at most $O(\sqrt{k})$ time per position in the text as at most $\sqrt{k}$ positions in the filter can match one character.
4. The total number of characters in the filter is $2k$ so a k-mismatch must match at least k positions in the filter.
5. As the total sum of all the counts is at most $n\sqrt{k}$ and each k-mismatch must have a count of at least k then there are at most $n/\sqrt{k}$ possible positions where a k-mismatch can occur.

Following the filtering stage, the number of mismatches at each remaining location in the text can be checked and an $O(k)$ time mismatch algorithm based on repeated constant time longest common extension (LCE) computations allows this to be performed efficiently. The LCE is classically implement using constant time LCA operations on the generalised suffix tree of p and t and will be run at at most $n/\sqrt{k}$ positions. If there were originally fewer than $\sqrt{k}$ characters that occur at least $\sqrt{2k}$ times in the pattern then an alternative stage of match counting is performed using fast FFT based cross-correlation calculations.

The difficulties that arise in this algorithm when considering don't care symbols are twofold. Even assuming that only one of the pattern and text contains don't care symbols, there is no known constant (or even sublinear) time LCE algorithm for strings that allows don't care symbols. Second, if we extend our problem to allow don't cares in both the pattern and text then the filtering process fails immediately as there is no way of knowing how many matches the filter should expect to find. A further minor obstacle is that to achieve the final time complexity of $O(n\sqrt{k\log k})$, certain periodicity properties of strings are required which also do not hold when don't cares are permitted.

5   The New Filtering Algorithm

In order to tackle the k-mismatch problem with don't cares in either the pattern or the text we generalise both the frequent/infrequent approach of Kosaraju and Abrahamson [1, 13] and the filtering approach to pattern matching. Here we call a symbol frequent if it appears at least d times in the pattern. We will separate our algorithm into two cases. The first case will be when we have fewer than f frequent symbols and the second case is when we have at least f frequent symbols. f and d are variables that we will set later on in order to minimise the overall running time of the algorithm.

5.1 Fewer Than f Frequent Symbols

Let m' be the number of solid symbols in the pattern. That is $m' = m$ minus the number of don't cares in the pattern. Instead of directly counting the number of mismatches we will count the number of matches of solid symbols and subtract this number from m' giving us the number of mismatches.

1. Count the number of matches involving frequent characters. As there are no more than f frequent characters this can be done with no more than f cross-correlation calculations. The time for this step is therefore $O(nf \log m)$.
2. Count the number of matches involving infrequent characters. For each position i in the text we can find all the positions j in the pattern such that $p_j = t_i$ and add 1 to the $i - j + 1$th position of an auxiliary array C. At the end of this step C will contain the number of matches involving infrequent characters for every location i in t. There can be no more than d matching positions for each i and so the total time for this step is $O(nd)$.

The overall running time when there are fewer than f frequent symbols is therefore $O(nd + nf \log m)$.

5.2 At Least f Frequent Symbols

When there are at least f frequent symbols we can longer afford to perform a cross-correlation for every different symbol. Instead we perform a filtering step to reduce the number of different possible positions in the text at which the pattern might have a k-mismatch. The following Theorem is the basis for the filtering method.

Theorem 1. *Assume there are at least f frequent symbols in the pattern, each of which occurs at least d times. If $fd > k$ then there are at most $\frac{nd}{fd-k}$ possible k-mismatch positions in the text.*

Proof. The first step is to construct an appropriate filter F and then show its use implies the result. For each of f different frequent symbols in p we choose d different locations where they occur in p. At each one of those locations j set $F_j = p_j$. We let $|F| = m$ and place don't care symbols at all other positions in F. In this way there are fd symbols from the alphabet in F and $m - fd$ don't care symbols which match any symbol in t. For each position i in t, look up the position of the at most d solid symbols in F that match $t[i]$. The location of each of these matches in the pattern is used to update an auxiliary array D of counts. Specifically, add 1 to each position $D[i - j + 1]$ where it is found that $F[j] = t[i]$.

The sum of all the counts in the array D can be no more than nd as we have added 1 at most d times for each position in t. However, for any given position of F in t the total number of solid matches is at most fd. This implies that if there is a k-mismatch at any position then the corresponding count in D must be at least $fd - k$. Therefore, the total number of positions for which there is a k-mismatch of the filter F is at most $nd/(fd - k)$. As the filter is a subpattern of p this implies the same upper bound for the number of k-mismatches of p in t.

Input: Pattern p, text t and an integer k
Output: $O[i] = HD_k(p, t[i, \ldots, i+m-1])$
if *number of frequent symbols is fewer than the threshold f* **then**
    Count "frequent" matches by performing cross-correlations;
    Count "infrequent" matches naively;
    $O[i]$ = number of solid symbols in $p[i, \ldots, i+m-1]$ minus sum of frequent
    and infrequent counts;
else
    Construct filter pattern F;
    Eliminate positions that can't match the filter F;
    Split pattern into ℓ contiguous sub-patterns P_j;
    At each remaining position, eliminate P_j which match exactly;
    $O[i]$ = sum of mismatches for each remaining P_j;
end

Algorithm 1. k-mismatch with don't cares

As before, we check each one of the at most $nd/(fd-k)$ positions where a k-mismatch can occur. Each position takes $O(km/\ell)$ time to check giving an overall time of $O(ndkm/(\ell(fd-k)))$ when there are at least f frequent symbols

Theorem 2. *The k-mismatch with don't cares problem can be solved using Algorithm 1 in $O(nm^{1/3}k^{1/3}\log^{2/3} m)$ time.*

Proof. The running time of Algorithm 1 is the maximum of the two cases described. This gives the running time as $O(\max(ndkm/(\ell(fd-k)), nd+nf\log m))$. By setting $\ell = f = d/\log m = \sqrt[3]{\frac{mk}{\log m}}$ we can derive the overall time complexity which is $O(nm^{1/3}k^{1/3}\log^{2/3} m)$.

6 Discussion

We have given the first filtering based algorithm for the k-mismatch problem when don't cares are allowed in either the pattern or the text. There is still a considerable gap between the fastest $\tilde{O}(n\sqrt{k})$ time k-mismatch algorithm without don't cares and our $\tilde{O}(nm^{1/3}k^{1/3})$ time solution which we conjecture can be at least partially closed. An even wider gap exists for LCE computation and it is another intriguing question whether a sublinear solution to this problem can be found when don't care characters are permitted.

References

[1] Abrahamson, K.: Generalized string matching. SIAM journal on Computing 16(6), 1039–1051 (1987)
[2] Akutsu, T.: Approximate string matching with don't care characters. Information Procesing Letters 55, 235–239 (1995)

[3] Amir, A., Lewenstein, M., Porat, E.: Faster algorithms for string matching with k mismatches. J. Algorithms 50(2), 257–275 (2004)

[4] Boyer, R.S., Moore, J.S.: A fast string matching algorithm. Communications of the ACM 20, 762–772 (1977)

[5] Clifford, P., Clifford, R.: Simple deterministic wildcard matching. Information Processing Letters 101(2), 53–54 (2007)

[6] Cole, R., Hariharan, R.: Verifying candidate matches in sparse and wildcard matching. In: Proceedings of the Annual ACM Symposium on Theory of Computing, pp. 592–601. ACM Press, New York (2002)

[7] Cormen, T.H., Leiserson, C.E., Rivest, R.L.: Introduction to Algorithms. MIT Press, Cambridge (1990)

[8] Durbin, R., Eddy, S.R., Krogh, A., Mitchison, G.: Biological Sequence Analysis: Probabilistic Models of Proteins and Nucleic Acids. Cambridge University Press, Cambridge (1999)

[9] Fischer, M., Paterson, M.: String matching and other products. In: Karp, R. (ed.) Proceedings of the 7th SIAM-AMS Complexity of Computation, pp. 113–125 (1974)

[10] Indyk, P.: Faster algorithms for string matching problems: Matching the convolution bound. In: Proceedings of the 38th Annual Symposium on Foundations of Computer Science, pp. 166–173 (1998)

[11] Kalai, A.: Efficient pattern-matching with don't cares. In: Proceedings of the 13th Annual ACM-SIAM Symposium on Discrete Algorithms, pp. 655–656, Philadelphia, PA, USA, Society for Industrial and Applied Mathematics (2002)

[12] Knuth, D.E., Morris, J.H., Pratt, V.B.: Fast pattern matching in strings. SIAM Journal of Computing 6, 323–350 (1977)

[13] Kosaraju, S.R.: Efficient string matching. Manuscript (1987)

[14] Landau, G.M., Vishkin, U.: Efficient string matching with k mismatches. Theoretical Computer Science 43, 239–249 (1986)

[15] Landau, G.M., Vishkin, U.: Efficient string matching in the presence of errors. In: Proc. 26th IEEE FOCS, p. 126 (1985)

Compact Set Representation for Information Retrieval

J. Shane Culpepper and Alistair Moffat

NICTA Victoria Laboratory
Department of Computer Science and Software Engineering
The University of Melbourne, Victoria 3010, Australia

Abstract. Conjunctive Boolean queries are a fundamental operation in web search engines. These queries can be reduced to the problem of intersecting ordered sets of integers, where each set represents the documents containing one of the query terms. But there is tension between the desire to store the lists effectively, in a compressed form, and the desire to carry out intersection operations efficiently, using non-sequential processing modes. In this paper we evaluate intersection algorithms on compressed sets, comparing them to the best non-sequential array-based intersection algorithms. By adding a simple, low-cost, auxiliary index, we show that compressed storage need not hinder efficient and high-speed intersection operations.

1 Introduction

Conjunctive Boolean queries are a fundamental operation in modern search engines. They are used for both traditional AND-mode querying, and also in ranked querying environments when dynamic pruning techniques are used, or when pre-computed static scores such as PageRank contribute to answer ordering [Zobel and Moffat, 2006].

In abstraction, a conjunctive query q is handled by performing a $|q|$-way intersection over $|q|$ ordered sets of integers, with each set being drawn from a pre-computed index and representing the documents containing one of the query terms. In this abstraction, any efficient algorithm can be used to compute the set intersection. But there is considerable tension between the desire to compress the index lists, and the need to process them using efficient intersection algorithms. In particular, the majority of set-versus-set and multi-way merging algorithms that have been described make use of non-sequential access to the elements of the set and are thus at odds with standard sequential decompression methods for compressed data.

In this paper we evaluate intersection algorithms on compressed sets, comparing them to the best array-based intersection algorithms. Because sequential decompression implies linear search, compressed methods seem fated to be slower than array-based alternatives. But, by adding a simple and low cost auxiliary index, we show that compressed storage need not hinder efficient and high-speed intersection operations.

2 Sets and Set Operations

Techniques for the manipulation of sets and set data have been a rich area of research for several decades. At the most basic level, set manipulation can be reduced to the classical *dictionary problem*, with three key operations needed:

N. Ziviani and R. Baeza-Yates (Eds.): SPIRE 2007, LNCS 4726, pp. 137–148, 2007.

INSERT(S, x)   Return $S \cup x$.
DELETE(S, x)   Return $S - x$.
MEMBER(S, x)   Return TRUE and a pointer to x if $x \in S$; otherwise return
                        FALSE.

Standard efficient structures supporting this group of operations include binary, balanced, and self-adjusting trees; hash tables; and so on. If the *universe* over which the
sets are drawn is dense, and readily mapped to the integers $1 \ldots u$, for some value u,
then direct-access structures such the *bitvector* can be used, in which the space required
is proportional to u rather than to n, the number of elements in the set. Different families
of operations may be needed in some applications. For example:

INTERSECT(S, T)   Return $S \cap T$.
JOIN(S, T)   Return $S \cup T$.
DIFFERENCE(S, T)   Return $S - T$.

These high-level set operations are often implemented using a number of more primitive
operations. In the next group, there is a notion of "the current element", and as the
sequence of operations unfolds, the locus of activity proceeds from one element to
another, and the current element migrates around the set:

PREDECESSOR(S)   Return a pointer to the element in S that immediately pre-
                        cedes the current one.
SUCCESSOR(S)   Return a pointer to the element in S that immediately fol-
                        lows the current one.
F-SEARCH(S, x)   Return a pointer to the least element $y \in S$ for which
                        $y \geq x$, where x is greater than the value of the current
                        element.

For example, intersection operations on sets of comparable size can be readily be implemented using the SUCCESSOR operation; and intersection and union operations on
sets of differing size can be implemented using the F-SEARCH operation, as described
in more detail in Section 3.

The three mid-level set operations can, in turn, be implemented on top of two basic
set operations:

RANK(S, x)   Return $|\{y \mid y \in S \text{ and } y \leq x\}|$.
SELECT(S, r)   Return a pointer to the r th largest element in S.

For example, SUCCESSOR(S) can be implemented as SELECT($S, 1 + $ RANK(*curr*)),
where *curr* is the value of the current item. In addition, once the RANK and SELECT
operations are available, the use of strategically chosen indices in a sequence of SELECT
operations can be used to provide an efficient non-sequential implementation of the
F-SEARCH operation.

In this paper we are primarily interested in INTERSECT operations, implementing
them via a sequence of SUCCESSOR and/or F-SEARCH calls. To set the scene for our
evaluation of techniques for implementing these operations, the next section briefly
summarizes several key data structures that can be used to represent sets.

3 Set Representations

There are several fundamentally different ways in which sets of integers can be stored, with different attributes, and different ways of handling the basic set operations.

Array of integers: A simple and efficient representation for a set is to store it in a sorted array of integers. For example, a set of $n = 15$ objects over the universe $1 \ldots u$, with (say) $u = 32$, can be stored in 15 words, or, via bit-packing techniques, in 15 five-bit binary numbers. More generally, a set of n items each in the range $1 \ldots u$ can be stored in $n \log u$ bits. Implementation of the operation SELECT in $O(1)$ time is immediate, via array indexing; and MEMBER, RANK, and F-SEARCH require $O(\log n)$ time, $O(\log n)$ time, and $O(\log d)$ time respectively, where d is the number of elements that the current position is shifted by. Large-scale operations such as JOIN and DIFFERENCE take $O(n_2)$ time, where $n_1 \leq n_2$ are the sizes of the two sets involved, because every member of the larger of the two sets might need to be listed in the output array. The final operation, INTERSECT, is considered in more detail in the next section, but can be implemented to require $O(n_1 \log(n_2/n_1))$ time in the worst case.

Bitvectors: Another classic set representation is as a bitvector – a u-bit sequence in which the x th bit is a 1 if and only if $x \in S$. Use of a bitvector shifts the cost balance of the various set operations. All of INSERT, DELETE, and MEMBER take $O(1)$ time; but JOIN, DIFFERENCE, and INTERSECT now take $O(u)$ time, if an output set in the same format is to be constructed. The F-SEARCH, RANK and SELECT operations are also expensive if unadorned bitvector representations are used. But in text querying applications, the output set need not be of the same data type, and can be generated as an array of integers. That means that $O(n_1)$-time intersection is possible via a sequence of MEMBER operations, where n_1 is the size of the smaller set.

A drawback of bitvectors is that their $O(u)$ space requirement is significantly more than the corresponding array representation when $n \ll u$. Also, if the application requires PREDECESSOR and SUCCESSOR query support then the basic bitvector representation may not be efficient.

Jacobson [1988] showed that the addition of a controlled amount of extra space allowed RANK and SELECT to be supported using $O(\log u)$ bit probes, and thus that SUCCESSOR, F-SEARCH and PREDECESSOR could also can be made efficient. Munro [1996] later showed that these operations can be supported in $O(1)$ time. Several further improvements to the original approach have been reported, including some that compress the bitvector by exploiting zones with low "1" densities (for example, [Clark, 1996, Pagh, 2001, Raman et al., 2002]). However, from a practical standpoint, the constant factors to implement the data structures described are high, and these succinct representations are driven by the desire for faster RANK and SELECT operations, neither of which is necessary when processing conjunctive Boolean queries.

Compressed representations: Compact representations of sets are almost all built around a simple transformation, which takes a sorted list of elements and converts them into a set of d-gaps. Any method for coding the gaps as variable length codes, including Huffman codes, Golomb codes, Elias γ and δ codes, and static byte- and nibble-based codes, can then be used to encode the transformed list. Examples of these approaches are presented by Witten et al. [1999] and Zobel and Moffat [2006].

Given that there are $C_n^u = u!/((u-n)!n!)$ ways of extracting an n-subset from a universe of u possibilities, compressed representations have as their target a cost of

$$\log C_n^u = \log \frac{u!}{(u-n)!n!} \approx n \left(\log \frac{u}{n} + 1.44 \right)$$

bits when $n \ll u$. Several of the codes listed in the previous paragraph attain, or come close to, this bound. It is also possible to outperform this worst-case bound if there is significant *clustering* within the set, and the n elements in the set are not a random subset of the available universe. One such code is the Interpolative code of Moffat and Stuiver [2000], which represents sets using binary codes, but in a non-linear sequence that makes it sensitive, or *adaptive*, to any non-uniformity. Adaptivity is explored further below.

The great drawback of most compressed representations is their inability to efficiently support any of the key set operations other than SUCCESSOR. In particular, none of F-SEARCH, RANK, and SELECT are efficient. Indeed, in the face of sequential compression techniques based on d-gaps, a search that shifts the current element by d positions requires $O(d)$ time. To regain faster F-SEARCH functionality, additional information can be added in to compressed set representation. For example, Moffat and Zobel [1996] explore adding periodic *skip* information into the compressed set, so that forward jumps can be taken. They suggest inserting such synchronization points every $O(\sqrt{n})$ positions, and demonstrate improved performance on conjunctive Boolean queries and pruned ranked queries, compared to sequential decoding.

More recently, Gupta et al. [2006] describe a two-level structure in which each of the levels is itself a searchable structure containing compressed information, extending earlier work by Blandford and Blelloch [2004]. In the Gupta et al. method, each block of elements at the lower level is a compressed sequential representation of a balanced binary search tree, stored using a pre-order traversal, and with a skip pointer inserted after each node so that its left subtree can be bypassed if the search is to proceed next into the right subtree. Sitting on top of these blocks is a further data structure that allows the correct tree to be quickly identified. By balancing the sizes and performance of the two structures, good performance is achieved.

4   Intersection Algorithms

This section reviews methods for intersecting sets in array-based implementations, before considering the cost of intersecting compressed sets.

Intersecting two sets: There is an interesting duality between set intersection techniques and integer compression methods. To explore that duality, consider the standard paradigm for calculating the intersection of two sets that is shown in Algorithm 1, where $n_1 = |S| \leq n_2 = |T|$, and each of the elements of the smaller set S is searched for in turn the larger set T, with the search always moving forward.

Assuming an array representation of T, there are a range of options for implementing F-SEARCH. One is to use a full binary search, taking $O(\log n_2)$ time per operation, and

Algorithm 1. Standard two-set intersection, INTERSECT(S, T)

```
1: without loss of generality assume that n₁ = |S| ≤ n₂ = |T|
2: set A ← { }
3: set x ← FIRST(S)
4: while x is defined do
5:     set y ← F-SEARCH(T, x)
6:     if x = y then
7:         add x to A
8:     set x ← SUCCESSOR(S)
```

$O(n_1 \log n_2)$ time overall. This approach is the dual of the $O(n \log u)$ cost of using binary codes to store an n-item set over the universe $1 \ldots u$. Another simple approach is to implement F-SEARCH as a linear search from the current location, so that forwards traversal over d elements, requires $O(d)$ time. This approach is the dual of storing a set using a Unary code, which in turn is equivalent to the use of a bitvector.

Better algorithms for array-based intersection also have dual codes. The Hwang and Lin [1973] intersection approach corresponds to the use of a Golomb code (see Witten et al. [1999]) to represent a subset. In the Golomb code, a gap of $d \geq 1$ is represented by coding $1 + (d-1)$ div b in Unary, and then $1 + (d-1)$ mod b in Binary, choosing parameter b as $(\ln 2) \cdot (u/n) \approx 0.69(u/n)$. Similarly, in the Hwang and Lin intersection algorithm a parameter $b = 0.69((n1 + n2)/n_1)$ is computed, and the F-SEARCH operations in Algorithm 1 are implemented by stepping b items at a time from the current location in T, and reverting to a binary search over a range of b once a straddling interval has been determined. When Algorithm 1 is coupled with this Golomb-Hwang-Lin searching method, the time required is $O(n_1 \log(n_2/n_1))$, which is worst-case optimal in an information-theoretic sense.

Other integer codes also have duals. The Elias γ code (see Witten et al. [1999]) is the dual of the exponential search mechanism of Bentley and Yao [1976] (referred to by some authors as "galloping" search), and has also been used as a basis for F-SEARCH operations. In the Elias γ code, the representation for a gap of d requires $1 + 2\lfloor \log d \rfloor = O(\log d)$ bits, and a subset of n of u elements requires at most $n (2 \log(u/n) + 1)$ bits. Similarly, the Baeza-Yates [2004] non-sequential intersection algorithm is the dual of the Interpolative code of Moffat and Stuiver [2000], mentioned above. In this method, the median of the smaller set is located in the larger set. Both sets are then partitioned, and two recursive subproblems handled.

Adaptive algorithms: There has also been interest in the *adaptive* complexity of codes, and hence (in the dual) of adaptive intersection methods. For example, if all elements in S happen to be smaller than the first element in T, then an implementation of F-SEARCH using linear search, or an exponential search, will execute in $O(n_1)$ time. On the other hand the Golomb-based searching approach is non-adaptive and still gives rise to an $O(n_1 \log(n_2/n_1))$ execution time, even for highly favorable arrangements.

Intersecting multiple sets: When there are multiple sets to be intersected, as is the situation in a text retrieval system handling multi-word queries, the operations can either

Algorithm 2. The Max Successor intersection algorithm

1: without loss of generality assume that $|S_1| \leq |S_2| \leq \cdots \leq |S_{|q|}|$
2: set $A \leftarrow \{\}$
3: set $x \leftarrow \text{FIRST}(S_1)$
4: **while** x is defined **do**
5:     **for** $i = 1$ to $|q|$ **do**
6:         set $y \leftarrow \text{F-SEARCH}(S_i, x)$
7:         **if** $x \neq y$ **then**
8:             set $x \leftarrow \max(y, \text{SUCCESSOR}(S_1))$
9:             **break**
10:        **else if** $i = |q|$ **then**
11:            add x to A
12:            set $x \leftarrow \text{SUCCESSOR}(S_1)$

be implemented as a sequence of binary set intersections, or as a single operation on multiple sets. Both approaches have their advantages and disadvantages. Set versus set methods (svs) start with the smallest set, and in turn intersect it against each of the others, in increasing order of size. Because the pivotal set of candidates can only get smaller, the worst-case cost of this approach in an array-based implementation is

$$\sum_{i=2}^{|q|} n_1 \log \frac{n_i}{n_1} \leq n_1(|q| - 1) \log \frac{n_{|q|}}{n_1},$$

where the ordering on the sets is such that $n_1 \leq n_2 \leq \cdots \leq n_{|q|}$. This method is both simple to implement, and also localized in its data access pattern – only two sets are in action at any given time. Either the standard intersection approach shown in Algorithm 1 can be used, or the adaptive Baeza-Yates method can be used.

The other approach is to process all sets simultaneously, and determine the elements in their intersection in an interleaved manner. The simplest approach is to take each element of the smallest set in turn, using it as an *eliminator*, and search for it in the other sets until either it is found in all of them, or is not found in one of them. If it is found, it is then part of the answer; if it is not found in one of the tests, it is eliminated, and the next item from the smallest set is taken.

Demaine et al. [2000] suggested that the set ordering should be dynamic, and be based at all times on the number of remaining elements, so that the cost of every operation is minimized in a greedy sense. As an alternative to the meta-cost of keeping the collection of sets ordered by their number of unprocessed values, Barbay and Kenyon [2002] suggested that the eliminator be chosen instead from the set that caused the previous eliminator to be rejected, so that all sets have the opportunity to provide the eliminator if they are the cause of a big "jump" in the locus of activity. Both of these modified approaches – referred to as adp and seq respectively in the results that appear below – are only of benefit if the input sets are non-uniform with respect to the universe. Barbay et al. [2006] provide a useful overview of how the different search and intersection techniques interact, and summarize a range of previous work.

Max Successor: We also tested another method, max. It takes eliminators from the smallest set, but when the eliminator is beaten, takes as the next eliminator the larger of

Original list of d–gaps

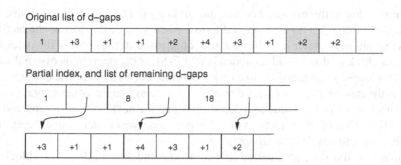

Fig. 1. Extracting every pth document number, and storing it in full in an auxiliary array. In this example $p = 4$, so every fourth d-gap from the list of d-gaps (top row, shaded entries) is extracted and expanded and stored in the auxiliary array (middle row), together with a byte offset into each block of $p - 1$ remaining d-gaps (bottom row).

the item that beat it, or the successor from the first list. Processing then starts at the first list again. Algorithm 2 describes this new approach.

5   Practical Indexing

To provide practical access to compressed sets, we return to a classic algorithmic theme, and build a partial index into the list of compressed d-gaps. Figure 1 sketches the proposed arrangement. Every pth d-gap is removed from the compressed index list, expanded into a document number, and stored in the auxiliary index. A bit offset (or byte offset, for byte-aligned codes) is also stored, as shown in the middle row in Figure 1. To search for a value x, the auxiliary index is first searched, to determine a containing block. Any of the available searching methods can be employed. Once the block that might contain x is identified, it is sequentially decoded and d-gaps resolved, starting at the access pointer. The cost of searching a set of n values is thus at most $O(\log(n/p))$ values accessed for a binary search in the auxiliary index, plus $O(p)$ values decoded during the linear search within the block. Taking $p = k \log n$ for some constant k gives search costs that are $O(\log n)$.

To compute the storage cost of the altered arrangement, suppose that the underlying compression method is an efficient one, and that the full set of n original d-gaps is stored in $n \log(u/n)$ bits. Removing every pth gap multiplies that by $(p-1)/p$. Each of the n/p entries in the auxiliary index requires $\log u$ bits for an uncompressed document number, and $\log(n \log(u/n)) < \log n + \log \log u$ bits for the access pointer, totaling

$$\frac{p-1}{p} n \log \frac{u}{n} + \frac{n}{p} (\log u + \log n + \log \log u)$$

bits. If we again take $p = k \log n$, this simplifies to

$$n \log \frac{u}{n} + \frac{n}{k} \left(2 + \frac{\log \log u}{\log n} \right).$$

When $n \geq \log u$, the overhead cost of the auxiliary index is thus $O(n/k)$ bits, with a search cost of $O(k \log n)$ time. In real terms, when $k = 1$, the cost of the auxiliary index is two bits per pointer in addition to the compressed storage cost of the index lists.

One drawback of this hybrid approach is that F-SEARCH operations over a distance of d are no longer guaranteed to take $O(\log d)$ time. For example, a search operation that shifts the current location forward by $d = \log n$ items in a list containing n pointers must, of necessity, involve sequential decoding within the next block of d-gaps, and thus $O(\log n)$ time. One of the objectives of our experiments was to determine the extent to which this issue affected practical operations.

Compared to the *skipped inverted lists* of Moffat and Zobel [1996], our blocks are much shorter, the auxiliary index is maintained separately to the main sequence of d-gaps rather than interleaved with it, and the auxiliary index is stored uncompressed. These differences add to the space requirement of the inverted index, but allow faster F-SEARCH operations, and thus faster INTERSECT computation. In recent independent work, Sanders and Transier [2007] also investigate two-level representations to improve intersection in compact sets. The main focus of their work is a variation on most significant bit tabling to create buckets of roughly uniform size. Sanders and Transier also consider the possibility of deterministic bucket sizes, in a method similar to the approach proposed here.

6   Experiments

This section describes the arrangements used to measure the execution cost of different set intersection techniques in an environment typical of text search engines.

Collection and queries: All of our results are based on experiments with a large set of queries, and the GOV2 collection of 426 GB of web data used in the TREC Terabyte Track (see http://trec.nist.gov). This collection contains just over 25 million documents, and about 87 million unique words. For our measurements, words that appeared only one or twice in the collection were assumed to be handled directly in the vocabulary rather than via index lists, and this meant that a total of 19,783,975 index lists were considered. Each list was an ordered sequence of document numbers in the range 1 to $u = 25,205,181$. Table 1 lists the cost of storing those lists using different representations. For example, stored as uncompressed 32-bit integers, the index requires 23 GB, compared to a combinatorial set cost, summed over all the lists, of 6 GB. Byte codes do not attain the latter target, nevertheless they provide an attractive space saving compared to uncompressed integers, and an even greater saving compared to bitvectors.

The queries used against this collection were derived from a set supplied by Microsoft as being queries for which at least one of the top three answer documents was in the .gov domain, as of early 2005. A total of 27,004 unique multi-word queries in the set had conjunctive Boolean matches in the GOV2 collection. Table 2 shows the distribution of query lengths in the query set, and the range of set sizes involved. Note how, even in two term queries, the most common term appears in more than 5% of the documents. The average query length tested was 2.73 which is near the expected average query length of 2.4 [Spink et al., 2001].

Table 1. Total space cost in gigabytes to store (in the center column) all of the inverted lists for the 426 GB TREC GOV2 collection, and (in the right column) the subset of the inverted lists referred to by the experimental query set.

Data Representation	TREC GOV2 19,783,975 words	Query Set 15,208 words
Bitvector	58,051.4	44.6
Integer (32-bit)	22.6	14.1
d-gaps, byte code, and auxiliary index, $k = 2$	8.5	4.4
d-gaps and byte code	7.4	3.8
Combinatorial cost	5.9	2.7

Table 2. Length distribution of the 27,004 queries. The average query length is 2.73 terms.

query length $|q|$	2	3	4	5	6	7	8	9	10+
number of queries	15,517	7,014	2,678	1,002	384	169	94	44	102
matches ('000)	124	78	56	41	27	15	10	11	3
average n_1 ('000)	338	325	348	356	288	248	165	226	112
average $n_{|q|}$ ('000)	1,698	5,725	10,311	14,317	15,927	17,365	17,958	18,407	19,236

Measurement: To carry out experiments, the index lists for the 15,208 words that appeared in the query set were extracted from the index into a separate file, as shown in the right-hand column of Table 1. Low frequency terms are rarely queried, and the cost of the bitvector representation drops dramatically. On the other hand, the relative fractions of the compressed representations suggest that more than half of the full index still needs to be manipulated.

The set of queries was then executed using the various intersection algorithms. To process one query, the set of lists pertinent to that query was read into memory while the execution clock was stopped; the clock was then monitored while the query was executed five times in a row to generate a list of answer document numbers in $1 \ldots u$; and then the CPU time taken by the five operations was added to a running total, according to the length of that query. For example, the time recorded for queries of length two is the average of $5 \times 15,517 = 77,585$ executions of 15,517 different queries. We also recorded the number of comparisons performed by each method, as a check against previous results, but report only CPU times here.

Array-based intersection: The first question was the extent to which the adaptive methods were superior to the standard ones. There are two levels at which adaptivity is possible – by using exponential search rather than the worst-case optimal Golomb search; and by performing all $|q| - 1$ merge operations in tandem with an enhanced choice of eliminator at each step, as described in the previous section. The results of these first experiments are shown in Figure 2, where it is assumed throughout that the sets are stored as arrays. In the left-hand graph, Golomb search is used with three multi-way methods, and the svs approach. The right-hand graph shows the same experiment, but using exponential search. In both arrangements the svs ordering outperforms the multi-set approaches, presumably as a consequence of the more tightly localized

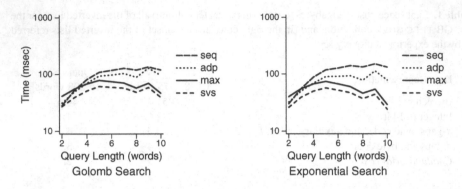

Fig. 2. Efficiency of different intersection algorithms with non-sequential search mechanisms, for different query lengths in the TREC GOV2 dataset, and an array set representation. Methods adp, seq, and max are multi-way methods; svs is the set-versus-set approach.

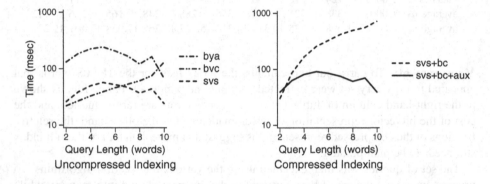

Fig. 3. Efficiency of intersection algorithms for different query lengths in the TREC GOV2 dataset on a 2.8 Ghz Intel Xeon with 2 GB of RAM: (a) fast methods, including the use of random access, where svs is set-vs-set, using exponential search, bya is set-vs-set using the Baeza-Yates method and binary search, and bvc is a bitvector-based evaluation; and (b) two methods that use compressed data formats, where svs+bc involves sequential processing of byte codes, and svs+bc+aux makes use of byte codes indexed by an auxiliary array with $k = 2$.

memory access pattern. Comparing the two graphs, the svs method benefits slightly from exponential search. Note also that execution time tends not to grow as more terms are added to the query – the cost is largely determined by the frequency of the rarest element, and long queries are likely to use at least one low-frequency term.

We also tested the binary search-based Baeza-Yates [2004] method, which is adaptive by virtue of the search sequence. It operates on two sets at a time, but has little locality of reference, and was slower than the svs approach.

Compressed indexing: Figure 3 compares uncompressed representations with two different compressed representations, in all cases using an underlying set-versus-set approach. In the left-hand graph the three best methods are shown – two using array

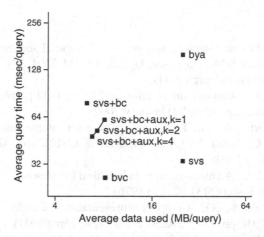

Fig. 4. Tradeoffs between index cost and query throughput, plotted as the average amount of data processed per query over all 27,004 queries, versus the average time taken per query

representations and non-sequential searching methods, and one (labeled bvc) based on bitvector operations. In the right-hand graph both lines refer to indexes stored in compressed form using byte codes. In the svs+bc approach, F-SEARCH operations are handled via sequential access and linear search; and in the svs+bc+aux method, through the use of an auxiliary index array. Use of the index array greatly increases processing speed on long queries, and allows intersections to be handled in times very close to the uncompressed svs cost in the left-hand graph.

Disk traffic: One aspect of our experiments that is not completely faithful to the operations of an information retrieval system is that we have not measured disk traffic as part of the query cost. Figure 4 shows data transfer volumes plotted against query time, in both cases averaged over the mix of 27,004 queries. The svs+bc+aux methods, using the auxiliary array, require only slightly more disk traffic than the fully-compressed svs+bc approach, and execute in as little as half the time. The indexed compressed methods are slower than the uncompressed svs+exp method, using exponential search, but the latter involves more disk traffic. The bvc bitvector approach also provides a surprising blend of data transfer economy (because most query terms are relatively common in the collection) and query speed (because most queries are short). It may be that hybrid approaches involving some terms stored as bitvectors and some using byte codes are capable of even faster performance, and we plan to explore this option next.

Acknowledgment. The second author was funded by the Australian Research Council, and by the ARC Center for Perceptive and Intelligent Machines in Complex Environments. National ICT Australia (NICTA) is funded by the Australian Government's Backing Australia's Ability initiative, in part through the Australian Research Council.

References

Baeza-Yates, R.: A fast set intersection algorithm for sorted sequences. In: Sahinalp, S.C., Muthukrishnan, S.M., Dogrusoz, U. (eds.) CPM 2004. LNCS, vol. 3109, pp. 400–408. Springer, Heidelberg (2004)

Barbay, J., Kenyon, C.: Adaptive intersection and t-threshold problems. In: Eppstein, D. (ed.) SODA 2002, pp. 390–399 (January 2002)

Barbay, J., López-Ortiz, A., Lu, T.: Faster adaptive set intersections for text searching. In: Àlvarez, C., Serna, M. (eds.) WEA 2006. LNCS, vol. 4007, pp. 146–157. Springer, Heidelberg (2006)

Bentley, J., Yao, A.C-C.: An almost optimal algorithm for unbounded searching. Information Processing Letters 5(3), 82–87 (1976)

Blandford, D.K., Blelloch, G.E.: Compact representations of ordered sets. In: Munro, J.I. (ed.) SODA 2004, pp. 11–19. ACM Press, New York (2004)

Clark, D.: Compact PAT trees. PhD thesis, University of Waterloo (1996)

Demaine, E.D., López-Ortiz, A., Munro, J.I.: Adaptive set intersections, unions, and differences. In: SODA 2000, pp. 743–752 (2000)

Gupta, A., Hon, W.-K., Shah, R., Vitter, J.S.: Compressed dictionaries: Space measures, data sets, and experiments. In: Àlvarez, C., Serna, M. (eds.) WEA 2006. LNCS, vol. 4007, pp. 158–169. Springer, Heidelberg (2006)

Hwang, F.K., Lin, S.: A simple algorithm for merging two disjoint linearly ordered list. SIAM Journal on Computing 1, 31–39 (1973)

Jacobson, G.: Succinct static data structures. PhD thesis, Carnegie Mellon University (1988)

Moffat, A., Stuiver, L.: Binary interpolative coding for effective index compression. Information Retrieval 3(1), 25–47 (2000)

Moffat, A., Zobel, J.: Self-indexing inverted files for fast text retrieval. ACM Transactions on Information Systems 14(4), 349–379 (1996)

Munro, J.I.: Tables. In: Chandru, V., Vinay, V. (eds.) STACS. LNCS, vol. 1180, pp. 37–42. Springer, Heidelberg (1996)

Pagh, R.: Low redundancy in static dictionaries with constant time query. SIAM Journal on Computing 31(2), 353–363 (2001), http://www.brics.dk/ pagh/ papers/dict-jour.pdf

Raman, R., Raman, V., Rao, S.S.: Succinct indexable dictionaries with applications to encoding k-ary trees and multisets. In: Munro, J.I. (ed.) SODA 2002. Society for Industrial and Applied Mathematics, pp. 233–242 (January 2002)

Sanders, P., Transier, F.: Intersection in integer inverted indices. In: ALENEX 2007, pp. 71–83 (January 2007)

Spink, A., Wolfram, D., Jansen, B.J., Saracevic, T.: Searching the web: The public and their queries. Journal of the American Society for Information Science 52(3), 226–234 (2001)

Witten, I.H., Moffat, A., Bell, T.A.: Managing Gigabytes: Compressing and Indexing Documents and Images, 2nd edn. Morgan Kaufmann, San Francisco (1999)

Zobel, J., Moffat, A.: Inverted files for text search engines. ACM Computing Surveys 38(2), 1–56 (2006)

Approximate Swap and Mismatch Edit Distance

Yair Dombb[1], Ohad Lipsky[1], Benny Porat[1], Ely Porat[1,2], and Asaf Tsur[1]

[1] Department of Computer Science,
Bar-Ilan University, Ramat-Gan, Israel
dombby@cs.biu.ac.il, ohadlipsky@yahoo.com, bennyporat@gmail.com,
zurasa@cs.biu.ac.il
Tel.: (972-3)531-8866
[2] Bar-Ilan University and Google Inc.
porately@cs.biu.ac.il

Abstract. There is no known algorithm that solves the general case of the *Approximate Edit Distance* problem, where the edit operations are: insertion, deletion, mismatch, and swap, in time $o(nm)$, where n is the length of the text and m is the length of the pattern.

In the effort to study this problem, the edit operations were analyzed independently. Karloff [10] showed an algorithm that approximates the edit distance problem with only the mismatch operation in time $O(\frac{1}{\epsilon^2} n \log^3 m)$. Amir et. al. [3] showed that if the only edit operations allowed are swap and mismatch, then the exact edit distance problem can be solved in time $O(n\sqrt{m}\log m)$.

In this paper, we discuss the problem of *approximate edit distance with swap and mismatch*. We show a randomized $O(\frac{1}{\epsilon^3} n \log n \log^3 m)$ time algorithm for the problem. The algorithm guarantees an approximation factor of $(1 + \epsilon)$ with probability of at least $1 - \frac{1}{n}$.

1 Introduction

Approximate string matching is a widely studied area in computer science. In approximate matching, one defines a distance metric between the objects (e.g. strings, matrices) and seeks all text location where the pattern matches the text by a pre-specified "small" distance.

The earliest and best known distance functions are Levenshtein's *edit distance* [13] and the *Hamming distance*. Let n be the text length and m the pattern length. Lowrance and Wagner [14,15] proposed an $O(nm)$ dynamic programming algorithm for the extended edit distance problem. In [12] the first $O(kn)$ algorithm was given for the edit distance with only k allowed edit operations. Cole and Hariharan [7] presented an $O(nk^4/m + n + m)$ algorithm for this problem. To this moment, however, there is no known algorithm that solves the general case of the extended edit distance problem in time $o(nm)$.

Since the upper bound for the edit distance seems very tough to break, attempts were made to consider the edit operations separately. If only mismatches are counted for the distance metric, we get the *Hamming distance*, which defines the *string matching with mismatches* problem. A great amount of work was done

N. Ziviani and R. Baeza-Yates (Eds.): SPIRE 2007, LNCS 4726, pp. 149–163, 2007.

on finding efficient algorithms for string matching with mismatches [1,11,5]. The most efficient deterministic worst-case algorithm for finding the Hamming distance of the pattern at every text location runs in time $O(n\sqrt{m}\log m)$. Karloff [10] presented an $O(\frac{1}{\epsilon^2}n\log^3 m)$ time algorithm that approximates the Hamming distance with a factor of $1 + \epsilon$. Attalah, Chyzak and Dumas [6] presented a randomized algorithm for estimating the number of matches in each text location in time $O(kn\log m)$ with variance of $(m - c_i)/k$. Isolating the swap edit operation yielded even better results [2,4], with a worst-case running time of $O(n\log m\log \sigma)$.

Amir, Eisenberg and Porat [3] faced the challenge of integration of the above two results. Integration has proven tricky since various algorithms often involve different techniques. For example, there are efficient algorithms for string matching with don't cares (e.g. [9]) and efficient algorithms for indexing exact matching (e.g. [16]), both are over 30 years old. Yet there is no known efficient algorithm for indexing with don't cares. In fact, sometimes the integration of two efficiently solvable operations ends up intractable. For example, Wagner [15] proves that edit distance with the *two* operations: *insertion* and *swap* in $\mathcal{NP}$-hard, while each one separately can be solved in polynomial time and the general edit distance – consisting of the *four* operations *insertion, deletion, mismatch* and *swap* – is also polynomially solvable.

In this context they [3] provided an efficient algorithm for edit distance with *two* operations: *mismatch* and *swap*. Their algorithm runs in time $O(n\sqrt{m}\log m)$.

In this paper we discuss the problem of approximating the edit distance with only swap and mismatch operations allowed. We think this result is essential for a complete analysis of the "swap and mismatch edit distance" problem. We present an $O(\frac{1}{\epsilon}n\log m)$ time algorithm for binary alphabet. For a small alphabet, one can use the binary algorithm $|\Sigma|^2$ times (consider each possible pair of symbols seperately). In general alphabets we show that the problem can be solved in $O(\frac{1}{\epsilon}n\log n\log^3 m)$ independently of the size of the alphabet Σ. We also show that the problem of (approximately) counting mismatches linearly reduces to (approximate) swap and mismatch edit distance. The current best time algorithm for approximately counting mismatches runs in time $O(\frac{1}{\epsilon^2}n\log^3 m)$ [10].

The techniques used by the algorithm are novel cases of overlap matching and convolutions as well as new bounded divide and conquer approach for alphabet size reduction.

2   Problem Definition

Definition 1. *Let $S = s_1 \ldots s_n$ be a string over alphabet Σ. An* edit operation *E on S is a function $E : \Sigma^n \to \Sigma^m$. Let $S = s_1 \ldots s_n$ and $T = t_1 \ldots t_\ell$ be strings over alphabet Σ. Let OP be a set of local operations. OP is called the* set of edit operations. *The* edit distance *of S and T is the minimum number k such that there exist a sequence of k edit operations $\langle E_1, ..., E_k \rangle$ for which $E_k(E_{k-1}(\cdots E_1(T)\cdots)) = S$.*

Example. The Lowrance and Wagner edit operations are: $\{INS_{i,\sigma}, DEL_i, REP_{i,\sigma}, \text{and } SWAP_i\}$, where

$INS_{i,\sigma}(s_1 \ldots s_n) = s_1 \ldots s_i, \sigma, s_{i+1} \ldots s_n,$

$DEL_i(s_1 \ldots s_n) = s_1 \ldots s_{i-1}, s_{i+1} \ldots s_n,$

$REP_{i,\sigma}(s_1 \ldots s_n) = s_1 \ldots s_{i-1}, \sigma s_{i+1} \ldots s_n, \text{ and}$

$SWAP_i(s_1 \ldots s_n) = s_1 \ldots s_{i-1}, s_{i+1}, s_i, s_{i+2} \ldots s_n.$

Definition 2. *Let* $T = t_1 \ldots t_n$ *be a text string, and* $P = p_1 \ldots p_m$ *be a pattern string over alphabet* Σ. *The edit distance problem of P in T is that of computing, for each* $i = 1, ..., n$, *the minimum edit distance of P and a prefix of* $t_i \ldots t_n$.

Lowrance and Wagner [14,15] give an $O(nm)$ algorithm for computing the edit distance problem with the above four edit operations. To date, no better algorithm is known for the general case. We consider the following problem.

Definition 3. *The* swap and mismatch edit distance problem *is the following.*

INPUT: Text string $T = t_1 \ldots t_n$ *and pattern string* $P = p_1 \ldots p_m$ *over alphabet* Σ.
OUTPUT: For each $i = 1, ..., n$, *compute the minimum edit distance of P and a prefix of* $t_i \ldots t_n$, *where the edit operations are* $\{REP_{i,\sigma}, SWAP_i\}$.

The following observation plays a role in our algorithm.

Observation 1. *Every swap operation can be viewed as two replacement operations.*

2.1   Convolutions

Convolutions are used for filtering in signal processing and other applications. A convolution uses two initial functions, t and p, to produce a third function $t \otimes p$. We formally define a discrete convolution.

Definition 4. *Let T be a function whose domain is* $\{0, ..., n-1\}$ *and P a function whose domain is* $\{0, ..., m-1\}$. *We may view T and P as arrays of numbers, whose lengths are n and m, respectively. The discrete convolution of T and P is the polynomial multiplication* $T \otimes P$, *where:*

$$(T \otimes P)[j] = \sum_{i=0}^{m-1} T[j+i]P[i], \quad j = 0, ..., n-m+1.$$

In the general case, the convolution can be computed by using the Fast Fourier Transform (FFT) [8] on T and P^R, the reverse of P. This can be done in time $O(n \log m)$, in a computational model with word size $O(\log m)$.

Important Property: The crucial property contributing to the usefulness of convolutions is the following. For every fixed location j_0 in T, we are, in essence, overlaying P on T, starting at j_0, i.e. $P[0]$ corresponds to $T[j_0]$, $P[1]$ to $T[j_0 + 1]$, ..., $P[i]$ to $T[j_0 + i]$, ..., $P[m-1]$ to $T[j_0 + m - 1]$. We multiply each element of P by its corresponding element of T and add all m resulting products. This is the convolution's value at location j_0.

Clearly, computing the convolution's value for every text location j, can be done in time $O(nm)$. The fortunate property of convolutions over algebraically close fields is that they can be computed *for all* n text locations in time $O(n \log m)$ using the FFT.

In the next few sections we will be using this property of convolutions to efficiently compute relations of patterns in texts. This will be done via *linear reductions* to convolutions. In the definition below $\mathcal{N}$ represents the natural numbers and $\mathcal{R}$ represents the real numbers.

Definition 5. *Let P be a pattern of length m and T a text of length n over some alphabet Σ. Let $R(S_1, S_2)$ be a relation on strings of length m over Σ. We say that the relation R holds between P and location j of T if $R(P[0] \cdots P[m-1], T[j]T[j+1] \cdots T[j+m-1])$.*

We say that R is linearly reduced *to convolutions if there exist a natural number c, a constant time computable function $f : \mathcal{N}^c \to \{0,1\}$, and linear time functions $\ell_1^m, ..., \ell_c^m$ and $r_1^n, ..., r_c^n$, $\forall n, m \in \mathcal{N}$, where $\ell_i^m : \Sigma^m \to \mathcal{R}^m$, $r_i^n : \Sigma^n \to \mathcal{R}^n$, $i = 1, ..., c$ such that R holds between P and location j in T iff $f(\ell_1^m(P) \otimes r_1^n(T)[j], \ell_2^m(P) \otimes r_2^n(T)[j], ..., \ell_c^m(P) \otimes r_c^n(T)[j]) = 1$.*

Let R be a relation that is linearly reduced to convolutions. It follows immediately from the definition that, using the FFT to compute the c convolutions, it is possible to find all locations j in T where relation R holds in time $O(n \log m)$.

Example. Let $\Sigma = \{a, b\}$ and R the equality relation. The locations where R holds between P and T are the locations j where $T[j+i] = P[i]$, $i = 0, ..., m-1$. Fischer and Patterson [9] showed that it can be computed in time $O(n \log m)$ by the following trivial reduction to convolutions.

Let $\ell_1 = \chi_a$, $\ell_2 = \chi_b$, $r_1 = \chi_{\bar{a}}$, $r_2 = \chi_{\bar{b}}$ where

$$\chi_\sigma(x) = \begin{cases} 1, & \text{if } x = \sigma; \\ 0, & \text{otherwise.} \end{cases} \quad \text{and} \quad \chi_{\bar{\sigma}}(x) = \begin{cases} 1, & \text{if } x \neq \sigma; \\ 0, & \text{otherwise.} \end{cases}$$

and where we extend the definition of the functions χ_σ to a strings in the usual manner, i.e. for $S = s_1 s_2 \ldots s_n$, $\chi_\sigma(S) = \chi_\sigma(s_1)\chi_\sigma(s_2) \ldots \chi_\sigma(s_n)$.

Let

$$f(x,y) = \begin{cases} 1, & \text{if } x = y = 0; \\ & \text{otherwise.} \end{cases}$$

Then for every text location j, $f(\ell_1(P) \otimes r_1(T)[j], \ell_2(P) \otimes r_2(T)[j]) = 0$ iff there is an exact matching of P at location j of T.

3   Algorithm for Binary Alphabets

For simplicity's sake we solve the problem for binary alphabets $\Sigma = \{0, 1\}$. We later show how to handle larger alphabets.

When considering a binary alphabet, a swap operation can be effective only in the cases where the text has a pair 10 aligned with an 01 in the pattern, or vice versa. Therefore, we are interested in analysing the case of alternating sequences of zeros and ones. We define this concept formally, since it is the key to the algorithm's idea.

An alternating segment of a string $S \in \{0,1\}^*$ is a substring alternating between 0s and 1s. A maximal alternating segment, or *segment* for short, is an alternating segment such that the character to the left of the leftmost character x in the alternating segment, if any, is identical to x, and similarly, the character to the right of the rightmost character y, if any, is identical to y.

Any string over $\Sigma = \{0,1\}$ can be represented as a concatenation of segments. We need to identify the cases where aligned text and pattern segments match via swap operations only and the cases where replacements are also necessary.

We now show the key property necessary to reduce swap matching to overlap matching. To this end we partition the text and pattern into segments.

Example. Let $P = 1010001110101001101001110101$.
P's segments are: 1010 0 01 1 101010 01 1010 01 1 101010.
The following lemma was proven at [2].

Lemma 1. *The pattern does not (swap-) match in particular alignment if and only if there exists a segment A in the text and a segment B in the pattern such that: 1. The characters of A and B misalign in the overlap. 2. The overlap is of odd-length.*

The conditions of the above lemma are also useful for our problem.

Lemma 2. *The number of mismatches that are not part of a swap, is exactly the number of the overlaps that implement condition 1. and 2. of lemma 1.*

Proof. We will examine all possibilities:

1. Condition 1. of the lemma does not hold. Then there is no misalignment of the text. Indeed it matches the pattern.
2. Condition 1. holds but condition 2. does not. According to lemma 1 there is a swap-match.
3. If the two conditions hold then either one of the two segments A and B is entirely contained in the other or the overlap is a real substring of both segments. For the first case we may assume, without loss of generality, that segment B of the pattern is contained in segment A of the text (the other case is treated in a similar fashion). The situation is that there is a misalignment and the overlap length is odd. Schematically, we have (with B and A boldfaced):
   Pattern: · · ·*a***abab** · · · ·**abaa** · ··
   Text· · · ·**ababa** · · ·**baba** · ··

Since swapping $B's$ edges will not help, the only swaps possible are internal to B. This means that there is exactly one element that remains mismatched after the swap.

The other situation is when the overlap is a real substring of both segments. We assume that B starts before A (the other case is handled in a similar fashion). The situation is:

Pattern: **abab** · · · ·**aba**a · ··
Text: · *b*ba · · · ·**ba**ba · ··

Again it is clear that the only possible swaps are internal to B, leaving one
element mismatched even after all possible swaps.   □

The outline of our algorithm is as follows:

1. For every text location i, count the number of mismatches m_i of the pattern
   starting at location i using two convolutions as described in [9].
2. Partition the text and pattern into segments by parity of starting and ending
   locations, and by length.
3. For every text location i, count the number r_i of odd-length misaligned overlaps,
   allowing an error of $\pm \epsilon m_i$. An odd-length misaligned overlap causes exactly one
   "real" (non-swap) mismatch.
4. The approximate number of swap-errors at location i is $s_i = (m_i - r_i)/2$.
5. The approximate number of swap and mismatch edit errors at location i is $r_i + s_i$.

3.1   Grouping Text Segments by Parity of Starting and Ending Location

We follow some of the implementation ideas of [2]. However, there it was only
necessary to check existence of odd-length mismatched overlaps, and we need
to count them as well. The main idea we use is to separate the segments of the
text and pattern into a small number of groups. In each of these groups it will
be possible to count the required overlaps in time $O(n\sqrt{m}\log m)$ using a limited
divide-and-conquer scheme based on polynomial multiplications (convolutions).
In the subsections that follow we handle the different cases. Some of these cases
necessitate new and creative uses of convolutions.

For checking the parity of the overlap, we need to know whether a text segment
ends at an odd or even text location. Consequently, we define new texts where
each text has *exactly* those segments of a given start and end parity, with all
other text elements defined as ϕ (*don't care*) which consequently never contribute
an error. (In the polynomial multiplication there will always be a 0 in these text
locations.) Henceforth, for short, we will talk of multiplying a text and pattern,
by which we mean a polynomial multiplication, where each of the text and
pattern is viewed as a vector of coefficients for the corresponding polynomials.

Definition. T^{oo} is a string of length n where for every location i, if t_i is in a
segment whose first element is in an odd location and whose last element is in an
odd location, $T^{oo}[i] = T[i]$. In all other locations j, $T^{oo}[j] = \phi$. T^{ee} is defined in
a similar fashion. We similarly define T^{oe} and T^{eo}, but for technical reasons we
need to further split these cases into $T_1^{oe}, T_2^{oe}, T_1^{eo}$ and T_2^{eo}. T_1^{oe} contains all the
odds *oe*-segments (the first *oe*-segment, the third *oe*-segment, etc.). T_2^{oe} contains
all the even *oe*-segments (the second *oe*-segment, the forth *oe*-segment, etc.). We
similarly define T_1^{eo} and T_2^{eo}.

Example. T = 10100011101010011010100111010101.
T segments are: 1010 0 01 1 101010 01 1010 01 1 1010101

$$T^{oo} \;=\; \phi\phi\phi1\phi\phi\phi\phi\phi\phi\phi\phi\phi\phi\phi\phi\phi\phi\phi1\phi\phi\phi\phi\phi\phi$$
$$T^{ee} \;=\; \phi\phi\phi\phi\phi\phi1\phi\phi\phi\phi\phi\phi\phi\phi\phi\phi\phi\phi\phi\phi\phi\phi\phi\phi$$
$$T^{oe}_1 \;=\; 1111\phi\phi\phi\phi\phi\phi\phi\phi\phi11\phi\phi\phi\phi11\phi\phi\phi\phi\phi\phi\phi$$
$$T^{oe}_2 \;=\; \phi\phi\phi\phi\phi\phi\phi111111\phi\phi1111\phi\phi\phi\phi\phi\phi\phi\phi\phi\phi\phi\phi$$
$$T^{eo}_1 \;=\; \phi\phi\phi\phi\phi11\phi\phi\phi\phi\phi\phi\phi\phi\phi\phi\phi\phi\phi\phi\phi\phi\phi\phi\phi\phi\phi\phi\phi\phi$$
$$T^{eo}_2 \;=\; \phi\phi\phi\phi\phi\phi\phi\phi\phi\phi\phi\phi\phi\phi\phi\phi\phi\phi\phi\phi\phi\phi\phi\phi\phi\phi\phi\phi\phi\phi\phi$$

Note that the segments of T^{oo} are exactly the segments of T that start and end at an odd location.

3.2   Grouping the Pattern Segments

The pattern segments are defined in exactly the same way as the text segments and we would like to group them in the same way. However, there is a difficulty here in "nailing down" the parity of a location, since the pattern is shifted and compared to every text location, i.e. the grouping of pattern segments needs to be related to the text parity in order to measure the parity of overlap. However, since the only property we have used in our grouping of text segments was the parity of the text location, it is clear that in all the pattern alignments the pattern segments that start in odd text locations are either all at odd pattern locations or all at even pattern locations. Similarly, these parities will be the same for all pattern alignments that start in even text locations.

We are now down to the combinations $T^{ti,tj}$ and $P^{pi,pj}$. This gives us 16 cases. Many cases are similar, though, so we need to handle separately only the following three types of cases:

1. $T^{ti,tj}$ and $P^{pi,pj}$ where either $ti = pi$ or $tj = pj$. (This type covers 12 cases.) These situations are handled in Section 4.
2. $T^{ti,tj}$ and $P^{pi,pj}$ where $ti, tj = oe$ and $pi, pj = eo$; or where $ti, tj = eo$ and $pi, pj = oe$. These cases are handled in Section 5.
3. $T^{ti,tj}$ and $P^{pi,pj}$ where $ti, tj = oo$ and $pi, pj = ee$; or where $ti, tj = ee$ and $pi, pj = oo$. These cases are handled in Section 6.

4   Segments with Equal Parity Start or End

Consider the case $T^{ti,tj}$ and $P^{pi,pj}$ where $ti = pi$.

Observation 2. *For every two segments, S_t in $T^{ti,tj}$, starting at location x, and S_p in $P^{pi,pj}$, starting at location y, $|x - y|$ is always even.*

We are interested in the number of odd overlaps. We now show a convolution for which the resulting value at location i is n exactly if there is n odd-length overlaps with the pattern starting at location i. (The convolution with T^{eo} (or T^{oe}, P^{eo}, P^{oe}) we need to do two convolutions, the first with T^{eo}_1, and the second with T^{eo}_2).

The Convolution: Pattern $P' = p'_1 \cdots p'_m$ is constructed as follows:

$$p'_i = \begin{cases} 0, & \text{if } P^{pi,pj}[i] = \phi; \\ 1, & \text{otherwise.} \end{cases}$$

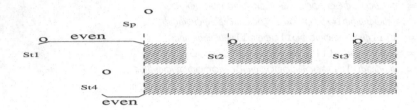

Fig. 1. The cases of both text and pattern segments starting in locations with the same parity

Text $T' = t'_1 \cdots t'_n$ is constructed by replacing every ϕ in $T^{ti,tj}$ by 0, and every segment in $T^{ti,tj}$ by a segment of alternating 1s and -1s, starting with 1. Then P' and T' are convolved.

Lemma 3. *Let $(T' \otimes P')[q]$ be the qth element in the result of the convolution. $(T' \otimes P')[q]$ is equal to the number of odd overlaps of the relevant text segments and relevant pattern segments.*

Proof. This follows from the definitions of convolutions, of T', and of P' and from the observation that for all cases where the starting location of a pattern segment is smaller than the starting location of a text segment and the pair overlaps the contribution to the result of the convolution will be 1 if the length of the overlap is odd and 0 if it is even (since every text segment starts with a 1 and then alternates between -1 and 1). Because of Observation 2, even when the text segment starts at a smaller location than the pattern segment, the difference between the starting locations has even length. Therefore in the area of the overlap, the text starts with a 1 and alternates between -1 and 1. Thus the convolution gives us the desired result. $\square$

Locations where $(T' \otimes P')[q] = 0$ are locations without odd-overlap between relevant text and pattern segments.

This solves all eight cases of $T^{ti,tj}$ and $P^{pi,pj}$ where $ti = pi$. For the additional four cases where $tj = pj$ we simply reverse the text and pattern to obtain the case considered above.

Note that this gives us the *exact* number of odd-length misaligned overlaps of segments with equal parity start or end.

5   The Odd-Even Even-Odd Segments

Consider the case T^{oe}_{1or2} and P^{eo}_{1or2} (the case of T^{eo}_{1or2} and P^{oe}_{1or2} is symmetric).

Terminology. Let S_t be a text segment whose starting location is s_t and whose ending location is f_t. Let S_p be a pattern segment being compared to the text at starting position s_p and ending position f_p. If $s_t < s_p < f_p < f_t$ then we say that S_t *contains* S_p. If $s_p < s_t < f_t < f_p$ then we say that S_p *contains*

S_t. If $s_t < s_p < f_t < f_p$ then we say that S_t has a *left overlap* with S_p. If $s_p < s_t < f_p < f_t$ then we say that S_t has a *right overlap* with S_p. We will sometimes refer to a left or right overlap as a *side overlap*.

Observation 3. *For every two segments, S_t in T^{oe}_{1or2} and S_p in P^{eo}_{1or2} if either S_p is contained in S_t or S_t is contained in S_p then the overlap is of even length. If the overlap is a left overlap or right overlap then it is of odd length. All possible cases are shown in figure 2 below.*

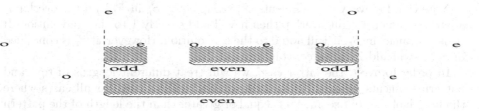

Fig. 2. The cases where the text segment starts at an odd location and ends at an even location, and the pattern segment does the opposite

The correctness of the observation is immediate. Segments of these types have even length. Thus, if one contains the other the overlap is necessarily of even length. Conversely, in case of a left or right overlap, which we call side overlaps, the overlap starting and ending locations have the same parity, making the length of the overlap odd. Remember our desire is to count all locations where there are segments which have odd overlap.

5.1   Grouping Text and Pattern Segments by Length

In addition to grouping the text and pattern segments by parity of their starting and ending indexes, we will also group them by their length. We will need the grouping by length for the convolutions presented in the next subsection.

Definition. For every length l we will define $T^{oo,l}$ to be the string of length n where for every location i if t_i is in a segment of length l which starts and ends in an odd location, $T^{oo,l}[i] = 1$. In all other locations j, $T^{oo,l}[j] = \phi$.

We will also define $T^{oo,\leq l}$ ($T^{oo,>l}$) to be the string of length n where for every location i if t_i is in a segment of length $\leq l$ ($> l$) which starts and ends in an odd location, $T^{oo,l}[i] = 1$. In all other locations j, $T^{oo,\leq l}[j] = \phi$ ($T^{oo,>l}[j] = \phi$).

$T^{ee,l}$, $T^{ee,\leq l}$ and $T^{ee,>l}$ are defined in a similar fashion. We similarly define $T^{t_i,t_j,l}$, $T^{t_i,t_j,\leq l}$ and $T^{t_i,t_j,>l}$ for $t_i \neq t_j$ but as before we will split these cases into $T^{t_i,t_j,l}_\alpha$, $T^{t_i,t_j,\leq l}_\alpha$ and $T^{t_i,t_j,>l}_\alpha$ for $\alpha \in \{1,2\}$.

As before, we will use convolutions. The desired property for such a convolution is as follows.

5.2   The Convolutions for the Odd-Even Even-Odd Segments Case

As we said earlier, for counting the 'real' mismatches in this case it is enough to count all side (left or right) overlaps. For that, we use the following convolutions:

Convolution 1. The Pattern $P' = p'_1 \cdots p'_m$ is constructed as follows:

$$p'_i = \begin{cases} 0, & \text{if } P^{pi,pj}[i] = \phi; \\ 1, & \text{otherwise.} \end{cases}$$

Text $T' = t'_1 \cdots t'_n$ is constructed by replacing every ϕ in $T^{ti,tj}$ by 0, and first and last place in every segment in T by 1, all the other places by 0.

Note that for every two segments, S_t in T^{oe}_{1or2} and S_p in P^{eo}_{1or2} if the overlap is a left overlap or a right overlap then it will add exactly 1 to the convolution. If S_p is contained in S_t, it will add 0 to the convolution. However, if S_t is contained in S_p, it will add 2 to the convolution.

In order to avoid the latter case, we will treat different lengths of text and pattern segments separately. This convolution will be used for all cases where the length of the text segments is equal or greater than the length of the pattern segments. For the cases where the pattern segments are longer than the text segments, we will use the following convolution:

Convolution 2. The Text $T' = t'_1 \cdots t'_n$ is constructed as follows:

$$t'_i = \begin{cases} 0, & \text{if } T^{ti,tj}[i] = \phi; \\ 1, & \text{otherwise.} \end{cases}$$

Pattern $P' = p'_1 \cdots p'_m$ is constructed by replacing every ϕ in $P^{pi,pj}$ by 0, and first and last place in every segment in P by 1, all the other places by 0.

Like in convolution 1, for every two segments, S_t in T^{oe}_{1or2} and S_p in P^{eo}_{1or2} if the overlap is a left overlap or a right overlap then it will add exactly 1 to the convolution. However, using this convolution, if S_t is contained in S_p, it will add 0 to the convolution, and if S_p is contained in S_t, it will add 2 to the convolution. Therefore, this convolution can be used to handle the cases where the length of the pattern segments is greater than the length of the text segments.

Using these convolutions for counting the exact number of real mismatches will require treating every length of pattern segments separately. Alas, this might take up to $O(\sqrt{m})$ convolutions, and so in order decrease the number of convolutions, we will group all segments of length $\frac{4}{\epsilon}$ or longer and treat them as one length. We will notate $T^L_{1or2} = T^{oe,\geq \frac{4}{\epsilon}}_{1or2}$, and similarly $P^L_{1or2} = P^{eo,\geq \frac{4}{\epsilon}}_{1or2}$.

For each length $l < L$ we will use convolution 1 to count real mismatches between $T^{oe,l}_{1or2}$ and $P^{eo,\leq l}_{1or2}$. Since we are comparing text segments only to shorter pattern segments, convolution 1 counts the exact number of mismatches. Also, we will use convolution 2 to count real mismatches between $T^{oe,<l}_{1or2}$ and $P^{eo,l}_{1or2}$. The total number of convolutions for this part is at most $\frac{8}{\epsilon}$.

In addition, we will use convolution 1 once to count all real mismatches between $T^{oe,L}_{1or2}$ and $P^{eo,<L}_{1or2}$, and convolution 2 once for counting all real mismatches between $T^{oe,<L}_{1or2}$ and $P^{eo,L}_{1or2}$.

Finally, we will use convolution 1 once more to count all real mismatches between $T_{1or2}^{oe,L}$ and $P_{1or2}^{eo,L}$. This convolution will count 2 mismatches for every occurrence where S_t is contained in S_p, but note that every such overlap (which is at least $\frac{4}{\epsilon}$ characters long) has at least $\frac{2}{\epsilon}$ swap errors, so the mistake when counting *swap and mismatch* errors will be at most ϵ.

6   The Odd-Odd Even-Even Segments

Consider the case T^{oo} and P^{ee} (the case T^{ee} and P^{oo} is symmetric).

Observation 4. *For every two segments, S_t in T^{oo} and S_p in P^{ee} if either S_p is contained in S_t or S_t is contained in S_p then the overlap is of odd length. If the overlap is a left overlap or right overlap then it is of even length. All possible cases are shown in figure 3 below.*

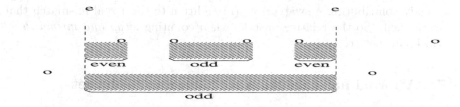

Fig. 3. Every containment has odd length; every side overlap has even length

The correctness of the observation is immediate. Segments of these types have odd lengths, thus if one contains the other then the overlap is necessarily of odd length. Conversely, in case of a left or right overlap, the overlap starting and ending locations have opposite parity, making the length of the overlap even.

6.1   The Convolutions for the Odd-Odd Even-Even Segments Case

The Convolution. Text $T' = t'_1 \cdots t'_n$ is constructed as follows:

$$t'_i = \begin{cases} 0, & \text{if } T^{ti,tj}[i] = \phi; \\ 1, & \text{otherwise.} \end{cases}$$

Pattern $P' = p'_1 \cdots p'_m$ is constructed by replacing every ϕ in $P^{pi,pj}$ by 0, and every segment in $P^{pi,pj}$ by a segment of alternating 1s and -1s, starting with 1. Then P' and T' are convolved.

Note that for every two segments, S_t in T^{oo} and S_p in P^{ee} if the overlap is a left overlap or a right overlap then it will add exactly 0 to the convolution. If S_p is contained in S_t, it will add 1 to the convolution. However, if S_t is contained in S_p, it will add -1 to the convolution.

In order to avoid miscalculations caused by some of the mismatches contributing 1 to the sum while others are contributing -1 to the sum, we will, like in the Odd-Even Even-Odd case, treat different sizes of text and pattern segments separately. In this case, we will notate $T^L = T^{oo, \geq \frac{2}{\epsilon}}$, and similarly $P^L = P^{ee, \geq \frac{2}{\epsilon}}$.

For each length $l < L$ we will use our convolution to count real mismatches between $T^{oo,l}$ and $P^{ee, \leq l}$. Since we are comparing text segments only to shorter pattern segments, our convolution counts the exact number of mismatches. Also, we will use our convolution to count real mismatches between $T^{oo,<l}$ and $P^{ee,l}$ (we will get a negative total, and take its absolute value). The total number of convolutions for this part is at most $\frac{4}{\epsilon}$.

In addition, we will use this convolution once to count all real mismatches between $T^{oo,L}$ and $P^{ee,<L}$, and again for counting all real mismatches between $T^{oo,<L}$ and $P^{ee,L}$ (taking the absolute value of the result, like before).

Note that we are not doing the convolution for the case where the text *and* pattern segments are longer than $\frac{2}{\epsilon}$. Is this case, real mismatches only occur in cases where text segments are contained in pattern segments (or vice versa), and since all such segments are of length of at least $\frac{2}{\epsilon}$ characters, they are already contributing $\frac{1}{\epsilon}$ swap errors (in addition to the 1 real mismatch that we overlooked). So the relative mistake when counting *swap and mismatch* errors will be at most ϵ.

7   Approximate Solution for General Alphabet

First, we note that in the general alphabet case, every two different symbols that appear in adjacent places somewhere in the pattern define a different *segment type*. To reduce the number of segment types we handle, we will project the segment types set into a smaller set of size $\frac{1}{\epsilon}$. Once the number of different segment types is bounded, we can use the binary algorithm for each type separately and sum up the results.

Let $S \subseteq \Sigma \times \Sigma$ be the set of different segment types in the pattern, Where a segment type determined only by the alternating two symbols. Note that $|S| \leq m$. We can construct an algorithm that approximates the swap mismatch edit distance in $O(|S|f(n, m, \epsilon) + mis(n, m, \epsilon))$ Where $f(n, m, \epsilon)$ is the time needed to $(1 + \epsilon)$-approximation of the binary alphabet swap and mismatch problem and $mis(n, m, \epsilon)$ is the time needed to approximate counting mismatches. The algorithm will work as follows:

Few Segment Types Algorithm:

1. For every text location $i = 1, \ldots, n-m+1$, approximately count the number of mismatches M_i of the pattern starting at location i [10].
2. For each segment type (σ_1, σ_2)
   (a) Replace all the symbols in segments of other types with a ϕ symbol.
   (b) Compute s_i the number of swap errors with segment type (σ_1, σ_2) for every text location $i = 1, \ldots, n - m + 1$, using the algorithm for the binary alphabet case.

(c) Add s_i to S_i For every $i = 1, \ldots, n - m + 1$.

3. The approximate swap mismatch edit distance at location i is $M_i - S_i$.

Lemma 4. *The algorithm* $(1+\epsilon)$*-approximates the swap mismatch edit distance and works in time* $O(|S| \frac{1}{\epsilon} n \log m + \frac{1}{\epsilon^2} n \log^3 m)$.

Proof. The inexact parts of this algorithm are the first step of counting mismatches and the use of the binary alphabet case algorithm. Both are guaranteed to have $(1 + \epsilon)$ approximation factor and approximation is kept under addition operations. We now discuss the time needed for each step. The first step is done using Karloff's algorithm in $O(\frac{1}{\epsilon^2} n \log^3 m)$. For each segment type computing s_i's can be done by counting the number of mismatches using convolutions (since there are only two symbols, this can be done in $O(n \log m)$ time) and running the algorithm for binary alphabet case which takes $O(\frac{1}{\epsilon} n \log m)$.

Observation 5. *Every pair of segments (in the text and in the pattern) that caused swap mistakes before the projection, will also cause swap mistakes after the projection. Pairs of segments that caused mismatch mistakes before the projection might cause mismatch mistakes or swap mistakes after the projection. For a given pair, the probability of the projection changing the type of the mistakes caused by that pair from mismatch mistakes to swap mistakes, is* ϵ.

The correctness of the observation is immediate, since every pair of segment types that caused swap mistakes before the projection, must have been a pair of two identical segment types. Therefore, such pair will be projected to a pair of two identical segment types and will continue to cause swap mistakes after the projection. Pairs of segment types that caused a mismatch mistake must have been a pair of two different segment types; The probability of such a pair to be projected to a pair of two identical segment types is exactly ϵ.

Many Segment Types Algorithm:

1. Randomly choose $\frac{1}{\epsilon} \log n$ projections $\Pi_j : S \to S'$ where $|S'| = \frac{1}{\epsilon}$.
2. For each Π_j Replace segments according to Π_j and approximately count swap and mismatch errors using the algorithm for few segment types.
3. Compute approximate swap and mismatch edit distance by taking the average between the different results.

Lemma 5. *The above algorithm correctly* $(1 + \epsilon)$*-approximates the swap and mismatch edit distance with error probability less than* $\frac{1}{n^3}$.

Proof. Using observation 5 we have that the swap errors before the projection are necessarily swap errors after the projections. Furthermore, other mismatch errors have the probability of less than ϵ to change to either matches or swap errors. We use $\frac{1}{\epsilon} \log n$ independent projections. Using Chernoff inequality we get that the probability that in more than $5 \log n$ projections a specific error will change is less than $\frac{1}{n^5}$. $Pr[X > 5 \log n] \leq (\frac{e^4}{(5)^5})^{\log n} < \frac{1}{n^5}$.

The above probability is for a specific location and error. Using union bound we get a total error probability of at most $\frac{1}{n^3}$. $5\log n$ is a 5ϵ fraction of $\frac{1}{\epsilon}\log n$ projections, taking $\epsilon' = \epsilon/5$ yields the desired approximation ratio.

Time complexity. For each projection we need $O(\frac{1}{\epsilon^2}n\log^3 m)$ time. In total our algorithm works in $O(\frac{1}{\epsilon^3}n\log n\log^3 m)$ time complexity.

8   Reduction from *Mismatch* to *Swap and Mismatch*

Theorem 1. *Let A_{sm} be an algorithm that solves the problem of swap and mismatch running in time $O(f(n))$. Then there is an algorithm A_m that solves mismatch in time $O(n + f(n))$.*

Proof. Let $T = t_1, t_2, \ldots, t_n$ be a text, and $P = p_1, p_2, \ldots, p_n$ be a pattern over alphabet Σ. We want to find the Hamming distance (i.e. the number of *mismatches*) between the pattern and every location in the text. Let $\psi \notin \Sigma$ and define $T' = t_1, \psi, t_2, \psi, \ldots, \psi, t_n$. In other words, T' is T with ψ inserted between every two adjacent characters of the original text T. Define P' the same way. Run the algorithm A_{sm} on T' and P' and return the result of the odd locations.

Observation. All the mistakes that A_{sm} finds are *mismatches*. This is true since between every two characters there is a ψ character which does not appear in Σ so it is impossible to have a *swap*. The additional ψs don't contribute any additional mismatches, since each ψ in the pattern is aligned with a ψ in the text when we consider only odd locations. $\square$

References

1. Abrahamson, K.R.: Generalized string matching. SIAM J. Comput. 16(6), 1039–1051 (1987)
2. Amir, A., Cole, R., Hariharan, R., Lewenstein, M., Porat, E.: Overlap matching. Inf. Comput. 181(1), 57–74 (2003)
3. Amir, A., Eisenberg, E., Porat, E.: Swap and mismatch edit distance. In: Albers, S., Radzik, T. (eds.) ESA 2004. LNCS, vol. 3221, pp. 16–27. Springer, Heidelberg (2004)
4. Amir, A., Lewenstein, M., Porat, E.: Approximate swapped matching. Inf. Process. Lett. 83(1), 33–39 (2002)
5. Amir, A., Lewenstein, M., Porat, E.: Faster algorithms for string matching with k mismatches. J. Algorithms 50(2), 257–275 (2004)
6. Atallah, M.J., Chyzak, F., Dumas, P.: A randomized algorithm for approximate string matching. Algorithmica 29(3), 468–486 (2001)
7. Cole, R., Hariharan, R.: Approximate string matching: A faster simpler algorithm. In: SODA, pp. 463–472. ACM Press, New York (1998)
8. Cormen, T.H., Leiserson, C.E., Rivest, R.L.: Introduction to Algorithms. MIT Press and McGraw-Hill (1992)
9. Fischer, M.J., Paterson, M.S.: String matching and other products. Complexity of Computation. In: Karp, R.M. (ed.) SIAM-AMS Proceedings, vol. 7, pp. 113–125 (1974)

10. Karloff, H.: Fast algorithms for approximately counting mismatches. Information Processing Letters 48(2), 53–60 (1993)
11. Landau, G.M., Vishkin, U.: Efficient string matching with k mismatches. Theoretical Computer Science 43, 239–249 (1986)
12. Landau, G.M., Vishkin, U.: Fast parallel and serial approximate string matching. Journal of Algorithms 10(2), 157–169 (1989)
13. Levenshtein, V.I.: Binary codes capable of correcting, deletions, insertions and reversals. Soviet Phys. Dokl. 10, 707–710 (1966)
14. Lowrance, R., Wagner, R.A.: An extension of the string-to-string correction problem. J. of the ACM, 177–183 (1975)
15. Wagner, R.A.: On the complexity of the extended string-to-string correction problem. In: Proc. 7th ACM STOC, pp. 218–223. ACM Press, New York (1975)
16. Weiner, P.: Linear pattern matching algorithm. In: Proc. 14 IEEE Symposium on Switching and Automata Theory, pp. 1–11 (1973)

Approximating Constrained LCS

Zvi Gotthilf and Moshe Lewenstein

Department of Computer Science, Bar-Ilan University, Ramat Gan 52900, Israel
{gotthiz,moshe}@cs.biu.ac.il

Abstract. The problem of finding the longest common subsequence (LCS) of two given strings A and B is a well-studied problem. The Constrained longest common subsequence (C-LCS) for three strings A, B and C is the longest common subsequence of A and B that contains C as a subsequence. The fastest algorithm solving the C-LCS problem has a time complexity of $O(mnk)$ where m, n and k are the lengths of A, B and C respectively. We propose to consider the approximate version of the LCS and the Constrained LCS. For LCS we propose a simple linear time approximation algorithm that yields an approximation ratio of $\frac{1}{|\Sigma|}$. For C-LCS we obtain the first two approximation algorithms. Our first algorithm has an approximation factor of $\frac{1}{\sqrt{\min(m,n)}}$ with an $O(mn)$ running time, while the second algorithm yields a $\frac{1}{\sqrt{\min(m,n)|\Sigma|}}$ approximation factor within a running time of $O(m + n)$.

1 Introduction

The problem of finding the longest common subsequence (LCS) of two given strings A and B is a well-studied problem, see [2,5,6,9]. The *constrained* longest common subsequence (C-LCS) for three strings A, B and C is the longest common subsequence of A and B that contains C as a subsequence. Tsai [8] gave a dynamic programming algorithm for the problem which runs in $O(n^2m^2k)$ where m, n and k are the lengths of A, B and C respectively. Improved dynamic programming algorithms were proposed in [1,3] which run in time $O(nmk)$.

Many problems in pattern matching are solved with dynamic programming solutions. Among the most prominent of these is the LCS problem. These solutions are elegant and simple, yet usually their running times are quadratic or more. It is a desireable goal to find algorithms which offer faster running times. One slight improvement, a reduction of a log factor, is the use of the Four-Russians trick, see [7]. However, in general, faster algorithms have proven to be rather elusive over the years (and perhaps it is indeed impossible).

To circumvent the NP-completeness of many natural, real-life problems, approximation algorithms have been suggested for many problems. Of course, the desire in this case is to provide some *polynomial-time* algorithm. When it comes to problems which have a polynomial time algorithm it only makes sense to suggest an approximation algorithm if the time is faster than the deterministic algorithm.

N. Ziviani and R. Baeza-Yates (Eds.): SPIRE 2007, LNCS 4726, pp. 164–172, 2007.

We propose to consider the approximate version of the LCS and the Constrained LCS. We believe this is an important venue of research that can shed light on many of the problems in pattern matching which suffer quadratic (or higher) running times.

In this paper, we propose a simple linear time approximation algorithm for LCS that yields an approximation ratio of $\frac{1}{|\Sigma|}$. For C-LCS we obtain the first two approximation algorithms. Our first algorithm has an approximation factor of $\frac{1}{\sqrt{\min(m,n)}}$ with an $O(mn)$ running time, while the second algorithm yields a $\frac{1}{\sqrt{\min(m,n)|\Sigma|}}$ approximation factor within a running time of $O(m+n)$.

2   Preliminaries

Let $A = \langle a_1, a_2, \ldots, a_m \rangle$, $B = \langle b_1, b_2, \ldots, b_n \rangle$ and $C = \langle c_1, c_2, \ldots, c_k \rangle$ be three strings. The LCS problem for two strings is finding their common subsequence of maximal length. The longest constrained subsequence (C-LCS, for short) of A, B and C is the longest common subsequence of A and B that contains C as a subsequence. First we define the approximation version of the LCS as follows: Let OPT_{lcs} be the optimal solution for the LCS problem and APP_{lcs} the result of an approximation algorithm such that APP_{lcs} is a common subsequence of A and B. The approximation factor of an algorithm APP will be the smallest ratio between $|APP_{lcs}|$ and $|OPT_{lcs}|$ over all possible input strings A, B and C. Similarly we define the approximation version of the C-LCS as follows: let OPT_{clcs} be the optimal solution for the C-LCS problem and APP_{clcs} the result of an approximation algorithm such that:

- APP_{clcs} is a common subsequence of A and B.
- C is a subsequence of APP_{clcs}.

The approximation factor of an algorithm APP will be the smallest ratio between $|APP_{clcs}|$ and $|OPT_{clcs}|$ over all possible input strings A, B and C.

2.1   Approximating LCS

In order to approximate the classical LCS problem, we propose the following simple linear time algorithm that yields an approximation factor of $\frac{1}{|\Sigma|}$, which will be used later on for approximating the C-LCS problem.

Algorithm 1. LCS approximation algorithm

1   **for** *every* $s \in \Sigma$ **do**
2       $Occ_A(s) \leftarrow$ number of occurrences of $s \in A$;
3       $Occ_B(s) \leftarrow$ number of occurrences of $s \in B$;
4   **for** *every* $s \in \Sigma$ **do** $Occ(s) \leftarrow \min(Occ_A(s), Occ_B(s))$
5   find the $s \in \Sigma$ with maximal $Occ(s)$ and return $s^{Occ(s)}$;

The following observation in regards to Algorithm 1 is easy to prove.

Observation 1. *Given two strings A and B over alphabet Σ, Algorithm 1 finds a common subsequence (of A and B) of length $\geq \frac{\min(m,n)}{|\Sigma|} \geq \frac{|OPT_{lcs}|}{|\Sigma|}$.*

3   Approximating Constrained LCS

In this section we present two approximation algorithms for the constrained LCS problem. First, we give a couple of definitions that are necessary to understand both algorithms.

Let S be a string. Denote with $S[i,j]$ the substring of S that starts at location i and ends at location j .

Denote with $start(a,i)$ and $end(a,i)$ the leftmost and rightmost locations in A such that $c_1, c_2, \ldots, c_i$ is a subsequence of $A[1, start(a,i)]$ and $c_i, c_{i+1}, \ldots, c_k$ is a subsequence of $A[end(a,i), m]$ (the same holds for $start(b,i)$ and $end(b,i)$). See figure 1. Recall that m, n and k are the lengths of A, B and C respectively.

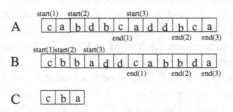

Fig. 1. An example of start(i) and end(i)

Let OPT be an optimal C-LCS solution. By definition, C must be a subsequence of OPT. Choose an arbitrary subsequence C of OPT and denote with $p_1, p_2, \ldots, p_k$ the positions of $c_1, c_2, \ldots, c_k$ in OPT (there may be many possible subsequences of C in OPT). We say that the positions $p_1, p_2, \ldots, p_k$ are an *assignment* of C over OPT.

The following Lemma and Corollary are instrumental in achieving our approximation algorithms.

Lemma 1. *Let k be the length of C and OPT the optimal C-LCS, then for any assignment of C over OPT and for every $1 \leq i \leq k-1$ the following statement holds:*

$$|LCS(A[start(a,i)+1, end(a,i+1)-1], B[start(b,i)+1, end(b,i+1)-1])| \geq |OPT[p_i+1, p_{i+1}-1]|.$$

Proof: Let us assume there is an assignment of C over OPT such that:

$|LCS(A[start(a,i)+1, end(a,i+1)-1], B[start(b,i)+1, end(b,i+1)-1])| < |OPT[p_i+1, p_{(i+1)}-1]|$. Note that $OPT[p_i+1, p_{(i+1)}-1]$ must be a common subsequence of two substrings of A and B, moreover those substrings must start at a location $\geq start(a,i) + 1$ and end at a location $\leq end(a, i+1) - 1$ (the same must hold for B). This contradicts the fact that the LCS of two substrings cannot be longer than the LCS of the original strings.   □

The next corollary follows from Lemma 1

Corollary 1. *Let k be the length of C and OPT the optimal C-LCS, then we can get a $\frac{1}{k}$-approximation algorithm for the C-LCS.*

Proof: Choosing the maximal LCS of $A[start(a,i) + 1, end(a, i+1) - 1]$ and $B[start(b,i) + 1, end(b, i+1) - 1]$ over $1 \leq i \leq k-1$ then by Lemma 1 we get a C-LCS of size $\frac{1}{k}|OPT|$.   □

3.1   An $O(mn)$ Time Approximation

In this section we show a $\sqrt{\min(m,n)}$-approximation algorithm with an $O(mn)$ running time. The motivation of this algorithm is derived from Corollary 1. We simply compute the exact LCS between $A[start(a,i) + 1, end(a, i+1) - 1]$ and $B[start(b,i) + 1, end(b, i+1) - 1]$ (for $1 \leq i \leq k-1$) using a dynamic method for updating LCS presented by Landau, Myers and Schmidt [4]. Given two strings A and B, let D be the traditional LCS dynamic programming matrix of A and B. By using the method of Landau, Myers and Schmidt we can compute in $O(|A|)$ time D', the dynamic programming matrix for the LCS of A and bB, where b is a single symbol.

Our algorithm has two cases:

1) $k \geq \sqrt{\min(m,n)}$.
2) $k < \sqrt{\min(m,n)}$.

Observation 2. *If $k \geq \sqrt{\min(m,n)}$ then any valid solution for the C-LCS problem yields an approximation factor of:* $\frac{\sqrt{\min(m,n)}}{\min(m,n)} = \frac{1}{\sqrt{\min(m,n)}}$.

Proof: Since C must be a subsequence of the C-LCS, if $k \geq \sqrt{\min(m,n)}$ then the length of any valid solution must be $\geq \sqrt{\min(m,n)}$. Therefore, using an approximation algorithm which simply returns C, we achieve an approximation factor of: $\frac{\sqrt{\min(m,n)}}{\min(m,n)} = \frac{1}{\sqrt{\min(m,n)}}$.   □

If $k < \sqrt{\min(m,n)}$ then we find a solution of length $\geq \frac{|OPT|}{\sqrt{\min(m,n)}}$. Thus, throughout this section we assume $k < \sqrt{\min(m,n)}$.

Our algorithm works as follows. First we compute the LCS of $A[a_1, end(a, 1) - 1]^R$ and $B[b_1, end(b, 1) - 1]^R$ then we scan both strings while dynamically updating the LCS matrix. We compare the lengths of the LCS of: $A[start(a,i), end(a, i+1)]^R$ and $B[start(b,i), end(b, i+1)]^R$ for every $1 \leq i \leq k-1$.

The algorithm performs one search of A and B (from left to right) using two pointers LocAStart and LocAEnd initialized with LocAStart $= 1$ and LocAEnd $= end(a, 1) - 1$ (symmetrically for B).

For every character we read, we update the LCS matrix of the reverse strings i.e. adding a character at the beginning of A or B. For every removed character we simply update the LCS matrix by deleting the last row or column.

See Algorithm 2 for a full description.

Algorithm 2. $O(mn)$ Time Approximation Algorithm

1 **if** $k \geq \sqrt{\min(m,n)}$ **then return** C and quit;
2 $tmpLCS \leftarrow$ LCS of $\langle a_1, \ldots, end(a,1) - 1 \rangle^R$ and $\langle b_1, \ldots, end(b,1) - 1 \rangle^R$;
3 $length \leftarrow length(tmpLCS)$;
4 $cLoc \leftarrow 0$;
5 **for** $i \leftarrow 1$ **to** $k - 1$ **do**
6   **while** $LocAEnd \neq end(a,i) - 1$ **do** read character of A and update the LCS matrix;
7   **while** $LocAStart \neq start(a,i) + 1$ **do** remove character of A and update the LCS matrix;
8   **while** $LocBEnd \neq end(b,i) - 1$ **do** read character of B and update the LCS matrix;
9   **while** $LocBStart \neq start(b,i) + 1$ **do** remove character of B and update the LCS matrix;
10   **if** $length(currentLCS) > length$ **then**
11       $tmpLCS \leftarrow currentLCS$;
12       $length \leftarrow length(currentLCS)$;
13       $cLoc \leftarrow i$;
14 **while** $LocAEnd \neq a_m$ **do** read character of A and update the LCS matrix;
15 **while** $LocAStart \neq start(a,k) + 1$ **do** remove character of A and update the LCS matrix;
16 **while** $LocBEnd \neq b_n$ **do** read character of B and update the LCS matrix;
17 **while** $LocBStart \neq start(b,k) + 1$ **do** remove character of B and update the LCS matrix;
18 **if** $length(currentLCS) > length$ **then**
19    $tmpLCS \leftarrow currentLCS$;
20    $cLoc \leftarrow k$;
21 **return** $\langle c_1, \ldots, c_{cLoc} \rangle \cdot \langle tmpLCS \rangle \cdot \langle c_{cLoc+1}, \ldots, c_k \rangle$;

Time and Correctness Analysis:
Let S be the output string of the algorithm. S fulfills the following conditions:

1) S is a common subsequence of A and B (while assuming C is a subsequence of both A and B).
2) S contains C as a subsequence (this happens either in the first step or in the step number 21).

The running time is $O(mn)$, since the first computation of the LCS takes $O(mn)$ and we perform no more than an additional $2(m+n)$ update operations (we insert and remove every character exactly once). Every update operation is done in a linear running time. Hence, the total running time is $O(mn)$.

Lemma 2. *The approximation ratio of the Algorithm 2 is* $\frac{1}{\sqrt{\min(m,n)}}$.

Proof: If our algorithm stops at the first line then the length of the output $\geq \sqrt{\min(m,n)}$. Moreover, $|OPT|$ must be $\leq \sqrt{\min(m,n)}$. Therefore, the approximation ratio is at least: $\frac{\sqrt{\min(m,n)}}{\min(m,n)} = \frac{1}{\sqrt{\min(m,n)}}$.

If our algorithm stops at the 21st line then $k < \sqrt{\min(m,n)}$. Say OPT consists of the following parts: $OPT[1, p_1 - 1], OPT_{P1}, OPT[p_1 + 1, p_2 - 1], OPT_{P2}, OPT$ $[p_2 + 1, p_3 - 1], \ldots, OPT[p_k + 1, p_l]$. Let $OPT[p_j + 1, p_{j+1}]$ be the part with the maximal length over all $0 \leq i \leq k + 1$ (denote p_0 as opt_1 and p_{k+1} as opt_l only for the scope of this proof).

According to Lemma 1, $|OPT[p_j + 1, p_{j+1} - 1]| \leq |LCS(A[start(a,j)+1, end$ $(a, j+1) - 1)], B[start(b,j) + 1, end(b, j+1) - 1])|$. Moreover the length of the C-LCS our algorithm returns in the 21st line

$$\geq |LCS(A[start(a,j) + 1, end(a, j+1) - 1], \tag{1}$$
$$B[start(b,j) + 1, end(b, j+1) - 1])| \tag{2}$$
$$\geq |OPT[p_j + 1, p_{j+1} - 1]| \tag{3}$$
$$\geq \frac{|OPT|}{k+1} \tag{4}$$
$$\geq \frac{|OPT|}{\sqrt{\min(m,n)}}. \tag{5}$$

Therefore, the approximation ratio of our algorithm is at least: $\frac{1}{\sqrt{\min(m,n)}}$. □

3.2   A Linear Time Approximation

In this section we present a linear time approximation algorithm for the C-LCS problem, which yields an approximation factor of $\frac{1}{\sqrt{\min(m,n)|\Sigma|}}$. During this algorithm we use the above mentioned Algorithm 1. Our goal is to find a linear time approximation algorithm with approximation factor as close as possible to the one presented in the previous section.

Recall that we desire to compute compute the LCS of $A[start(a,i)+1, end(a, i+1) - 1]$ and $B[start(b,i) + 1, end(b, i+1) - 1]$. However, since we are looking for a linear time algorithm, it is clear that we cannot compute the LCS even for a specific i, since the LCS computation may cost us $O(mn)$ running time.

Therefore, we want to use Algorithm 1 to approximate the LCS of those sections ($1 \leq i \leq k$). However, if we perform $\sqrt{\frac{\min(m,n)}{|\Sigma|}}$ iterations ($1 \leq i \leq k$) of Algorithm 1 we can still get a running time of $O(\min(m,n)^{1.5})$. Therefore, we use a more efficient method in order to find a $\frac{1}{|\Sigma|}$ approximation for the LCS problem. Our method is similar to counting sort and works as follows:

Our first goal is to find the maximal approximated LCS of $A[a_i + 1, end(a, i+1) - 1]$ and $B[b_i + 1, end(b, i+1) - 1]$ ($1 \leq i \leq k - 1$). Afterwards we set this maximal approximate LCS in its appropriate position in C and then return it as an approximate C-LCS. In order to do so in linear time, we maintain a sorted array of number of occurrences of every symbol $\in \Sigma$, since the number of occurrences of every symbol must be $\leq \min(m, n)$ we can simply maintain all the needed information in linear time and space. The algorithm performs one search of A and B using two pointers such that insertion and deletion operations are done exactly twice for every character.

Our algorithm has two cases:

1) $k \geq \sqrt{\frac{\min(m,n)}{|\Sigma|}}$.

2) $k < \sqrt{\frac{\min(m,n)}{|\Sigma|}}$.

If $k \geq \sqrt{\frac{\min(m,n)}{|\Sigma|}}$ then similarly to observation 2 any valid solution for the C-LCS problem must yield an approximation factor of $\frac{1}{\sqrt{\min(m,n)|\Sigma|}}$.

If $k < \sqrt{\frac{\min(m,n)}{|\Sigma|}}$ then we find a solution of length $\geq \frac{|OPT|}{\sqrt{\min(m,n)|\Sigma|}}$.

Thus, throughout this section we assume $k < \sqrt{\frac{\min(m,n)}{|\Sigma|}}$.

See Algorithm 3 for a full description.

During the algorithm we use the following definitions: for every $\sigma \in \Sigma$, denote with $C_A(\sigma, i, j)$ the number of σ's in $A[i, j]$ (the same for $C_B(\sigma, i, j)$). Moreover, $l[i, j, p, q](\sigma) = \min(C_A(\sigma, i, j), C_B(\sigma, p, q))$ and $l^*(i, j, p, q) = \max m[i, j, p, q](\sigma)$.

With the use of $C_A(\sigma, i, j)$, $C_B(\sigma, i, j)$ and some additional tables, the following lemma can be straightforwardly be seen to be true.

Lemma 3. $l^*(i + 1, j, p, q), l^*(i, j + 1, p, q), l^*(i, j, p + 1, q)$ and $l^*(i, j, p, q + 1)$ can be computed from $l^*(i, j, p, q)$ in $O(1)$ time, given $O(n)$ space.

Similar to the previous algorithm, we perform one search of A and B from left to right (can be done using two pointers for every string).

Time and Correctness Analysis: Let S be the output string of the algorithm. S fulfill both conditions:

1) S is common subsequence of A and B (where C is a subsequence of both A and B).

2) S contains C as a subsequence (see either first step or step number 23).

The running time is $O(m+n)$, since the computation of $l^*(1, end(a, 1) - 1, 1, end(b, 1) - 1)$ takes $O(m+n)$ and we perform no more than additional $2(m+n)$ update operations (we insert and delete every character exactly once). According to Lemma 3 every update operation is done in $O(1)$ time then the total running time remains linear.

Lemma 4. *The approximation ratio of the Linear Time C-LCS Approximation Algorithm is* $\frac{1}{\sqrt{\min(m,n)|\Sigma|}}$.

Algorithm 3. Linear Time Approximation Algorithm

1 if $k \geq \sqrt{\frac{\min(m,n)}{|\Sigma|}}$ **then return** C and quit;

2 compute $l^*(1, end(a,1) - 1, 1, end(b,1) - 1)$

3 set: $Symbol \leftarrow MaxSym$;

4 $Occ \leftarrow MaxOcc$;

5 $cLoc \leftarrow 0$;

6 **for** $i \leftarrow 1$ **to** $k - 1$ **do**

7   compute $l^*(start(a,i) + 1, end(a, i+1) - 1, start(b,i) + 1, end(b, i+1) - 1)$

8   **if** $MaxOcc > Occ$ **then**

9 $Symbol \leftarrow MaxSym$;

10 $Occ \leftarrow MaxOcc$;

11 $cLoc \leftarrow i$;

12 compute $l^*(start(a,k) + 1, m, start(b,k) + 1, n)$

13 **if** $MaxOcc > Occ$ **then**

14 $Symbol \leftarrow MaxSym$;

15 $Occ \leftarrow MaxOcc$;

16 $cLoc \leftarrow k$;

17 **return** $C[c_1, c_{cLoc}] \cdot \langle Symbol^{Occ} \rangle \cdot C[c_{cLoc+1}, c_k]$;

Proof: If our algorithm stops at the first line then the length of the output $\geq \sqrt{\frac{\min(m,n)}{|\Sigma|}}$. Moreover $|OPT|$ must be $\leq \sqrt{\min(m,n)}$. Therefore, the approximation ratio is at least: $\frac{\sqrt{\frac{\min(m,n)}{|\Sigma|}}}{\min(m,n)} = \frac{1}{\sqrt{\min(m,n)|\Sigma|}}$.

If our algorithm stops at the 23rd line then $k < \sqrt{\frac{\min(m,n)}{|\Sigma|}}$.

OPT consists of the following parts: $OPT[1, p_1 - 1], OPT_{P1}, OPT[p_1 + 1, p_2 - 1], OPT_{P2}, OPT[p_2 + 1, p_3 - 1], \ldots, OPT[p_k + 1, p_l]$. Let $OPT[p_i + 1, p_{i+1}]$ be the part with the maximal length over all $0 \leq i \leq k + 1$ (denote p_0 as opt_1 and p_{k+1} as opt_l only for the scope of this proof).

According to Lemma 1, $|OPT[p_j + 1, p_{(j+1)} - 1]| \leq |LCS(A[start(a,j) + 1, end(a, j+1) - 1], B[start(b,j) + 1, end(b, j+1) - 1])|$. Moreover the length of the C-LCS our algorithm returns in the 23rd line

$$\geq \frac{1}{|\Sigma|} |LCS(A[start(a,j) + 1, end(a, j+1) - 1], \tag{6}$$

$$B[start(b,j) + 1, end(b, j+1) - 1])| \tag{7}$$

$$\geq \frac{1}{|\Sigma|} |OPT[p_j + 1, p_{j+1} - 1]| \tag{8}$$

$$\geq \frac{1}{|\Sigma|} \frac{|OPT|}{k + 1} \tag{9}$$

$$\geq \frac{1}{|\Sigma|} \frac{|OPT|}{\sqrt{\frac{\min(m,n)}{|\Sigma|}}} \tag{10}$$

$$= \frac{|OPT|}{\sqrt{\min(m,n)|\Sigma|}} \tag{11}$$

Therefore, the approximation ratio of our algorithm is at least: $\frac{1}{\sqrt{\min(m,n)|\Sigma|}}$. □

4   Open Questions

In our opinion a central question is whether it is possible to find a linear time algorithm approximate the classical LCS problem with an approximation factor better than $\frac{1}{|\Sigma|}$. Moreover, one may ask whether there are better approximation algorithms for the C-LCS problem, which improves either the approximation factors or the running times.

Moreover, an interesting question raised is regarding the more general variant in which there are more than two strings or more than one constraint.

References

1. Arslan, A.N., Egecioglu, O.: Algorithms for the constrained longest common subsequence problems. Int. J. Found. Comput. Sci. 16(6), 1099–1109 (2005)
2. Bergroth, L., Hakonen, H., Raita, T.: A survey of longest common subsequence algorithms. In: SPIRE 2000, IEEE Computer Society Press, Washington, DC, USA (2000)
3. Chin, F.Y.L., De Santis, A., Ferrara, A.L., Ho, N.L., Kim, S.K.: A simple algorithm for the constrained sequence problems. Inf. Process. Lett. 90(4), 175–179 (2004)
4. Myers, E.W., Landau, G.M., Schmidt, J.P.: Incremental string comparison. SIAM J. Comput 27(2), 557–582 (1998)
5. Hirschberg, D.S.: A linear space algorithm for computing maximal common subsequences. Commun. ACM 18(6), 341–343 (1975)
6. Hirschberg, D.S.: Algorithms for the longest common subsequence problem. J. ACM 24(4), 664–675 (1977)
7. Masek, W.J., Paterson, M.S.: A faster algorithm computing string edit distances. J. Comput. Syst. Sci. 20, 18–31 (1980)
8. Tsai, Y.-T.: The constrained longest common subsequence problem. Inf. Process. Lett. 88(4), 173–176 (2003)
9. Ullman, J.D., Aho, A.V., Hirschberg, D.S.: Bounds on the complexity of the longest common subsequence problem. J. ACM 23(1), 1–12 (1976)

Tuning Approximate Boyer-Moore
for Gene Sequences*

Petri Kalsi, Leena Salmela, and Jorma Tarhio

Helsinki University of Technology
{pkalsi,lsalmela,tarhio}@cs.hut.fi

Abstract. Recently a new variation of approximate Boyer-Moore string matching was presented for the k-mismatch problem. This variation was developed for gene sequences. We further tuned this algorithm gaining speedups in both preprocessing and search times. Our preprocessing has lower time complexity than the previous algorithm and our experiments show that our algorithm is over 30% faster than the previous one. We also present two variations of the algorithm for the k-difference problem.

1 Introduction

We consider two variations of approximate string matching, the k-mismatch problem and the k-difference problem. In both of the problems, we have a pattern $p = p_0, \ldots, p_{m-1}$ of m characters drawn from an alphabet Σ of size σ and a text $t = t_0, \ldots, t_{n-1}$ of n characters over the same alphabet. We need to find all such substrings of the text that the distance between the substring and the pattern is at most k. In the k-difference problem the distance between two strings is the standard edit distance where mismatches, deletions and insertions are allowed. The k-mismatch problem is a more restricted one using the Hamming distance where only mismatches are allowed.

Several algorithms [12] for both variations of approximate string matching have been presented. Many of the algorithms have been developed with text data in mind and these algorithms do not necessarily work well with a small alphabet. Recently developing algorithms for small alphabets has attracted attention as approximate searching of large volumes of gene sequences has become common. One example of such a biological problem is the gene sequence acquisition problem in which a collection of gene sequences and a primer is given and we need to extract all those sequences that contain the primer with at most k mismatches.

The approximate Boyer-Moore (ABM) algorithm [14] is an adaptation of the Boyer-Moore-Horspool algorithm [8] to approximate matching. ABM performs well on moderately large alphabets and low error levels. ABM was originally not designed for small alphabets and in fact it performs rather poorly on them. Liu et al. [9] tuned the k-mismatch version of ABM for smaller alphabets. Their

---

* Work supported by Academy of Finland.

algorithm, called FAAST, has a stronger shift function which makes it faster than ABM.

In this paper we introduce improvements to the FAAST algorithm gaining considerable speedups in both preprocessing and search times. The preprocessing we present is simpler having a lower time complexity than that of FAAST. While the FAAST algorithm can only handle the k-mismatch problem, we show that with the simpler preprocessing the algorithm can be modified to also handle the k-difference problem.

2   Previous Work

So far many algorithms have been developed based on Boyer-Moore string matching [5] for the k-mismatch problem. Here we consider mainly ABM [14] and FAAST [9], but two other variations developed by Baeza-Yates & Gonnet [3] and El-Mabrouk & Crochemore [6] are worth mentioning. The shift function of the Baeza-Yates-Gonnet algorithm is based on the triangular inequality, whereas the El-Mabrouk-Crochemore algorithm applies the Shift-Add approach [2]. Three [6,9,14] of these four algorithms have been shown to be sublinear on the average. E.g. the average case complexity of ABM (without preprocessing) is $O(nk(1/(m-k)+k/\sigma))$.

Typically algorithms of Boyer-Moore type have two phases: preprocessing of the pattern and searching of its occurrences in the text. ABM uses the bad character rule for shifting and is thus a direct generalization of the Boyer-Moore-Horspool algorithm [8]. Instead of stopping at the first mismatch in the matching loop, the control stops at the $k+1^{\text{st}}$ mismatch or when an occurrence of the whole pattern is found. The shift is calculated considering $k+1$ characters currently aligned with the end of the pattern. The shift is the minimum of the precomputed shifts for those $k+1$ characters. After shifting, at least one of the these characters will be aligned correctly with the pattern.

FAAST is an improved variation of ABM for small alphabets using a variation of the Four-Russians technique [1,10,15] to speed up the search. Instead of minimizing $k+1$ shifts during search, it uses a precomputed shift table for a $(k+x)$-gram aligned with the end of the pattern, where $x \geq 1$ is a parameter of the algorithm. The shift table is calculated so that after the shift at least x characters are aligned correctly. It is obvious that this stronger requirement leads to longer shifts in most situations, when $x > 1$ holds, and the shift is never shorter than the shift of ABM. Note that for $x = 1$ the length of shift is the same for both the algorithms, but the shift is minimized during preprocessing only in FAAST. So the algorithms are different even for $x = 1$. The optimal value of x for maximum searching speed depends on other problem parameters and the computing platform. However, an increment of x makes the preprocessing time grow. FAAST presents a clear improvement on solving the k-mismatch problem on DNA data as compared to the ABM algorithm. The preprocessing phase of FAAST is advanced because it includes the minimization step of ABM. The preprocessing time of FAAST is $O((k+x)((m-k)\sigma^{k+x}+m))$.

3 Algorithm for the k-Mismatch Problem

Our aim is to develop a faster algorithm for DNA data based on FAAST which uses a $(k+x)$-gram for shifting. We make two major changes to FAAST. We implement a simpler and faster preprocessing phase based on dynamic programming. FAAST counts the number of mismatches in the $(k+x)$-gram aligned with the end of the pattern during the searching phase. Our approach makes it possible to compute this number during preprocessing, which improves the searching speed.

The preprocessing phase computes the Hamming distance between an arbitrary $(k+x)$-gram and each $(k+x)$-gram of the pattern using dynamic programming. The first row and column of the dynamic programming table are initialized to 0, and the rest of the table can be filled with a simple iteration:

$$D[i,j] = D[i-1, j-1] + \alpha \quad \text{where } \alpha = \begin{cases} 0 & \text{if } t_{i-1} = p_{j-1}, \\ 1 & \text{otherwise} \end{cases}$$

Note that all $(k+x)$-grams of the pattern are handled in the same table. As an example, let us consider a situation where a pattern $p =$ "ggcaa" has been aligned with the text string "gcata", and $k = x = 2$ holds. The reference $(k+x)$-gram is now "cata", and the corresponding Hamming distance table of size $(k+x+1) \times (m+1)$, calculated during preprocessing, is shown in Fig. 1. First of all, we see that the last cell $D[k+x, m] = 3 > k$, and therefore it is not possible to find a match at this position, as already the suffix of the aligned text string contains too many mismatches. Otherwise, we would have to check for a match by examining the amount of mismatches in the beginning of the aligned string.

We will also look at the bottom row of the table, and find the rightmost cell $D[k+x, j]$ with a value $h \le k$, except for the last cell $D[k+x, m]$. This is the next possible candidate for aligning the pattern with the text with less than k mismatches and the correct shift is equal to $m - j$. In our example, the cell $D[k+x, 2] = 2$, and we would shift the pattern by $5 - 2 = 3$ positions to get the next alignment.

We do not need the whole table to obtain this information, so we just store the calculated Hamming distance for each generated $(k+x)$-gram in a table M which is indexed by a number obtained by transforming the $(k+x)$-gram to an integer. The precalculated shifts are stored in a table D_{kx}. During the searching

$$D$$

$i \backslash j$		g	g	c	a	a
	0	1	2	3	4	5
0	0	0	0	0	0	0
c 1	0	1	1	0	1	1
a 2	0	1	2	2	0	1
t 3	0	1	2	3	3	1
a 4	0	1	2	3	3	3

Fig. 1. The Hamming distance table D of size $(k+x+1) \times (m+1)$ for k-mismatch problem $(k = 2,\ x = 2)$. The pattern is "ggcaa" and the reference $(k+x)$-gram is "cata".

phase we convert the last $(k + x)$-gram of the aligned text string into an index y to the tables, and check for an occurrence if $M[y] \leq k$. Note that if the text is not pure DNA data, we need to check the whole aligned text string against the pattern as there might be some indeterminate characters. Finally, we shift the pattern according to $D_{kx}[y]$.

We can improve the preprocessing time by applying the technique used previously by Fredriksson and Navarro [7] for approximate matching and Navarro et al. [13] for indexed approximate matching. If the $(k + x)$-grams are generated in the lexicographical order, the dynamic programming table differs only by the last few rows. Therefore we can speed up the preprocessing if we only recalculate the last rows of the table at each step, starting from the first changed character.

This can be implemented by traversing the trie built of all $(k + x)$-grams in depth first order. Nodes at the ith level of the trie correspond to strings of length i. Thus there are σ^i nodes on level i and the total number of nodes in the trie is

$$\sum_{i=1}^{k+x} \sigma^i = \sigma \frac{\sigma^{k+x} - 1}{\sigma - 1} = O(\sigma^{k+x}).$$

If we have the dynamic programming table for a node in the trie, the tables for the children nodes can be obtained by calculating one more row to the dynamic programming table taking $O(m)$ time per child so calculating the dynamic programming tables for all nodes in the trie takes $O(\sigma^{k+x}m)$ time. At the leaf nodes we have the dynamic programming table for the corresponding $(k+x)$-gram and we need to figure out the number of mismatches entered to table M and the shift value entered to table D_{kx} which takes $O(m)$ time. The extra calculation needed at leaf nodes is thus $O(\sigma^{k+x}m)$ because there are σ^{k+x} leaf nodes. Therefore the time complexity of the preprocessing phase is $O(2\sigma^{k+x}m) = O(\sigma^{k+x}m)$. Note that if we implement the traversing of the trie by recursion, we actually do not need to explicitly build the trie.

We call this algorithm for the k-mismatch problem Algorithm 1. The shift behaviors of Algorithm 1 and FAAST are exactly the same. In FAAST the number of mismatches in the last $(k + x)$-gram of an alignment is computed during the searching phase whereas in Algorithm 1 this is fetched from a table. However, we still need to read the $(k + x)$-gram and thus the time complexity of the search phase of Algorithm 1 is the same as in FAAST.

Implementation note. For maximum performance it is crucial how the value of a $(k + x)$-gram is computed during searching. We mapped the ASCII values of DNA characters to integers $\{0, 1, 2, 3\}$ and used a shift-or loop to construct a bit representation of a $(k + x)$-gram.

4   Algorithms for the k-Difference Problem

Algorithm 1 can be easily modified to solve the k-difference problem. We initialize the dynamic programming table as in the k-mismatch case, but now we apply the traditional equations for the k-difference problem

$$D[i,j] = \min \left\{ \begin{array}{l} D[i-1,j-1] + \alpha, \\ D[i-1,j] + 1, \\ D[i,j-1] + 1 \end{array} \right\} \quad \text{where } \alpha = \left\{ \begin{array}{ll} 0 & \text{if } t_{i-1} = p_{j-1}, \\ 1 & \text{otherwise} \end{array} \right.$$

As before we construct the $(k+x+1) \times (m+1)$ table during preprocessing for each possible text string, and obtain the tables $M[y]$ and $D_{kx}[y]$ by checking the bottom row of the constructed table. The searching phase starts by aligning the pattern against the text prefix ending at position $m-k-1$. When examining an alignment ending at position s all matches ending before that position have been reported. At each alignment we have to construct a full $(m+k+1) \times (m+1)$ edit distance table D with the currently aligned text $t_{s-(m+k)+1} \ldots t_s$ against the pattern, if $M[t_{s-(k+x)+1} \ldots t_s] \leq k$. A match will be reported, if $D[m+k,m] \leq k$. After this operation we will shift the pattern according to D_{kx}. In order to observe correctly an occurrence of the pattern in the beginning of the text, we assume that $t_{-k}, \ldots, t_{-1}$ hold a character not in the pattern. The modification of Algorithm 1 for the k-difference problem is called Algorithm 2.

Example tables for the k-difference problem are shown in Fig. 2, using a pattern "ggcaa", a text string "aggcata" and parameters $k = x = 2$. We can see from the first table that $D_{kx}[\text{"cata"}] = 5-4 = 1$ and $M[\text{"cata"}] = D_0[k+x,m] = 1$. Therefore, we would construct a table D, and find that $D[m+k,m] = 1 \leq k$, and report a match at position s. We would continue the search by shifting the pattern by 1.

In the k-mismatch problem we did not need to reread the last $k+x$ characters from the text alignment when checking for an occurrence. Instead we had stored the number of mismatches in the table M and we could extend the match based on that information. For the k-difference problem the situation is not quite as simple because we need to compute the dynamic programming table to check for an occurrence. The problem with Algorithm 2 is that when checking for an occurrence the aligned text is read forward while during the preprocessing phase we have generated the dynamic programming table for the last characters of the pattern. In order to use that information and avoid rereading the last $k + x$ characters we need to invert the calculation of the dynamic programming table so that we start building the table from the end of the pattern and the text string.

First we will explain how the inverted table is built and then show how that information is used to speed up the checking of an occurrence. The initialization of the inverted table is different, as we set $D[0,j] = j$ and $D[i,0] = i$ for $i \in [0, k+x], j \in [0, m]$, instead of 0. We have to read the pattern and text in reverse, and therefore we get a new condition for α:

$$\alpha = \left\{ \begin{array}{ll} 0 & \text{if } t_{k+x-i} = p_{m-j}, \\ 1 & \text{otherwise} \end{array} \right.$$

This inverted table gives equivalent results when it comes to calculating the actual edit distance between the pattern and the aligned text string, but we still need to obtain the tables D_{kx} and M from a normal table. When the inverted edit distance table has been finished, we have to search for a match at the

D_0

i\j	g	g	c	a	a
	0	1	2	3	4	5
0	0	0	0	0	0	0
c 1	0	1	1	0	1	1
a 2	0	1	2	1	0	1
t 3	0	1	2	2	1	1
a 4	0	1	2	3	2	1

D

i\j	g	g	c	a	a
	0	1	2	3	4	5
0	0	0	0	0	0	0
a 1	0	1	1	1	0	0
g 2	0	0	1	2	1	1
g 3	0	0	0	1	2	2
c 4	0	1	1	0	1	2
a 5	0	1	2	1	0	1
t 6	0	1	2	2	1	1
a 7	0	1	2	3	2	1

D_{inv}

i\j	a	a	c	g	g
	0	1	2	3	4	5
0	0	1	2	3	4	5
a 1	1	0	1	2	3	4
t 2	2	1	1	2	3	4
a 3	3	2	1	2	3	4
c 4	4	3	2	1	2	3
g 5	5	4	3	2	1	2
g 6	6	5	4	3	2	1
a 7	7	6	5	4	3	2

Fig. 2. Normal and inverted edit distance tables for k-difference problem ($k = 2$, $x = 2$) with the pattern "ggcaa" and the aligned text "aggcata". Sizes of the tables are $(k + x + 1) \times (m + 1)$ for D_0 and $(m + k + 1) \times (m + 1)$ for D and D_{inv}.

last column. To be exact, we need to check $2k + 1$ different cells of the table for a possible match, because the match can contain up to k insert or delete operations, and the match length can therefore vary. All possible matches that end in the character t_s will be found in the last cells of the last column of the inverted table. We can either report the first match with less than $k + 1$ differences, or search for the match with the minimum differences. The current alignment $t_{s-(m+i)+1} \ldots t_s$ matches the pattern $p_0 \ldots p_{m-1}$ with less than $k + 1$ differences, if

$$D_{inv}[m + i, m] \le k, i \in -k \ldots k$$

If we have an edit distance table calculated for the text suffix $t_{s-(k+x)+1} \ldots t_s$, we can check for a complete occurrence by filling the rest of the table rows from $t_{s-(k+x)}$ down to $t_{s-(m+k-1)}$. We can therefore store the last row of the inverted table $D_{inv}[k + x, j]$, $j \in [0, m]$ for each $(k + x)$-gram during the preprocessing phase. This row can then be used to fill up the rest of the table by dynamic programming during the search phase, when the aligned text needs to be checked for an occurrence, and we do not need to run the dynamic programming for the whole table every time. We modify Algorithm 2 to use the inverted table during the search phase, and we also store the last row of the inverted tables generated during the preprocessing phase. The new algorithm is called Algorithm 3, and its pseudo code is given in Fig. 3. For simplicity, the preprocessing part of the pseudo code does not use the optimization of generating the $(k + x)$-grams in lexicographic order and recalculating the dynamic programming table only for those rows that have changed.

The preprocessing phase of Algorithm 2 has the same time complexity as that of Algorithm 1. In Algorithm 3, we need to calculate both the original dynamic programming table and the reversed one. Because a $(k + x)$-gram is read in opposite directions when calculating these two tables we have to enumerate the $(k+x)$-grams twice. However the asymptotic time complexity remains the same.

preprocess (p, m, k, x)

1.   **for** $(i \in 0 \ldots k + x)$
2. $D[i, 0] \leftarrow 0$
3. $D_{inv}[i, 0] \leftarrow i$
4.   **for** $(j \in 0 \ldots m)$
5. $D[0, j] \leftarrow 0$
6. $D_{inv}[0, j] \leftarrow j$
7.   **for** $(t = t_0 \ldots t_{k+x-1} \in \Sigma^{k+x})$
8.   **for** $(i \in 1 \ldots k + x, \; j \in 1 \ldots m)$
9. $D[i, j] \leftarrow \min \left\{ \begin{array}{l} D[i-1, j-1] + \alpha, \\ D[i-1, j] + 1, \\ D[i, j-1] + 1 \end{array} \right\}, \alpha = \left\{ \begin{array}{l} 0 \text{ if } t_{i-1} = p_{j-1}, \\ 1 \text{ otherwise} \end{array} \right.$
10. $D_{inv}[i, j] \leftarrow \min \left\{ \begin{array}{l} D_{inv}[i-1, j-1] + \alpha, \\ D_{inv}[i-1, j] + 1, \\ D_{inv}[i, j-1] + 1 \end{array} \right\}, \alpha = \left\{ \begin{array}{l} 0 \text{ if } t_{k+x-i} = p_{m-j}, \\ 1 \text{ otherwise} \end{array} \right.$
11. $M[t] \leftarrow D[k + x, m]$
12. $lastRow[t] \leftarrow D_{inv}[k + x]$
13.   **for** $(j \in [m - 1, 0])$
14.   **if** $D[k + x, j] < k$
15. $D_{kx}[t] \leftarrow m - j$
16.   **break**

search (t, n, k, x)

1.   **for** $(i \in 0 \ldots m + k)$
2. $D_{inv}[i, 0] \leftarrow i$
3.   **for** $(j \in 0 \ldots m)$
4. $D_{inv}[0, j] \leftarrow j$
5. $s \leftarrow m - k - 1$
6.   **while** $(s < n)$
7.   **if** $(M[t_{s-(k+x)+1} \ldots t_s] \leq k)$ /* possible occurrence */
8. $D_{inv}[k + x] \leftarrow lastRow[t_{s-(k+x)+1} \ldots t_s]$
9.   **for** $(j \in 1 \ldots m, \; i \in k + x + 1 \ldots m + k)$
10. $D_{inv}[i, j] \leftarrow \min \left\{ \begin{array}{l} D_{inv}[i-1, j-1] + \alpha, \\ D_{inv}[i-1, j] + 1, \\ D_{inv}[i, j-1] + 1 \end{array} \right\}, \alpha = \left\{ \begin{array}{l} 0 \text{ if } t_{s-i+1} = p_{m-j}, \\ 1 \text{ otherwise} \end{array} \right.$
11.   **if** $(D_{inv}[m + i, m] \leq k, i \in -k \ldots k)$
12.   **Report match at** $t_{s-(m+i)+1} \ldots t_s$ **with** $D_{inv}[m + i, m]$ **differences**
13. $s \leftarrow s + D_{kx}[t_{s-(k+x)+1} \ldots t_s]$

Fig. 3. Algorithm 3 preprocessing and search phases

The shifts in the searching phase of Algorithm 2 and 3 are somewhat shorter than in Algorithm 1 because the probability of two strings matching with distance less than k is higher when using the standard edit distance than when using the Hamming distance.

5   Experimental Results

Tests were run on an Intel Pentium D 2.8 GHz dual core CPU with 1 gigabyte of memory. This processor contains 24+32 kilobytes of L1 cache, and 1024 kilobytes of L2 cache. Algorithms were implemented in C and compiled with gcc 4.0.2, using optimization level -O2 as 32-bit binaries. All the algorithms were run three times for the same patterns, and the listed search and preprocessing times are the average values observed between all runs. For comparison in the k-mismatch case we use the original ABM algorithm and our implementation of FAAST. The Myers algorithm [11], the algorithm by Baeza-Yates and Perleberg (BYP) [4] and a version of ABM are used for the k-difference problem. All the results are shown with the x-value gaining the fastest searching speed in FAAST and our new algorithms if otherwise is not stated. The best x-value is generally the same for our algorithms and for FAAST. The other algorithms do not utilize the x-value.

The searched text is a two megabytes long sequence of the fruit fly genome. The test patterns have been extracted randomly from the text. Each pattern set consists of 200 different patterns of the same length, and they are searched sequentially.

Table 1 shows the search times for the original ABM, FAAST and Algorithm 1 in the k-mismatch problem. Algorithm 1 is generally 30–50% faster than FAAST in the k-mismatch case for $k \in [1, 3]$. Also, the preprocessing phase of Algorithm 1 is 10 to 30 times faster than that of FAAST.

Experimental results for the k-difference problem are shown in Table 2, and Fig. 4 further illustrates the results with $k = 2$. In the k-difference problem, our new algorithms compare well against the Myers, BYP and ABM algorithms. Algorithms 2 and 3 are up to 50% faster than the BYP algorithm with $k = 1$, as it is shown in Table 2. For $k = 2$ Algorithm 3 is faster than BYP for short patterns but the BYP algorithm takes the lead for longer patterns. If we allow more differences, Myers is the fastest for short patterns and BYP for long ones. The basic version of the Myers algorithm is limited by the 32-bit word size, and it cannot handle patterns where $m > 32$. The modifications in Algorithm 3 decrease search time by 20-30%, when compared to Algorithm 2.

Table 1. Search times in seconds for k-mismatch, using best observed x-values. Preprocessing times are in parentheses.

m	$k = 1$ ABM	FAAST	Alg.1	$k = 2$ ABM	FAAST	Alg.1
15	7.28 (0.04)	1.17 (0.48)	0.64 (0.03)	15.65 (0.04)	2.17 (1.76)	1.21 (0.16)
20	7.28 (0.07)	0.92 (0.65)	0.54 (0.03)	15.59 (0.08)	1.68 (2.58)	0.98 (0.14)
25	7.24 (0.09)	0.78 (0.87)	0.44 (0.04)	15.63 (0.09)	1.47 (3.13)	0.81 (0.22)
30	7.22 (0.15)	0.68 (0.98)	0.40 (0.06)	15.71 (0.10)	1.30 (3.70)	0.69 (0.20)
35	7.34 (0.18)	0.60 (1.22)	0.36 (0.05)	15.65 (0.16)	1.22 (4.16)	0.53 (0.24)
40	7.31 (0.24)	0.53 (1.42)	0.33 (0.05)	15.69 (0.19)	1.11 (4.73)	0.54 (0.27)

Table 2. Search times in seconds for k-difference, using best observed x-values

m	$k = 1$					$k = 2$				
	ABM	Myers	BYP	Alg.2	Alg.3	ABM	Myers	BYP	Alg.2	Alg.3
15	8.82	7.35	2.85	1.98	1.65	38.58	7.33	6.90	6.70	5.04
20	8.27	7.41	2.74	1.63	1.44	27.24	7.36	4.50	5.75	4.53
25	7.99	7.34	2.69	1.41	1.34	19.49	7.37	3.79	5.58	4.09
30	8.07	7.37	2.67	1.32	1.15	14.80	7.37	3.89	5.61	4.03
35	8.07	-	2.62	1.29	1.13	12.48	-	3.73	5.77	4.00
40	7.99	-	2.63	1.23	1.05	11.08	-	3.94	5.95	4.04

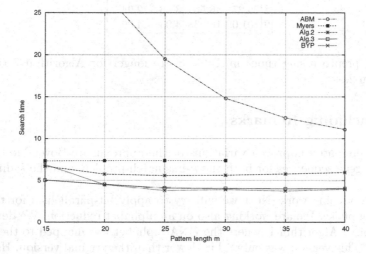

Fig. 4. Search times for k-difference with $k = 2$

We also ran some preliminary tests to compare Algorithm 3 and the algorithm by Fredriksson and Navarro [7]. We used the version of their algorithm that reads the window backwards. In these tests Algorithm 3 was faster for pattern lengths up to 18 when $k = 1$ and up to pattern length 15 when $k = 2$. For longer patterns the algorithm by Fredriksson and Navarro was faster.

The effect of increasing the precalculated edit distance table size, and thus increasing preprocessing time with a large x-value is shown in Table 3. With small values of x, the search time decreases as the amount of preprocessing increases, but after a certain limit increasing the x-value will begin to slow down the search. For these pattern lengths and k-values the optimal x-value was typically 4 for the k-mismatch problem and 6 for the k-difference problem.

In the implementation of Algorithm 2, preprocessing is optimized by generating the $(k + x)$-grams in lexicographic order and recalculating the dynamic programming table only for those characters that differ form the previous $(k+x)$-gram while Algorithm 3 needs to do this recursion twice, once to generate the normal dynamic programming table and once to calculate the reversed one.

Table 3. Preprocessing times and search times for k-difference, with different x-values
($k = 1$, $m = 20$)

x	Preprocessing		Search	
	Alg. 2	Alg. 3	Alg.2	Alg.3
1	<0.01	<0.01	977.30	724.61
2	0.01	0.01	213.43	144.53
3	0.02	0.05	45.57	28.92
4	0.10	0.18	11.64	7.08
5	0.37	0.71	3.94	2.44
6	1.59	2.76	1.84	1.44
7	6.38	11.35	1.63	1.51
8	25.27	46.50	3.06	2.94
9	101.09	188.38	4.03	4.06

Thus the preprocessing times in Table 3 are longer for Algorithm 3 than for
Algorithm 2.

6   Concluding Remarks

We have presented improved variations of the approximate Boyer-Moore algorithm for gene sequences for both the k-mismatch problem and the k-difference
problem.

This in ongoing work. Next we will try to apply bit-parallelism for the preprocessing phase. We are working also on an alphabet reduction. We developed
a variation of Algorithm 1, where the DNA alphabet was mapped to the binary
alphabet. This version was only a bit slower than the original version. However,
for short DNA texts the total time (preprocessing + searching) was the best with
the alphabet reduction. The alphabet reduction also extends the applicability of
our precomputed shift to larger alphabets.

Acknowledgments. We thank Janne Auvinen for implementing a part of the
algorithms.

References

1. Arlazarova, V., Dinic, E., Kronrod, M., Faradzev, I.: On economic construction
   of the transitive closure of a directed graph. Doklady Academi Nauk SSSR 194,
   487–488 (1970) (in Russian) (English translation in Soviet Mathematics Doklady
   11, 1209–1210 (1975))
2. Baeza-Yates, R., Gonnet, G.: A new approach to text searching. Communications
   of the ACM 35(10), 74–82 (1992)
3. Baeza-Yates, R., Gonnet, G.: Fast string matching with mismatches. Information
   and Computation 108(2), 187–199 (1994)
4. Baeza-Yates, R.A., Perleberg, C.H.: Fast and practical approximate string matching. Information Processing Letters 59(1), 21–27 (1996)

5. Boyer, R., Moore, J.: A fast string searching algorithm. Communications of the ACM 10(20), 762–772 (1977)
6. El-Mabrouk, N., Crochemore, M.: Boyer-Moore strategy to efficient approximate string matching. In: Hirschberg, D.S., Meyers, G. (eds.) CPM 1996. LNCS, vol. 1075, pp. 24–38. Springer, Heidelberg (1996)
7. Fredriksson, K., Navarro, G.: Average-optimal single and multiple approximate string matching. ACM Journal of Experimental Algorithmics 9, 1–47 (2004)
8. Horspool, N.: Practical fast searching in strings. Software Practice & Experience 10, 501–506 (1980)
9. Liu, Z., Chen, X., Borneman, J., Jiang, T.: A fast algorithm for approximate string matching on gene sequences. In: Apostolico, A., Crochemore, M., Park, K. (eds.) CPM 2005. LNCS, vol. 3537, pp. 79–90. Springer, Heidelberg (2005)
10. Masek, W., Paterson, M.: A faster algorithm for computing string edit distances. Journal of Computer and System Sciences 20, 18–31 (1980)
11. Myers, G.: A fast bit-vector algorithm for approximate string matching based on dynamic programming. Journal of the ACM 46(3), 395–415 (1999)
12. Navarro, G.: A guided tour to approximate string matching. ACM Computing Surveys 33(1), 31–88 (2001)
13. Navarro, G., Sutinen, E., Tanninen, J., Tarhio, J.: Indexing text with approximate q-grams. In: Giancarlo, R., Sankoff, D. (eds.) CPM 2000. LNCS, vol. 1848, pp. 350–363. Springer, Heidelberg (2000)
14. Tarhio, J., Ukkonen, E.: Approximate Boyer-Moore string matching. SIAM Journal on Computing 22, 243–260 (1993)
15. Wu, S., Manber, U., Myers, E.: A subquadratic algorithm for approximate limited expression matching. Algorithmica 15(1), 50–67 (1996)

Optimal Self-adjusting Trees for Dynamic String Data in Secondary Storage

Pang Ko and Srinivas Aluru

Department of Electrical and Computer Engineering
Iowa State University
{kopang,aluru}@iastate.edu

Abstract. We present a self-adjusting layout scheme for suffix trees in secondary storage that provides optimal number of disk accesses for a sequence of string or substring queries. This has been an open problem since Sleator and Tarjan presented their splaying technique to create self-adjusting binary search trees in 1985. In addition to resolving this open problem, our scheme provides two additional advantages: 1) The partitions are slowly readjusted, requiring fewer disk accesses than splaying methods, and 2) the initial state of the layout is balanced, making it useful even when the sequence of queries is not highly skewed. Our method is also applicable to PATRICIA trees, and potentially to other data structures.

1 Introduction

Suffix tree is an important data structure in information retrieval, text processing, and computational biology. It is especially suitable for indexing biological sequences where predefined boundaries such as words, phrases, and sentences are absent. Some of the best implementations of suffix trees take 10 bytes for each character to be indexed on average [10]. The large amount of data coupled with unpredictable access patterns make it necessary to improve the performance of suffix trees in secondary storage. Currently, there are three main areas of focus in attempting to improve the performance of suffix trees in secondary storage:

1. Reduce the memory footprint of suffix trees with succinct representations [1,8] or heuristics [6,10,11].
2. Develop heuristics or algorithms to reduce the number of disk accesses needed while constructing the suffix tree [2,7,9].
3. Organize the nodes of a suffix tree on disk, so that number of disk access for a query in the worst case is bounded [4,9].

Like most indexing structures, suffix trees are built with the intention that they will be queried many times. Therefore, it is very important to devise algorithms that not only guarantee the worst case performance of a single query, but also provide good performance for a large sequence of queries collectively. In 1985, Sleator and Tarjan [12] created the self-adjusting binary tree by using a "splay"

N. Ziviani and R. Baeza-Yates (Eds.): SPIRE 2007, LNCS 4726, pp. 184–194, 2007.

process, and proved that it produces the optimal number of disk accesses for a large sequence of queries. Since the publication of their ground breaking paper, the splaying technique has received wide attention.

However, the splaying process involves promoting a node in the tree to be the new root of the tree, and therefore is not suitable for suffix trees. Indeed, Sleator and Tarjan had left the development of a self-adjusting data-structure for text data as an open question. This open question has been partially resolved by Ciriani *et al.* [5], who provided a randomized algorithm that achieves the optimal number of disk accesses with high probability. Their method utilizes self-adjusting skip lists on top of a static suffix array. The problem of developing a deterministic algorithm for static and dynamic texts remained open.

In this paper, we resolve this open problem by designing a self-adjusting suffix tree layout that optimizes the total number of disk accesses for a sequence of queries. The main difficulty is that while a number of valid alternative topologies exist for binary search trees, allowing continual adjustment suited to the flow of queries, the suffix tree topology is fixed and unbalanced to begin with. We overcome this limitation by proposing a layout scheme that creates a mapping of suffix tree nodes to disk blocks. While the tree topology remains fixed, the mapping of the layout can be adjusted to the sequence of queries, producing the desired performance bound. We begin with the layout scheme we proposed in [9], which balances the number of disk accesses required for any root to leaf path. We make the following contributions:

1. We present the first self-adjusting organization of the suffix tree and show that it provides an optimal bound for the number of disk accesses for a large[1] sequence of queries. Our layout scheme is deterministic and allows for insertion and deletion of strings, thus answering the open question posed by Sleator and Tarjan in [12].
2. We show that a "slow" moving promotion of the nodes works as well as the dramatic promotion to the root, which is a radical departure from existing self-adjusting algorithms and data structures, and can potentially be applied to develop more efficient "splaying" heuristics.
3. In practice, the self-adjusting data structures do not perform as well as balanced trees except in cases where a few of the leaves are accessed significantly more frequently than others [3,13]. Our layout scheme is balanced in its initial state, thus combining the advantages of both types of data structures.
4. Besides suffix trees, our scheme can also be used for PATRICIA trees and potentially other data structures where the topology of the data structure cannot be altered.
5. Because of the topology of the suffix tree, our layout has the ability to reduce the number of disk accesses needed for a set of non-identical queries that share the same prefix.

The rest of the paper is organized as follows: In Section 2 through Section 4 we present our self-adjusting layout and show its performance in secondary storage.

---

[1] A sequence of queries is considered to be "large" if the number of queries is greater than the number of leaves in the suffix tree.

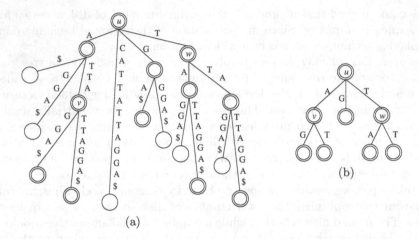

(a)

(b)

Fig. 1. (a) The suffix tree of the string CATTATTAGGA$. The nodes of an example partition are drawn as concentric circles. Braching nodes of the partition are labeled as u, v, w. (b) The skeleton partition tree of the example partition.

Some benefits of our algorithm are discussed in Section 5, and we conclude the paper in Section 6.

2   Self-adjusting Suffix Tree

Let s be a string of length n, and $s[i]$ denote the ith character of s, for $1 \le i \le n$. All characters of s are drawn from the alphabet Σ except the last character $s[n]$, which is a special character $\$ \notin \Sigma$. We denote the substring of s starting from character $s[i]$ and ending at $s[j]$ as $s[i..j]$, where $0 \le i < j \le n$. Suffix s_i is defined to be the substring $s[i..n]$, and can be uniquely identified by its starting position i. The suffix tree of s is a compacted trie of all suffixes of s. In the suffix tree each leaf represents exactly one suffix of s, and each suffix of s is represented by exactly one leaf. Figure 1(a) shows an example of the suffix tree.

Unlike a binary tree where each node represents an independent element or key, whose parent/child relationship can be modified, the parent/child relationship of a pair of nodes in the suffix tree cannot be altered. Thus a direct application of the "splaying" method introduced in [12] is not possible. Instead, we describe a partition scheme for suffix trees and show that optimal performance can be achieved by slowly altering the partitions.

Definition 1. *The **score** of a node v, denoted as $score_v$, is one plus the number of times v is accessed in a sequence of queries, if v is a leaf in the suffix tree. If v is an internal node, then the score is defined as the sum of the scores of all the leaves in the subtree under v.*

Note that when a leaf v is added to the suffix tree, its score is one by this definition.

Definition 2. *Let v be a node in a suffix tree. The **rank** of v is defined as $rank_v = \lfloor \log_B(score_v) \rfloor$.*

Definition 3. *Two nodes u and v are said to be in the same **partition** if and only if there is a path between u and v such that all nodes on the path have the same rank. The rank of a partition is the rank of the nodes in the partition, and is denoted as $rank_{\mathcal{P}}$.*

Note that a partition is a connected component of the suffix tree, and all partitions of a suffix tree are disjoint. Furthermore a partition can be viewed as a tree. Therefore for any partition $\mathcal{P}$, we define the root and leaves of the partition as follows:

Definition 4. *A node u in partition $\mathcal{P}$ is the **root** of $\mathcal{P}$, if and only if u's parent in the suffix tree is not in $\mathcal{P}$, or equivalently, if the rank of u's parent differs from u's rank.*

Definition 5. *A node u in partition $\mathcal{P}$ is a **leaf** of $\mathcal{P}$ if and only if none of u's children in the suffix tree is a part of $\mathcal{P}$, or u is a leaf of the suffix tree.*

However, while every internal node of a suffix tree has at least two children, this is not true for a partition of the suffix tree. So all the nodes of a partition can be divided into two types based on this property.

Definition 6. *If u is a node in partition $\mathcal{P}$ such that at least two of u's children are also in $\mathcal{P}$, then we refer to u as a **branching** node. Otherwise it is referred to as a **non-branching** node.*

Definition 7. *For a partition $\mathcal{P}$, its **skeleton partition tree**, $\mathcal{T}_{\mathcal{P}}$, contains the root, all the leaves and branching nodes of $\mathcal{P}$. If a node u is an ancestor of node v in $\mathcal{P}$ and no node on the path between u and v is in $\mathcal{T}_{\mathcal{P}}$, then there is an edge connecting u to v, and u is said to be the **parent** of v in $\mathcal{T}_{\mathcal{P}}$, while v is referred to as a **child** of u in $\mathcal{T}_{\mathcal{P}}$.*

Figure 1(a) shows an example suffix tree with the nodes of one of its partitions drawn as concentric circles, while Figure 1(b) shows the corresponding skeleton partition tree.

The number of nodes in a partition $\mathcal{P}$ can be as few as one, and as many as $O(B^{i+1})$ where i is the rank of $\mathcal{P}$, but the number of nodes in a skeleton partition tree is bounded by $O(B)$.

Lemma 1. *The number of nodes in a skeleton partition tree for any partition is at most $O(B)$.*

Proof. Suppose $V = \{v_1, v_2, \ldots, v_k\}$ is the set of all leaves of partition $\mathcal{P}$ with rank ℓ. Let r be the root of the partition. Then

$$B^{\ell+1} - 1 \geq score_r = \sum_{i=1}^{k} score_{v_i} \geq kB^\ell \Rightarrow B^{\ell+1} - 1 \geq kB^\ell \Rightarrow B > k$$

Therefore the number of leaves in a skeleton partition tree is at most $B - 1$. Since all nodes in the skeleton partition tree have at least two children except the root, there are at most $2B - 2 = O(B)$ number of nodes.   □

From Lemma 1 we know that the skeleton partition tree can be stored in a constant number of disk blocks depending on the choice of B.

Definition 8. *Let $C = v_1, v_2, \ldots, v_k$ be a path in $\mathcal{P}$, such that v_1 is a child of the root of $\mathcal{P}$ or a branching node, and v_k is a leaf of $\mathcal{P}$, C is referred to as a* **component** *of $\mathcal{P}$.*

By Definition 8 it is clear that any partition $\mathcal{P}$ can be divided into disjoint components such that each node is included in exactly one component. Since each component is a path in the suffix tree it can be stored as a linked list and packed in disk pages, such that at most $O(\frac{\ell}{B})$ disk pages are accessed if the ℓ-th node is needed. Let u be a node in $\mathcal{T}_\mathcal{P}$. There are two copies of u, one in $\mathcal{P}$ and another in $\mathcal{T}_\mathcal{P}$. We denote the copy of u stored in the skeleton partition tree as $u_\mathcal{T}$ and the copy of u stored in one of the components of $\mathcal{P}$ as $u_\mathcal{P}$. For consistency purposes, if a node v is not a part of $\mathcal{T}_\mathcal{P}$ we will still refer to the only copy of v in one of the components as $v_\mathcal{P}$. We use u to denote the node u in the conceptual suffix tree.

Let $u_\mathcal{T}$ be a node of the skeleton partition tree $\mathcal{T}_\mathcal{P}$ of partition $\mathcal{P}$. The following information is stored in $u_\mathcal{T}$.

1. The string depth u in the suffix tree.
2. For each child v of u in the suffix tree, the leading character of the edge label between u and v is stored.
3. If a child v of u is in $\mathcal{P}$ and also in $\mathcal{T}_\mathcal{P}$, then a pointer to $v_\mathcal{T}$ is stored.
4. If a child v of u is in $\mathcal{P}$ but is not a part of $\mathcal{T}_\mathcal{P}$, let v' be the nearest descendant of v such that v' is a part of $\mathcal{T}_\mathcal{P}$. A pointer to $v'_\mathcal{T}$ is stored. A pointer to $v_\mathcal{P}$ in one of the components is also stored.
5. If a child v of u does not belong to $\mathcal{P}$, then it is the root of another partition $\mathcal{Q}$. A pointer to $v_\mathcal{Q}$ is stored.
6. A representative suffix s_i such that the leaf representing s_i is a leaf in the subtree under u in the suffix tree.

To search for a pattern p in the suffix tree, the algorithm starts from the partition containing the root of the suffix tree, and moves from one partition to another. At each partition, the skeleton partition tree is first loaded into the memory. The search starts from the root of $\mathcal{T}_\mathcal{P}$.

1. Suppose the search is at node $v_\mathcal{T}$, and let d be the string depth of $v_\mathcal{T}$. Character $p[d + 1]$ is used to identify the correct child of $v_\mathcal{T}$ to move to. If the child is in $\mathcal{P}$, then the search moves to node $u_\mathcal{T}$, the child of $v_\mathcal{T}$ in $\mathcal{T}_\mathcal{P}$. Otherwise, the representative suffix is loaded and compared to p.
2. Let $W = \{w_\mathcal{T}^1, w_\mathcal{T}^2, \ldots, w_\mathcal{T}^h\}$ be the series of nodes the algorithm in Step 1 passes through. Let ℓ be the number of characters matched from the comparisons of the representative suffix and p. Identify $w_\mathcal{T}^k \in W$ such that the

string depth of w_T^k is less than or equal to ℓ, and the string depth of w_T^{k+1} is greater than ℓ.

3. Let v be the child of w_T^k whose leading character of the edge label is $p[d+1]$ where d is the string depth of w_T^k.

   (a) If v is in $\mathcal{P}$ then use the pointer stored at w_T^k to visit $v_\mathcal{P}$ in one of the components of $\mathcal{P}$, and continue down the component until $v_\mathcal{P}'$ is encountered such that $v_\mathcal{P}'$'s string depth is less than or equal to ℓ, while the next node in the component has a string depth greater than ℓ. Let the string depth of $v_\mathcal{P}'$ be d. If the first character of the edge label leading to one of the children of v' is the same as $p[d+1]$, then that partition is processed next. If no such child can be found then pattern p cannot be found in the suffix tree.

   (b) If v is not in $\mathcal{P}$ then the partition containing v is loaded and processed.

   (c) If no such v is found then pattern p cannot be found in the suffix tree.

Note that in Step 1 of the above algorithm, not all characters of p and the representative suffix need to be compared because some of the beginning characters of p have already been compared with the representative suffixes from previous partitions.

Lemma 2. *Given a pattern p of length $|p|$, the search algorithm takes $O(\frac{|p|}{B} + \log_B N)$ number of disk accesses, where N is the score of the root of the suffix tree.*

Proof. The rank of the root of the suffix tree is $\lfloor \log_B N \rfloor$, therefore on any path from the root of the suffix tree to a leaf of the suffix tree there are at most $\log_B N$ number of partitions. Each character of p is used at most twice, first in traversing down the skeleton partition tree, and then again in comparing with the representative suffix, so $O(\frac{|p|}{B})$ number of disk accesses are needed for all partitions. For each partition some of the non-branching nodes in one of its components are needed, but the total number of non-branching nodes that need to be accessed is at most $|p|$. Since non-branching nodes from the same component can be packed into disk pages, at most $O(\frac{|p|}{B} + \log_B N)$ disk accesses are needed. $\square$

3   Updating the Partitions

Lemma 3. *When leaf v is accessed or added, the rank of a node u may change only if u is an ancestor of v and it is the root of some partition $\mathcal{P}$.*

Proof. Suppose leaf v is accessed. If a node u is not an ancestor of v then its score and rank does not change. If a node u is an ancestor of v but is not the root of some partition $\mathcal{P}$, let $r_\mathcal{P}$ be the root of $\mathcal{P}$ and k be the rank of u and $r_\mathcal{P}$. Then before the access we have $B^{k+1} - 1 \geq score_{r_\mathcal{P}} > score_u$. So after the access of leaf v, $rank_u$ will remain unchanged. $\square$

As the score of each node changes, the rank of these nodes may change as well. When the rank of the root of a partition increases, then this node will be moved out of the current partition, either joining its parent in the suffix tree if their ranks are the same, or becoming a singleton partition, i.e., a partition containing only one node. After the removal of the root of a partition, the partition itself will split into multiple partitions. It is enough to divide the skeleton partition tree into multiple trees where each child of the old root will be a new root of the new partition. The components are not changed except the one containing the old root.

Lemma 4. *When a leaf v is removed from a suffix tree, then the rank of a node u will not change if u is not an ancestor of v, or it is a branching node in its partition, or the ancestor of a branching node in its partition.*

Proof. Suppose leaf v is removed from the suffix tree. For any node u that is not an ancestor of v in the suffix tree, u's score and rank are not changed. If $u_\mathcal{P}$ is a branching node and an ancestor of v, then the score and rank of at least one child of $u_\mathcal{P}$ also in $\mathcal{P}$ is not affected, and since $u_\mathcal{P}$ must have a rank equal to or greater than any of its children, $u_\mathcal{P}$'s rank is unaffected. If $u_\mathcal{P}$ is the ancestor of a branching node $w_\mathcal{P}$ and an ancestor of v, $w_\mathcal{P}$'s rank is not affected by v's deletion, so $u_\mathcal{P}$'s rank must not change either.   □

When the rank of a node v in a partition, say $\mathcal{Q}$, decreases, it is removed from the current partition. If the new rank of the node is the same as some of its children, then it and all its children that have the same rank will be moved to form a new partition $\mathcal{P}$. If none of the children has the same rank as the removed node v, then it will become a singleton partition. To merge the partitions the skeleton partition trees are merged and the node v is added as the new root. This node is also added to one of the components, while the other components remain unchanged.

Theorem 1. *The amortized number of disk accesses for each query, insertion, or deletion of a leaf in the suffix tree is $O(\frac{|p|}{B} + \log_B N)$ if the maximum number of children per node is constant, i.e. the size of alphabet is constant.*

Proof. The cost of querying is $O(\frac{|p|}{B} + \log_B N)$ as shown in Lemma 2. During an insertion, if the root of partition $\mathcal{P}$, $u_\mathcal{P}$, is promoted to partition $\mathcal{Q}$, then we put k credits in u's account, where k is the size of the alphabet. During a deletion at most k children will be moved to u's partition and we will deduct k from its account. Since any node must be first promoted then demoted, the account is alway positive. The total number of credits used is at most $k \log_B N$, so when k is constant the amortized cost for insertion and deletion is $O(\frac{|p|}{B} + \log_B N)$.   □

Although the above theorem shows that our layout scheme is limited for suffix tree indexing of strings drawn from a constant size alphabet, we claim that we can delay the maintenance of the skeleton partition tree and the partitions and slowly perform them in the future, so that we can accommodate strings drawn from alphabets of any size. The idea is to store all the child pointers of each internal node in a consecutive sequence of disk blocks. A pointer is then stored in each internal node pointing to the root of the tree.

4 Self-adjusting Performance

Given a sequence of N queries, we assume without loss of generality that all queries are successful and end at one of the leaf nodes of the suffix tree. If this sequence of queries is known beforehand and the suffix tree is partitioned accordingly, then the number of disk accesses needed to answer any of the queries will be $O\left(\frac{|p|}{B} + rank(r) - rank(v)\right) = O\left(\frac{|p|}{B} + \log_B \frac{N}{score(v)}\right)$ where $|p|$ is the length of the query, r is the the root node of the suffix tree, and v is the leaf where the query ends. Let $p_1, p_2, \ldots, p_N$ be the sequence of queries, and let $\{v_1, v_2, \ldots, v_M\}$ be the set of leaves in the suffix tree. Over the entire sequence of queries, the performance of our layout is

$$O\left(\sum_{i=1}^{N} \frac{|p_i|}{B} + \sum_{j=1}^{M} score(v) \log_B \frac{N}{score(v)}\right) \tag{1}$$

This is the optimal performance for a given sequence of queries for any data structure indexing strings [5]. We now show that this worst case performance can be achieved even if the sequence of queries is not known beforehand.

Theorem 2. *Let $P = p_1, p_2, \ldots, p_N$ be a sequence of N queries, and S be the set of numbers $\{s_1, s_2, \ldots, s_M\}$ such that s_i is the number of times leaf v_i is accessed by patterns in P. Then*

$$O\left(\sum_{i=1}^{N} \frac{|p_i|}{B} + \sum_{j=1}^{M} s_j \log_B \frac{N}{s_j}\right)$$

number of disk accesses are needed to answer all queries in P.

Proof. Since the sequence of queries is not known beforehand, we calculate how many more disk accesses are needed than the ideal scenario. Consider a suffix tree in our layout where all leaves have an initial score of one. Let v_i be a leaf node which will be accessed s_i times in the sequence of N queries. The ideal number of disk accesses is $O(\frac{|p|}{B} + \log_B N - \log_B s_i)$ for each of the s_i times v is accessed. For the first B queries made that end at v_i the number of disk accesses is $O\left(\frac{|p|}{B} + rank(r) - rank(v)\right) = O\left(\frac{|p|}{B} + \log_B N\right)$ which requires $O(\log_B s_i)$ more disk accesses than ideal. For the next $B^2 - B$ number of queries, the number of disk accesses is $O(\log_B s_i - 1)$ more than the ideal number of disk accesses, and so on. The sum of this telescoping series is

$$B \log_B k + (B^2 - B)(\log_B s_i - 1) + \ldots = B + B^2 \ldots + B^{\log_B s_i} = O(s_i)$$

Therefore the total number of disk accesses for all of the s_i times that v_i is accessed is $O\left(s_i \frac{|p|}{B} + s_i \log_B \frac{N}{s_i} + s_i\right)$, and for the sequence of queries P, the number of disk accesses needed is

$$O\left(\sum_{i=1}^{N} \frac{|p_i|}{B} + \sum_{j=1}^{M} s_j \left(\log_B \frac{N}{s_j} + \Theta(1)\right)\right) = O\left(\sum_{i=1}^{N} \frac{|p_i|}{B} + \sum_{j=1}^{M} s_j \log_B \frac{N}{s_j}\right) \quad \square$$

So even with the slow promotion of the leaf, the optimal disk access bound can be achieved.

5   Discussion

Let q_i and q_j be two queries from the sequence of N queries, that share a common prefix p. Let v_i and v_j be the two leaves where q_i and q_j will end, respectively. Then the score of the lowest common ancestor of v_i and v_j, say v, is at least the sum of the scores of v_i and v_j. Therefore even if q_i and q_j are infrequent queries compared to others queries in the sequence, v could still have a high rank, thus potentially reducing the number of disk accesses needed to find v_i and v_j. This is also true for a sequence of such queries.

For a balanced tree there will always be $\Theta\left(\frac{|p|}{B} + \log_B M\right)$ number of disk accesses, where M is the number of leaves in the tree. For any self-adjusting data structure for strings in secondary storage, suppose that every leaf is accessed equal number of times, then the worst case performance of the self-adjusting data structure is the same as the balanced tree, because $\frac{N}{k} = M$. But if the self-adjusting data structure is not balanced to start with, then it will take some time for it to became as efficient as the balanced tree. Therefore self-adjusting data structure will not be as effective as the balanced tree.

From Equation 1 we can also observe that if a leaf is accessed B times more frequently than other leaves in the tree in the sequence of N queries, it will only save a total of B disk accesses compared to the balanced tree. So if we assume that the self-adjusting data structure is not balanced in the beginning, then it would require a very skewed data set to offset the initial inefficiency of the self-adjusting data structure.

This observation provides an explanation for the results in [3,13], where the authors were surprised that self-adjusting data structures do not perform nearly as well as balance trees, except for very skewed data sets. However, if the suffix layout is built with our layout scheme, then it will be a balanced layout, potentially avoiding the initial inefficiencies. But it should be noted that all self-adjusting data structures will incur an overhead for the adjustment which will also affect their performance.

6   Conclusion

In this paper we presented a self-adjusting layout scheme for suffix trees in secondary storage, which allows insertion and deletion of strings. We showed that our layout scheme is optimal for a sequence of queries. This settles the question first proposed by Sleator and Tarjan [12] of constructing an optimal self-adjusting data structure for strings, which is partially solved by Ciriani *et al.* [5] for the static case with high probability to achieve the optimal bound. Our layout scheme can also be applied to PATRICIA trees and possibly to many other data structures whose topology is uniquely determined by the data being indexed and cannot be altered. However, we also call the benefit of non-balanced self-adjusting data structures for strings into question. We argued that due to the overhead needed to readjust itself, self-adjusting data structures are not likely to perform as well as balanced trees except for very skewed data sets. But since the initial state of our layout scheme is a balanced layout, and it only readjusts itself very infrequently it may perform well in practice. Thus, our scheme combines the benefits of both types of data structures. Another added benefit of our layout is that, if multiple non-identical queries shares the same prefix, then the node representing this prefix will have a higher rank even if the queries are infrequent by themselves. This reduces the number of disk accesses needed for all of them.

References

1. Barbay, J., Golynski, A., Munro, J.I., Rao, S.S.: Adaptive searching in succinctly encoded binary relations and tree-structured documents. In: Lewenstein, M., Valiente, G. (eds.) CPM 2006. LNCS, vol. 4009, pp. 24–35. Springer, Heidelberg (2006)
2. Bedathur, S., Haritsa, J.: Search-optimized suffix-tree storage for biological applications. In: Proc. 12th IEEE International Conference on High Performance Computing, pp. 29–39. IEEE Computer Society Press, Los Alamitos (2005)
3. Bell, J., l Gupta, G.: An evaluation of self-adjusting binary search tree techniques. Software - Practice and Experience 23(4), 369–382 (1993)
4. Brodal, G.S., Fagerberg, R.: Cache-oblivious string dictionaries. In: Proc. 17th Annual ACM-SIAM Symposium on Discrete Algorithms, pp. 581–590. ACM Press, New York (2006)
5. Ciriani, V., Ferragina, P., Luccio, F., Muthukrishnan, S.: Static optimality theorem for external memory string access. In: Proc. 43rd Annual Symposium on Foundations of Computer Science, pp. 219–227 (2002)
6. Clark, D.R., Munro, J.I.: Efficient suffix trees on secondary storage. In: Proc. 7th Annual ACM-SIAM Symposium on Discrete Algorithms, pp. 383–391. ACM Press, New York (1996)
7. Farach-Colton, M., Ferragina, P., Muthukrishnan, S.: On the sorting-complexity of suffix tree construction. Journal of the ACM 47(6), 987–1011 (2000)
8. Grossi, R., Vitter, J.S.: Compressed suffix arrays and suffix trees with applications to text indexing and string matching. In: Proc. 32nd Annual ACM Symposium on Theory of Computing, pp. 397–406. ACM Press, New York (2000)
9. Ko, P., Aluru, S.: Obtaining provably good performance from suffix trees in secondary storage. In: Lewenstein, M., Valiente, G. (eds.) CPM 2006. LNCS, vol. 4009, pp. 72–83. Springer, Heidelberg (2006)
10. Kurtz, S.: Reducing the space requirement of suffix trees. Software - Practice and Experience 29(13), 1149–1171 (1999)

11. Munro, J.I., Raman, V., Rao, S.S.: Space efficient suffix trees. J. Algorithms 39(2), 205–222 (2001)
12. Sleator, D.D., Tarjan, R.E.: Self-adjusting binary search trees. Journal of the ACM 32(3), 652–686 (1985)
13. Williams, H.E., Zobel, J., Heinz, S.: Self-adjusting trees in practice for large text collections. Software - Practice and Experience 31(10), 925–939 (2001)

Indexing a Dictionary for Subset Matching Queries

Gad M. Landau[1,2], Dekel Tsur[3], and Oren Weimann[4]

[1] Department of Computer Science, University of Haifa, Haifa - Israel
[2] Department of Computer and Information Science, Polytechnic University,
New York - USA
landau@cs.haifa.ac.il
[3] Department of Computer Science, Ben-Gurion University, Beer-Sheva - Israel
dekelts@cs.bgu.ac.il
[4] Computer Science and Artificial Intelligence Laboratory, MIT,
Cambridge MA - USA
oweimann@mit.edu

Abstract. We consider a subset matching variant of the *Dictionary Query* problem. Consider a dictionary D of n strings, where each string location contains a set of characters drawn from some alphabet Σ. Our goal is to preprocess D so when given a query pattern p, where each location in p contains a single character from Σ, we answer if p matches to D. p is said to match to D if there is some $s \in D$ where $|p| = |s|$ and $p[i] \in s[i]$ for every $1 \leq i \leq |p|$.

To achieve a query time of $O(|p|)$, we construct a compressed trie of all possible patterns that appear in D. Assuming that for every $s \in D$ there are at most k locations where $|s[i]| > 1$, we present two constructions of the trie that yield a preprocessing time of $O(nm + |\Sigma|^k n \lg(\min\{n, m\}))$, where n is the number of strings in D and m is the maximum length of a string in D. The first construction is based on divide and conquer and the second construction uses ideas introduced in [2] for text fingerprinting. Furthermore, we show how to obtain $O(nm + |\Sigma|^k n + |\Sigma|^{k/2} n \lg(\min\{n, m\}))$ preprocessing time and $O(|p| \lg \lg |\Sigma| + \min\{|p|, \lg(|\Sigma|^k n)\} \lg \lg(|\Sigma|^k n))$ query time by cutting the dictionary strings and constructing two compressed tries.

Our problem is motivated by haplotype inference from a library of genotypes [14,17]. There, D is a known library of genotypes ($|\Sigma| = 2$), and p is a haplotype. Indexing all possible haplotypes that can be inferred from D as well as gathering statistical information about them can be used to accelerate various haplotype inference algorithms. In particular, algorithms based on the "pure parsimony criteria" [13,16], greedy heuristics such as "Clarks rule" [6,18], EM based algorithms [1,11,12,20,26,30], and algorithms for inferring haplotypes from a set of Trios [4,27].

1 Introduction

In the *Dictionary Query* problem, one is given a set D of strings $s_1, \ldots, s_n$ and subsequent queries ask whether a given query pattern p appears in D. In [7],

N. Ziviani and R. Baeza-Yates (Eds.): SPIRE 2007, LNCS 4726, pp. 195–204, 2007.
© Springer-Verlag Berlin Heidelberg 2007

this paradigm was broaden to allow a bounded number of mismatches, or allow a bounded number of "don't care" characters. We further extend dictionary queries to support a restricted version of *subset matching*. In subset matching, the characters are subsets of some alphabet Σ. A pattern p is said to match a string s of the same length if $p[i] \subseteq s[i]$ for every $1 \le i \le |p|$. The subset matching problem of finding all occurrences of a pattern string p in a text string t was solved in $O(n \lg^2 n)$ deterministic time [8] and $(n \lg n)$ randomized time [22], where n is the sum of sizes of the sets in p and t.

In this paper we consider the problem of indexing a dictionary for subset matching queries. We focus on a relaxed version of subset matching requiring that the query pattern is over single characters from Σ rather than subsets of Σ. Formally, the problem we consider is defined as follows. We are given a dictionary D of strings $s_1, \ldots, s_n$ where each string character is a subset of some alphabet Σ. A query p is a string over the alphabet Σ, and we say that p matches to s_i if $|p| = |s_i|$ and $p[j] \in s_i[j]$ for every $1 \le j \le |p|$. Our goal is to preprocess D for queries of the form "does p match to a string in D?".

Let m denote the length of the longest string in D and let D' be the set of all strings that match to a string in D. For example, if D contains two strings, $ab\{c, d\}$ and $ab\{c, d\}g\{a, b, c\}ad$, then $D' = \{abc, abd, abcgaad, abcgbad, abcgcad, abdgaad, abdgbad, abdgcad\}$. Notice that a compressed trie of D' supports a query time of $O(|p|)$ for a pattern p. Such a trie can be naively constructed in $O(|\Sigma|^k nm)$ time and $O(|\Sigma||D'|)$ space, assuming every $s \in D$ has at most k locations in which $|s[i]| > 1$. The techniques of Cole et al. [7] can be used to solve the problem with $O(nm \lg(nm) + n(c_1 \lg n)^{k+1}/k!)$ preprocessing time, and $O(m + (c_2 \lg n)^k \lg \lg n)$ query time (c_1 and c_2 are some constants). For small $|\Sigma|$, this approach is less efficient than the trie approach.

In Sections 2 and 3 we present two faster constructions of the trie. The first construction is based on divide and conquer and requires $O(nm + |\Sigma|^k n \lg n)$ preprocessing time. The second construction uses ideas introduced in [2] for text fingerprinting and requires $O(nm + |\Sigma|^k n \lg m)$ preprocessing time. The space complexity is $O(|\Sigma||D'|)$, and it can be reduced to $O(|D'|)$ by using suffix tray [9] ideas. This comes at the cost of $O(|p| + \lg \lg |\Sigma|)$ query time. In Sections 4 we show that by cutting the dictionary strings and constructing two tries we can obtain $O(nm + |\Sigma|^k n + |\Sigma|^{k/2} n \lg(\min\{n, m\}))$ preprocessing time at the cost of $O(|p| \lg \lg |\Sigma| + \min\{|p|, \lg |D'|\} \lg \lg |D'|) = O(|p| \lg \lg |\Sigma| + \min\{|p|, \lg(|\Sigma|^k n)\} \lg \lg(|\Sigma|^k n))$ query time.

An important feature of our first two trie constructions is that they can calculate the number of appearances in D of each pattern in D' (i.e., which is most common? which is least common? etc.). This feature is useful in the application of *Haplotype Inference* that we next describe according to the presentation of Gusfield [13].

1.1 A Haplotype Trie from a Genotype Dictionary

In diploid organisms such as humans, there are two non-identical copies of each chromosome (except for the sex chromosome). A description of the data from a single copy is called a *haplotype* while a description of the conflated (mixed) data on the two copies is called a *genotype*. The underlying data that forms a haplotype is either the full DNA sequence in the region, or more commonly the values of only DNA positions that are *Single Nucleotide Polymorphisms* (SNP's). A SNP is a position in the genome at which exactly two (of four) nucleotides occur in a large percentage of the population. If we consider only the SNP positions, each position can have one of two nucleotides and a haplotype can thus be represented as a $0/1$ vector. A genotype can be represented as a $0/1/2$ vector, where 0 means that both copies contain the first nucleotide, 1 means that both copies contain the second nucleotide and 2 means that the two copies contain different nucleotides (but we don't know which copy contains which nucleotide).

The next high-priority phase of human genomics will involve the development and use of a full *Haplotype Map* of the human genome [21]. Unfortunately, it is prohibitively expensive to directly determine the haplotypes of an individual. As a result, almost all population data consists of genotypes and the haplotypes are currently inferred from raw genotype data. The input to the haplotype inference problem consists of n genotypes ($0/1/2$ vectors), each of length m. A solution to the problem associates every genotype with a pair of haplotypes (binary vectors) as follows. For any genotype g, the associated binary vectors v_1, v_2 must both have value 0 (respectively 1) at any position where g has value 0 (respectively 1); but for any position where g has value 2, exactly one of v_1, v_2 must have value 0, while the other has value 1.

In our settings, the dictionary D corresponds to the library of genotypes, where every genotype location that has the value 2 is replaced by the set $\{0, 1\}$. This way, $|\Sigma| = 2$ and D' consists of all the possible haplotypes that can be part of a pair inferred from D. Our trie stores all haplotypes in D' and we can calculate the number of appearances in D of each such haplotype while constructing the trie. The trie can then be used to accelerate haplotype inference algorithms based on the "pure parsimony criteria" [13,16], greedy heuristics such as "Clarks rule" [6,18], EM based algorithms [1,11,12,20,26,30], and algorithms for inferring haplotypes from a set of Trios [4,27].

2   An $O(nm + |\Sigma|^k n \lg n)$ Time Construction

In this section we present an $O(nm + |\Sigma|^k n \lg n)$ time construction for the compressed trie of D'. To simplify the presentation, for the rest of the paper we assume w.l.o.g. that all strings in D have the same length m.

We first describe an algorithm for merging two compressed tries T_1 and T_2.

1. If one of the tries T_1 or T_2 has a single vertex, then return a copy of the other trie.
2. If both the roots of T_1 and T_2 have degree 1, and the labels of the edges leaving the roots of T_1 and T_2 have a common first letter, then find the longest common prefix (LCP) p of these labels. Remove the string p from T_1, that is, if the label of the edge e that leaves the root of T_1 is equal to p,

remove the edge e and the root from T_1, and otherwise remove p from the label of e. Additionally, remove p from T_2.

Next, recursively merge the two modified tries T_1 and T_2, and let T be the result of the merge. Add a new root r to T and connect it by an edge to the old root of T, where the label of the edge is p.

3. If the two cases above do not occur, then split the trie T_1 as follows. For every edge $e = (r, v)$ that leaves the root r of T_1, create a new trie that contains r and all the descendents of v in T_1. This trie will be denoted T_1^a, where a is the first letter in the label of e. Similarly, split the trie T_2 and create tries $\{T_2^a\}_{a \in \Sigma}$.

For each letter $a \in \Sigma$, recursively merge the tries T_1^a and T_2^a if these two tries exist. Finally, merge the roots of the merged tries.

If the LCP of two edge labels can be obtained in $O(1)$ time, then the time complexity of this algorithm is $O(|T_1| + |T_2|)$, where $|T|$ denotes the number of vertices in the compressed trie T. Next, we present the algorithm for building a compressed trie of D'.

1. For every string in D, replace every character that is a set of size greater than one with a new symbol ϕ.
2. Build a generalized suffix tree $\hat{T}$ containing all suffixes of strings in D.
3. Build compressed tries $T_1, \ldots, T_n$, where T_i is a compressed trie containing all the patterns that match s_i (recall that $D = \{s_1, \ldots, s_n\}$).
4. Repeat $\lceil \lg n \rceil$ times:
   (a) Partition the compressed tries into pairs, except at most one trie.
   (b) Merge each pair of tries into a single trie.

Constructing $\hat{T}$ requires $O(nm)$ time. Each edge label b in some trie that is built during the algorithm, matches a substring $s_i[j..j+|b|-1]$ of some string s_i in D. It is important to notice that $|s_i[l]| = 1$ for every $j+1 \leq l \leq j+|b|-1$. Using the suffix tree $\hat{T}$, computing the longest prefix of two edge labels takes $O(1)$ time. Therefore, the merging of two compressed tries in the algorithm is performed in linear time. In each iteration of line 4, the total work is linear in the total sizes of the current tries, which is $O(|\Sigma|^k n)$. Thus, the overall time complexity of the algorithm is $O(nm + |\Sigma|^k n \lg n)$.

3   An $O(nm + |\Sigma|^k n \lg m)$ Time Construction

In this section we present an $O(nm + |\Sigma|^k n \lg m)$ time construction for the compressed trie of D'. Consider the lexicographical ordering of all the strings in D'. Notice that if we knew this ordering and the length of the LCP of every adjacent strings in this ordering, then we could construct the trie in $O(|D'|) = O(|\Sigma|^k n)$ time by adding the strings in order. We next describe how to obtain the required ordering and LCP information in $O(nm + |\Sigma|^k n \lg m)$ time.

We assign a unique name to every string in D' using fingerprinting techniques [2,10,25]. A *naming table* of some $p \in D'$ is a labeled complete binary tree whose leaves are the characters of p (without loss of generality $|p|$ is a power of two[1]). The naming table has therefore $1 + \lg |p|$ rows and the cells in the last row are named by the characters of p. For example, if $ab\{a,c\}b\{b,c\}cab \in D$ then $p = abcbbcab \in D'$ and the naming table of p might look like this:

25							
9			17				
1		2	3		1		
a	b	c	b	b	c	a	b

We assign integer names to the rest of the table using the *naming technique* [3,24], which is a modified version of the Karp, Miller and Rosenberg algorithm [23]. The names are assigned bottom-up, such that when we assign a name to a cell we look at the pair of cells below it. If this pair appeared before then we give the same name that was given before, otherwise we give a new name. Finally, the unique name assigned to p is in the root of the naming table.

The following property is what makes the naming technique appealing in our settings. Consider two strings (over the alphabet Σ) p and q that both match the same string in D. If p and q differ in one location then the naming table of p differs from the naming table of q only in $1 + \lg |p|$ cells (these cells are in bold in the following example where $p = ababbcab$ and $q = abcbbcab$).

37							
13			17				
1		**1**	3		1		
a	b	**a**	b	b	c	a	b

Consider all the strings that match a specific string $s \in D$. It is possible to enumerate these strings in an order $s^{(1)}, s^{(2)}, \ldots$ in which two consecutive strings differ in exactly one location. This means that we can compute names for these strings in $O(m + |\Sigma|^k \lg m)$ time as follows. We first build the naming table of $s^{(1)}$ from bottom to top, using a two-dimensional table B to store the names given so far. More precisely, $B[a,b]$ is the name given for the pair (a,b), if the pair (a,b) was named. Since checking whether a pair of names appeared before takes constant time, the time it takes to build the naming table is linear in the number of cells in the table, which is $m + m/2 + m/4 + \cdots + 1 = 2m - 1$. Next, we build the naming table of $s^{(2)}$ by updating $1 + \log m$ cells in the table of $s^{(1)}$, which takes $O(\log m)$ time. Then, we build the naming table of $s^{(3)}$ using the naming table of $s^{(2)}$, and so on.

Applying the naming procedure to all strings in D takes $O(nm + |\Sigma|^k n \lg m)$ time. The space complexity is $O((nm + |\Sigma|^k n \lg m)^2)$ due to the table B. The space complexity can be reduced to $O(nm + |\Sigma|^k n \lg m)$ as shown in [10]. The algorithm of [10] uses a different order of filling the naming tables. In the first step, the algorithm computes the names in the second row from the bottom of

---

[1] Otherwise, we can extend p until $|p|$ is a power of two by concatenating to p a string of a repeated new character.

the naming tables of all strings in D'. This is done by taking all pairs of names encountered in the first row of each naming table, lexicographically sorting these pairs, and then naming the pairs. In the second step, the algorithm computes the names in the third row from the bottom of the naming tables of all strings in D', and so on.

In addition to the reduction in space, the algorithm of [10] has the following property. For every two strings $s, s' \in D'$ with names a and a' respectively, $a < a'$ if and only if s is lexicographically smaller than s'.

After naming all strings in D', we sort these strings according to their names. As noted above, this gives the lexicographical ordering of D'. Furthermore, the LCP of any two strings in D' can be computed in $O(\lg m)$ time by comparing their naming tables top-down as noticed in [25]. Therefore, we can compute the length of the LCP of every two consecutive strings in the lexicographic ordering of D' in $O(|\Sigma|^k n \lg m)$ time, and then construct the trie in $O(|D'|) = O(|\Sigma|^k n)$ time by adding the strings in lexicographical order.

4 An $O(nm + |\Sigma|^k n + |\Sigma|^{k/2} n \lg(\min\{n, m\}))$ Time Construction

In this section we present a different approach for solving the dictionary query problem. Instead of building one trie, we build two tries. This reduces the construction time, but gives a penalty in the query time.

Let S be a set of elements with keys from some set $\mathcal{U}$. For every $x \in \mathcal{U}$, the *successor* of x in S is the element $y \in S$ such that $\text{key}(y) \geq x$ and $\text{key}(y)$ is minimal. A *successor data-structure* for the set S supports answering queries of the form "Given a value $x \in \mathcal{U}$, what is the successor of x in S?". For $\mathcal{U} = \{1, \ldots, U\}$, a successor data-structure for a set S can be built in $O(|S|)$ time and space such that successor queries are answered in $O(\lg \lg U)$ time (such a construction is obtained, for example, by combining the van Emde Boas data-structure [29] with the static dictionary of Hagerup et al. [15]).

In order to build a dictionary query data-structure, we split every string in D into two parts. For each $s_i \in D$ define s'_i to be the longest prefix of s_i that contains at most $\lceil k/2 \rceil$ sets of size greater than 1. Also, define s''_i to be the prefix of s_i^R (i.e. the string s_i reversed) of length $|s_i| - |s'_i| = m - |s'_i|$. For example, if $k = 2$ and $s_1 = ab\{c, d\}g\{a, b, c\}ad$ then $s'_1 = ab\{c, d\}g$ and $s''_1 = da\{a, b, c\}$.

Let $D_1 = \{s'_1, \ldots, s'_n\}$ and $D_2 = \{s''_1, \ldots, s''_n\}$. For $i = 1, 2$, let D'_i be the set of all strings that match to one of the strings in D_i. We wish to reduce the problem of matching a string p against the dictionary D to matching a prefix p' of p against D_1, and matching a prefix p'' of p^R against D_2, with $|p''| = |p| - |p'|$. However, there are two issues that need to be addressed: (1) It is possible that p' matches a string s'_i, while p'' matches to a string s''_j with $i \neq j$. This of course does not imply that p matches to a string in D. (2) We do not know the length of p', so we need to check all prefixes of p that match to a string in D_1.

Let T_1 be a compressed trie for D'_1 and T_2 be a compressed trie for D'_2. For each vertex of T_2 assign a distinct integer from the set $\{1, \ldots, |T_2|\}$. The integer assigned to a vertex v is denoted id(v). The string that *corresponds* to a vertex v in a trie is the concatenation of the edge labels in the path from the root to v. The *depth* of a vertex v in a trie is the length of the strings that corresponds to v. We say that the vertices $v \in T_1$ and $w \in T_2$ are *paired* if the sum of their depths is m. For a vertex v in T_1 (respectively T_2) whose corresponding string is s, let L_v be the set of all indices i such that s matches to s_i' (respectively s_i''). For a vertex $v \in T_1$, let S_v be the set containing every vertex $w \in T_2$ that is paired with v and for which $L_v \cap L_w \neq \emptyset$.

The data-structure for the dictionary query problem consists of the tries T_1 and T_2, and each vertex $v \in T_1$ has a successor data-structure that stores the set S_v. The key of an element $w \in S_v$ is id(w).

Answering a query is done as follows. First find the longest path P_1 in T_1 that corresponds to a prefix of the query pattern p, and the longest path P_2 in T_2 that corresponds to prefix of p^R. For a vertex $v \in P_1$, if a vertex $w \in P_2$ is paired with v then by definition the depth of w is equal to m minus the depth of v. Since there is at most one vertex on P_2 with a given depth, we conclude that there is at most one vertex $w \in P_2$ that is paired with v. Moreover, it is possible to find all paired vertices $v \in P_1, w \in P_2$ in $O(|P_1| + |P_2|) = O(m)$ time by traversing P_1 from top to bottom, while concurrently traversing P_2 from bottom to top. To answer the query p, we just need to check whether $w \in S_v$ for some paired vertices $v \in P_1$ and $w \in P_2$. Checking whether $w \in S_v$ for some fixed v and w is done by performing a successor query on the successor data-structure of v. Answering a dictionary query requires at most $|P_1| \leq m$ searches in the successor structures, where each such search takes $O(\lg \lg |D'|)$ time. Therefore, the time to answer a query is $O(m \lg \lg |D'|)$.

We now discuss the time complexity of building the tries. The tries T_1 and T_2 are built using the algorithms in Sections 2 and 3 in $O(nm + |\Sigma|^{k/2} n \lg(\min(n, m)))$ time. In order to build the sets S_v for all v, we compute the intersections $L_v \cap L_w$ for all v and w. This is done as follows. For each i from 1 to n, go over all vertices $v \in T_1$ such that $i \in L_v$. For each such v, go over all $w \in T_2$ such that $i \in L_w$, and add the pair (id(w), i) to a list I_v that is stored at v. Then, for each $v \in T_1$, lexicographically sort the list I_v and obtain all the intersections involving v. Therefore, computing all the intersections and building the successor data-structures takes $O(|\Sigma|^k n)$ time. The total preprocessing time is $O(nm + |\Sigma|^k n + |\Sigma|^{k/2} n \lg(\min\{n, m\}))$.

In order to speed up the query time, we use the technique of fractional cascading [5]. Using a variant of this technique that is described in the next section, we can preprocess T_1 such that searching for a key x in all the successor data-structures of the vertices of some path P in T_1 is done in $O(|P| \lg \lg |\Sigma| + \lg \lg |D'|)$ time. Recall that in order to answer a query, we need to locate id(w) in the successor data-structures of v for every paired vertices $v \in P_1$ and $w \in P_2$. In order to use the fractional cascading speedup, we need to decrease the number of integers assigned to the vertices of P_2. Note that we can assign the same integer to several vertices of T_2 if their corresponding strings have different lengths. Thus, we partition the vertices of T_2 into paths $Q_1, \ldots, Q_r$ using heavy path decomposition [19].

This decomposition has the property that a path from some vertex of T_2 to the root passes through at most $\lg |T_2|$ different paths in the decomposition. We now define id(w) to be index i of the path Q_i that contains w.

Now, locating id(w) in the successor data-structures of v for every paired vertices $v \in P_1, w \in P_2$ is done by grouping all vertices $w \in P_2$ with id$(w) = i$, and performing a search for i in all the successor data-structures of the vertices of some sub-path of P_1 (the subpath that contains all the vertices in P_1 between the highest and lowest vertices of P_1 that are paired to vertices $w \in P_2$ with id$(w) = i$). We have that there are at most $\min\{m, \lg |T_2|\} = O(\min\{m, \lg |D'|\})$ different integers assigned to the vertices of P_2, and the different subpaths of P_1 that correspond to the different integers in P_2 are disjoint. Therefore, the time to answer a query is $O(m \lg\lg |\Sigma| + \min\{m, \lg |D'|\} \lg\lg |D'|)$.

4.1   Fractional Cascading

Let T be a rooted tree of maximum degree d. Each vertex v of T has a set $C_v \subseteq \{1, \ldots, U\}$. The goal is to preprocess T in order to answer the following queries "given a connected subtree T' of T and an integer x, find the successor of x in C_v for every $v \in T'$". The fractional cascading technique of [5] gives search time of $O(|T'| \lg d + \lg\lg U)$, with linear time preprocessing. We now present a variant of fractional cascading that gives better search time (our construction is similar to the one in [28]).

The preprocessing of T is as follows. For each vertex v of T construct a list A_v whose elements are kept in a non-decreasing order (the order of constructing the A_v lists is from the leaves up). For a leaf v, A_v contains exactly the elements of C_v. For an internal vertex v, A_v contains all the elements of C_v. Additionally, for every child w of v, A_v contains every second element of A_w Each element of A_v stores a pointer to its successor in the set C_v. An element of A_v which came from a set A_w keeps a pointer to its copy in A_w. This pointer is called a w-bridge.

Handling a query T', x is done by finding the successor of x in each set A_v for $v \in T'$. Then, using the successor pointers, the successor of x in each set C_v is obtained. Suppose we have found the successor y of x in A_v and we now wish to find the successor y' of x in A_w, where w is a child of v. If we know the first element that appears after y in A_v and has a w-bridge, then we can follow the bridge to A_w and y' is either the element at the end of the bridge or the element preceding it in A_w.

In order to efficiently find the first w-bridge after some element of A_v, perform additional preprocessing: Partition the elements of each list A_v into blocks $B_v^1, B_v^2, \ldots, B_v^{\lceil |A_v|/d \rceil}$ of d consecutive elements each (except perhaps the last block). Let $w_1, \ldots, w_{d'}$ be the children of v. For each block B_v^i build an array L_v^i, where $L_v^i[j]$ is the location of the first w_j-bridge that appear in the blocks $B_v^{i+1}, B_v^{i+2}, \ldots, B_v^{\lceil |A_v|/d \rceil}$. Moreover, for all j, build a successor data-structures $S_v^{i,j}$ that contains all the elements of the block B_v^i that have a w_j-bridge. The key of an element in $S_v^{i,j}$ is its rank in the block B_v^i.

Given an element y of A_v, finding the first element of A_v starting from y that has a w_j-bridge is done in $O(\lg \lg d)$ time. Therefore, the overall search time is $O(|T'| \lg \lg d + \lg \lg U)$.

5 Conclusion and Open Problems

We have shown two solutions for the subset dictionary query problem: one based on building a trie for D' and one based on building two tries. We conjecture that the trie of D' can be built in $O(nm + |\Sigma|^k n)$ time.

References

1. Abecasis, G.R., Martin, R., Lewitzky, S.: Estimation of haplotype frequencies from diploid data. American Journal of Human Genetics 69(4 Suppl. 1), 114 (2001)
2. Amir, A., Apostolico, A., Landau, G.M., Satta, G.: Efficient text fingerprinting via parikh mapping. Journal of Discrete Algorithms 1(5-6), 409–421 (2003)
3. Apostolico, A., Iliopoulos, C.S., Landau, G.M., Schieber, B., Vishkin, U.: Parallel construction of a suffix tree with applications. Algorithmica 3, 347–365 (1988)
4. Brinza, D., He, J., Mao, W., Zelikovsky, A.: Phasing and missing data recovery in family trios. In: Sunderam, V.S., van Albada, G.D., Sloot, P.M.A., Dongarra, J.J. (eds.) ICCS 2005. LNCS, vol. 3515, pp. 1011–1019. Springer, Heidelberg (2005)
5. Chazelle, B., Guibas, L.J.: Fractional cascading: I. a data structuring technique. Algorithmica 1(2), 133–162 (1986)
6. Clark, A.G.: Inference of haplotypes from pcr-amplified samples of diploid population. Molecular Biology and Evolution 7(2), 111–122 (1990)
7. Cole, R., Gottlieb, L., Lewenstein, M.: Dictionary matching and indexing with errors and don't cares. In: Proceedings of the 36th annual ACM Symposium on Theory Of Computing (STOC), pp. 91–100. ACM Press, New York (2004)
8. Cole, R., Hariharan, R.: Verifying candidate matches in sparse and wildcard matching. In: Proceedings of the 34th annual ACM Symposium on Theory Of Computing (STOC), pp. 592–601. ACM Press, New York (2002)
9. Cole, R., Kopelowitz, T., Lewenstein, M.: Suffix trays and suffix trists: structures for faster text indexing. In: Bugliesi, M., Preneel, B., Sassone, V., Wegener, I. (eds.) ICALP 2006. LNCS, vol. 4051, pp. 358–369. Springer, Heidelberg (2006)
10. Didier, G., Schmidt, T., Stoye, J., Tsur, D.: Character sets of strings. Journal of Discrete Algorithms (to appear)
11. Excoffier, L., Slatkin, M.: Maximum-likelihood estimation of molecular haplotype frequencies in a diploid population. Molecular Biology and Evolution 12(5), 921–927 (1995)
12. Fallin, D., Schork, N.J.: Accuracy of haplotype frequency estimation for biallelic loci, via the expectation-maximization algorithm for unphased diploid genotype data. American Journal of Human Genetics 67(4), 947–959 (2000)
13. Gusfield, D.: Haplotype inference by pure parsimony. In: Baeza-Yates, R.A., Chávez, E., Crochemore, M. (eds.) CPM 2003. LNCS, vol. 2676, pp. 144–155. Springer, Heidelberg (2003)
14. Gusfield, D., Orzack, S.H.: Haplotype inference. In: Aluru, S. (ed.) CRC handbook on bioinformatics (2005)

15. Hagerup, T., Miltersen, P.B., Pagh, R.: Deterministic dictionaries. J. of Algorithms 41(1), 69–85 (2001)
16. Hajiaghayi, M.T., Jain, K., Konwar, K., Lau, L.C., Mandoiu, I.I., Vazirani, V.V.: Minimum multicolored subgraph problem in multiplex pcr primer set selection and population haplotyping. In: Alexandrov, V.N., van Albada, G.D., Sloot, P.M.A., Dongarra, J.J. (eds.) ICCS 2006. LNCS, vol. 3991, pp. 758–766. Springer, Heidelberg (2006)
17. Halldórsson, B.V., Bafna, V., Edwards, N., Lippert, R., Yooseph, S., Istrail, S.: A survey of computational methods for determining haplotypes. In: Istrail, S., Waterman, M.S., Clark, A. (eds.) Computational Methods for SNPs and Haplotype Inference. LNCS (LNBI), vol. 2983, pp. 26–47. Springer, Heidelberg (2004)
18. Halperin, E., Karp, R.M.: The minimum-entropy set cover problem. In: Díaz, J., Karhumäki, J., Lepistö, A., Sannella, D. (eds.) ICALP 2004. LNCS, vol. 3142, pp. 733–744. Springer, Heidelberg (2004)
19. Harel, D., Tarjan, R.E.: Fast algorithms for finding nearest common ancestors. SIAM Journal of Computing 13(2), 338–355 (1984)
20. Hawley, M.E., Kidd, K.K.: Haplo: A program using the em algorithm to estimate the frequencies of multi-site haplotypes. Journal of Heredity 86, 409–411 (1995)
21. Helmuth, L.: Genome research: Map of human genome 3.0. Science 5530(293), 583–585 (2001)
22. Indyk, P.: Faster algorithms for string matching problems: Matching the convolution bound. In: Proceedings of the 39th annual Symposium on Foundations of Computer Science (FOCS), pp. 166–173 (1998)
23. Karp, R.M., Miller, R.E., Rosenberg, A.L.: Rapid identification of repeated patterns in strings, trees and arrays. In: Proceedings of the 4th annual ACM Symposium on Theory Of Computing (STOC), pp. 125–136. ACM Press, New York (1972)
24. Kedem, Z.M., Landau, G.M., Palem, K.V.: Parallel suffix-prefix-matching algorithm and applications. SIAM Journal of Computing 25(5), 998–1023 (1996)
25. Kolpakov, R., Raffinot, M.: New algorithms for text fingerprinting. In: Lewenstein, M., Valiente, G. (eds.) CPM 2006. LNCS, vol. 4009, pp. 342–353. Springer, Heidelberg (2006)
26. Long, J.C., Williams, R.C., Urbanek, M.: An E-M algorithm and testing strategy for multiple-locus haplotypes. American Journal of Human Genetics 56(2), 799–810 (1995)
27. Marchini, J., Cutler, D., Patterson, N., Stephens, M., Eskin, E., Halperin, E., Lin, S., Qin, Z.S., Munro, H.M., Abecasis, G., Donnelly, P.: The International HapMap Consortium.A comparison of phasing algorithms for trios and unrelated individuals. American Journal of Human Genetics 78, 437–450 (2006)
28. Shi, Q., JáJá, J.: Novel transformation techniques using q-heaps with applications to computational geometry. SIAM Journal of Computing 34(6), 1471–1492 (2005)
29. van Emde Boas, P.: Preserving order in a forest in less than logarithmic time and linear space. Information Processing Letters 6(3), 80–82 (1977)
30. Zhang, P., Sheng, H., Morabia, A., Gilliam, T.C.: Optimal step length em algorithm (oslem) for the estimation of haplotype frequency and its application in lipoprotein lipase genotyping. BMC Bioinformatics 4(3) (2003)

Extending Weighting Models with a Term Quality Measure

Christina Lioma and Iadh Ounis

University of Glasgow, Scotland G12 8QQ, U.K.

Abstract. Weighting models use lexical statistics, such as term frequencies, to derive term weights, which are used to estimate the relevance of a document to a query. Apart from the removal of stopwords, there is no other consideration of the quality of words that are being 'weighted'. It is often assumed that term frequency is a good indicator for a decision to be made as to how relevant a document is to a query. Our intuition is that raw term frequency could be enhanced to better discriminate between terms. To do so, we propose using non-lexical features to predict the 'quality' of words, before they are weighted for retrieval. Specifically, we show how parts of speech (e.g. nouns, verbs) can help estimate how informative a word generally is, regardless of its relevance to a query/document. Experimental results with two standard TREC[1] collections show that integrating the proposed term quality to two established weighting models enhances retrieval performance, over a baseline that uses the original weighting models, at all times.

1 Introduction

The aim of an Information Retrieval (IR) system is to retrieve relevant documents in response to a user need, which is usually expressed as a query. The retrieved documents are returned to the user in decreasing order of relevance, which is typically determined by weighting models. Most weighting models use term statistics, such as term frequency, to assign weights to individual terms, which represent the contribution of the term to the document content. These term weights are then used to estimate the similarity between queries and documents [18].

The underlying idea of most weighting models is to boost the weight of terms that occur frequently in a document and rarely in the rest of the collection of documents [18]. Various extensions have been applied on top of this, such as normalising term frequency according to document length [1,5,14], or using the term frequency in specific fields of structured documents (e.g. title, abstract) [8,13,15]. Further extensions include integrating query-independent evidence (e.g. PageRank [2]) to the weighting model in the form of prior probabilities [4,6,12] ('prior' because they are known before the query is issued). For example, assuming that a document's PageRank indicates its quality, integrating PageRank priors to the

---

[1] Text REtrieval Conference: http://trec.nist.org/

N. Ziviani and R. Baeza-Yates (Eds.): SPIRE 2007, LNCS 4726, pp. 205–216, 2007.

weighting model consists in using information about the document quality when computing how relevant that document is to a query.

We propose a measure of term quality, which is similar to the notion of document quality, in that it is known prior to a query. In addition, our proposed term quality is known prior to the document as well, because it represents how informative a term is generally in language, not with respect to a query or its contribution to the content of a specific document. Hence, this is an intrinsic notion of term quality. The intuition behind using it in IR is that integrating it into the term weighting process may enhance retrieval performance, similarly to the way document quality (in the form of priors, for example) can improve performance when integrated to document ranking [4,6,12].

We derive the proposed term quality in an empirical way from part of speech (POS) n-grams. POS n-grams are n-grams of parts of speech, extracted from POS tagged text. Specifically, we extend the work of [7] who estimate how informative a word sequence corresponding to a POS n-gram can be. We use this to derive a quality score for each term separately, not as a sequence, which we then integrate to the weighting model. Our intuition is that the term frequency statistics used by weighting models could be enhanced to better discriminate between terms, using our proposed measure of term informativeness. Hence, our goal is to assist the lexical features used by these models (e.g. term frequency), which are query/document-dependent, with a non-lexical feature (our proposed term quality), which is query/document-independent.

We evaluate the impact of integrating our proposed term quality into weighting models upon retrieval performance, using the original weighting model as a baseline. Experiments on two established weighting models and two standard TREC collections, show that term quality improves retrieval performance sometimes considerably, and consistently at all times.

The remainder of this paper is organised as follows. Section 2 presents related studies. Section 3 details our methodology for deriving term quality and for extending weighting models with it. Section 4 discusses the evaluation, and Section 5 summarises our conclusions.

2   Related Studies

We propose to extend weighting models with a measure of term quality. The notion of an intrinsic term quality is new in IR (to our knowledge), but not in linguistics. Word commonness, which measures how common a word is generally in language, is used in theoretical and practical linguistics, e.g. in quantitative linguistics, lexicography, and language teaching [16]. Mikk [9] suggests a 'corrected term frequency' based on word commonness, which predicts the complexity of a document's content. Our proposed term quality is based on the same intuition, namely that raw term frequency could be enhanced to better discriminate between terms. Unlike previous studies, we specifically apply this intuition to IR, and ask whether we can extend models that are tailored to process term frequencies, with the proposed term quality, so as to improve retrieval performance.

There are two novel contributions in this work. First, even though previous work has extended weighting models with various types of evidence, as briefly mentioned in Section 1 [4,6,8,15], to our knowledge, no study has reported integrating term evidence that is both query- and document-independent to the weighting model. The closest to this notion is removing stopwords before retrieval, in the sense that words from a standard list are removed regardless of the query/document. The second novelty of our approach is that we use POS n-grams to derive a measure of term quality. POS n-grams were first used in POS tagging, to determine the probability of occurrence of POS tags [3]. More recently, POS n-grams were used to estimate the quality of word sequences in IR [7]. This work is a continuation of the latter, because it uses the quality of POS n-grams to derive a measure of quality for individual terms.

3   Methodology

3.1   Deriving Term Quality from Parts of Speech

We derive a term quality measure from POS n-grams as follows. We use a POS tagger to POS tag a collection of documents. Any POS tagger and any large collection of documents can be used. We extract POS n-grams from each POS-tagged sentence in each document. For example, for a sentence ABCDEF, where parts of speech are denoted by the single letters A, B, C, D, E, F and where POS n-gram length = 4, the POS n-grams extracted are ABCD, BCDE, and CDEF. Then, we compute the quality of each POS n-gram, using the *content load* (*cl*) estimator described in [7]. This estimator considers nouns, verbs, adjectives and participles more informative than other parts of speech, with nouns the most informative, following evidence given in [11,19]. The formula is:

$$cl = \frac{C_N + C_{AVP} \cdot \varrho}{n} \tag{1}$$

where C_N = number of nouns in the POS n-gram, C_{AVP} = number of adjectives and/or verbs and/or participles in the POS n-gram, n = POS n-gram length, and ϱ = penalising variable applied to adjectives and/or verbs and/or participles. The value of ρ is automatically derived from collection statistics [7]. Using Equation (1), the content load of a POS n-gram is between 0 and 1, where 0 and 1 are the minimum and maximum values, respectively. This content load for POS n-grams approximates how important any of the word sequences that correspond to a POS n-gram can be. We extend the IR system's inverted file with information about the POS n-grams that are associated with each term, and their content load. The inverted file of an IR system contains statistics on term frequencies in each document and in the whole collection.

Based on Equation (1), we propose to compute the quality score for each term (*tqs*) as follows:

$$tqs = \frac{\sum cl_t}{f_{POSngram_t}} \tag{2}$$

Table 1. Example term quality for TREC query 451 (stemmed & no stopwords)

term	quality score	term	quality score	term	quality score	term	quality score
bengal	0.46	catteri	0.42	includ	0.19	program	0.29
breed	0.31	characterist	0.31	item	0.33	refer	0.25
breeder	0.42	club	0.35	name	0.25	relev	0.24
carri	0.20	discus	0.35	onli	0.16	tiger	0.39
cat	0.35	discuss	0.19	origin	0.22		

where cl_t is the content load of a POS n-gram that contains[2] term t (computed using Equation (1)), and $f_{POSngram_t}$ is the number of POS n-grams that contain term t. Note that we consider the content load of all POS n-grams, regardless of which documents they occur in, because our goal is to derive a global, as opposed to document-centric, estimation of term quality. Using Equation (2), the term quality score is between 0 and 1, where 0 and 1 are the minimum and maximum scores, respectively. The term quality score approximates how important a term generally is, based on its part of speech and the POS n-grams in which it occurs in the collection.

Table 1 gives an example of the quality score given to terms from TREC query number 451 (queries are presented in Section 4.). Terms are stemmed and stopwords are removed in this example. We see that `bengal`, `breeder` and `catteri` have the highest *tqs*, while `desc`, `includ` and `onli` have the lowest. Even though *tqs* is derived from POS n-grams, and specifically a formula that rewards nouns, slightly penalises adjectives, verbs and participles, and ignores everything else, the term quality score seems to discriminate between terms on the basis of more than just their POS class. Hence, the highest scoring term is an adjective (`bengal`), not a noun, in this query. Similarly, while both `name` and `tiger` are nouns, they have different scores (0.25 and 0.39, respectively). Overall, the main point to remember here is that the quality scores assigned to these query terms have been derived from POS, not lexical, statistics, extracted from a whole collection. Hence, these term quality scores are completely document independent.

POS tagging, extracting POS n-grams and computing term quality take place once at indexing time, with negligible overhead.

3.2   Integrating Term Quality to Term Weighting

Section 3.1 introduced a general quality measure for terms, which is document-independent. More simply, the proposed term quality measures how informative a term generally is, and not how relevant a term is to another. In order for such a general quality measure to be used in retrieval, it needs to be integrated with relevance weighting, i.e. classical term weighting that determines the relevance of a term to a document. We present how we integrate the proposed term quality to term weighting in this section.

---

[2] POS n-grams contain POS tags, not terms. By term 'contained' in a POS n-gram we mean a term that, when tagged, has its POS tag captured in a POS n-gram.

We estimate the relevance $R(d, Q)$ between a document d and a query Q, as:

$$R(d, Q) = \sum_{t \in Q} w(t, d) \cdot qtw \qquad (3)$$

where t is a term in Q, $w(t, d)$ is the weight of term t for a document d, and qtw is the query term weight. $w(t, d)$ can be computed by different weighting models in different ways [1,6,14]. All these models however use the frequency of a term in a document (tf) one way or another. For example, for BM25 [14]:

$$w(t, d) = w^{(1)} \cdot \frac{(k_3 + 1) \cdot qtf}{k_3 + qtf} \cdot tfn \qquad (4)$$

where k_3 is a parameter, qtf is the query term frequency, and tfn is the normalised term frequency in a document, given by:

$$tfn = \frac{(k_1 + 1) \cdot tf}{tf + k_1 \cdot (1 - b + b \cdot \frac{l}{avg\_l})} \qquad (5)$$

where k_1 & b are parameters, and l (avg_l) is the document length (average document length in the collection).

In Equation (4), $w^{(1)}$ is the weight of a term in the query, given by:

$$w^{(1)} = log \cdot \frac{N - n + 0.5}{n + 0.5} \qquad (6)$$

where N is the number of all documents in the collection, and n is the number of documents containing term t. Note that $w^{(1)}$ is the inverse document frequency (idf) component of BM25.

To recapitulate, tf is used as an integral part of BM25 to compute the relevance of a document to a query (Equation (5)).

Our aim is to show how $w(t, d)$ can be altered to include our proposed term quality. We extend weighting models with term quality (tqs), computed using Equation (2), by altering term frequency (tf) in the $w(t, d)$ component (see Equation (3)) of each weighting model, as follows:

$$tf_q = tf \cdot \frac{1}{1 - tqs} \qquad (7)$$

where tf_q is the term frequency that is altered by the term quality score, tf is the original term frequency of term t in a document, and tqs is the proposed term quality. The idea here is to boost the discriminative effect of term frequency with knowledge about how informative a term generally is in language. The reason why we use 1 / (1 - tqs), instead of raw tqs, is explained at the end of this section. The main point to remember here is that by using term quality to alter term frequency, we integrate it into the weighting model as part of the $w(t, d)$ component, and not externally (e.g. as a prior). Note that the integration proposed (Equation (7)) is one among several possible and potentially more

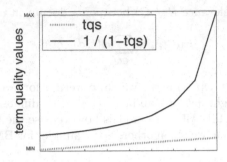

Fig. 1. *tqs* (computed using Equation (2)) across its value range ($min - max$=0-1)

sophisticated ways of integrating term quality to the term weighting model. Our focus is to initially test if term quality works for retrieval, and not yet to optimise its integration into term weighting.

We have shown how we integrate term quality to the weighting model by multiplying it to term frequency before the relevance of a document to a query is 'weighted'. In fact, this integration takes place even before term frequency is normalised with respect to document length. We do not know what the effect of normalisation will be. In this work, we assume that normalising term frequency should not affect the integration of term quality into the model. However, this assumption is worth testing in the future.

Why do we use 1 / (1 - *tqs*) instead of *tqs*? We know that term quality can be between 0 and 1. Figure 1 plots term quality for values within this range. Note that we plot *tqs* as a simple function here, meaning the x-axis is simply the arguments of the function. The distribution of term quality across its value range is the dotted line. We see that a simple transformation of *tqs*, namely 1 / (1 - *tqs*), widens the value range considerably (solid line). We assume that widening the value range of the proposed term quality will render it more discriminative, and this is why we prefer it over the raw *tqs*, when integrating it to the weighting model.

Implementation-wise, integrating term quality to the weighting model takes place when documents are matched to queries, and consists of a simple look-up of POS n-gram statistics in the IR system's inverted file. This is done simultaneously to the usual term statistics look-up, with negligible overhead.

4   Evaluation

We aim to test whether integrating the proposed term quality score into weighting models can enhance retrieval performance. We use two standard TREC collections, namely WT10G (TREC 2000-2001), which contains 1.69 million Web documents (10GB), and Disks 4&5 (TREC 2004), which contain 528 thousand mainly newswire documents (2GB). We remove the Congressional Record from Disks 4&5, according to TREC 2003-2004 settings. We use topics 451-550

Table 2. Weighting model parameters (b for BM25, c for PL2)

WT10G		Disk 4&5	
default	optimal	default	optimal
b=0.75, c=1.00	b=0.27, c=13.13	b=0.75, c=1.00	b=0.34, c=12.00

(WT10G) and topics 301-450 & 601-700 (Disks 4&5) to retrieve relevant documents. We use short topics (title-only) because they are more representative of real user queries. For indexing and retrieval, we use the Terrier IR platform [10], and apply stopword removal and Porter's stemming during indexing.

To compute the term quality score, we POS tag WT10G, using the TreeTagger, because it is fast ($\sim$10,000 tokens/sec) and has a low error rate ($\sim$3.8%) [17]. Following [7], we extract POS n-grams of length $n= 4^3$ from the collection, and compute their content load using Equation (1), with $\rho = 0.17$.

To match documents to query terms, we use BM25 [14] and PL2 from the Divergence From Randomness (DFR) framework [1]. Each of these models treats the matching process differently, giving us a varied setting for our experiments. Each model has a term frequency normalisation parameter (b for BM25, and c for PL2). These parameters can be tuned according to query/collection statistics, and can affect retrieval performance considerably [1].

To evaluate the impact of integrating term quality to the weighting model upon retrieval performance, with the original weighting models as a baseline, we conduct three series of experiments. Throuhgout, we use the mean average precision (MAP) to evaluate retrieval performance. 1) We set all weighting model parameters to default/recommended values. (See [14] for default b values; c values are recommended at [1] ^{4}.) 2) To test the effect of our approach on a stronger baseline than that of default values, we optimise all weighting model tf normalisation parameters for MAP, by training using data sweeping and simulated annealing over a large range of values. We optimise the baseline for MAP and use the same parameter for the weighting model with the term quality (i.e. we assume that the optimal b value for BM25 will also be optimal for BM25 with term quality). All parameter values used are shown in Table 2. 3) To raise even more the baseline, we use optimal values and query expansion (QE), which is an automatic performance-boosting technique that extracts the most relevant terms from the top retrieved documents, and uses them to expand the initial query. The expanded query is then used to retrieve documents anew. For query expansion, we use the Bo1 DFR weighting model [1], and extract the 30 most relevant terms from the top 5 retrieved documents, which is the recommended setting [1]. Table 3 displays the evaluation results.

Table 3 shows that, at all times, the proposed term quality improves MAP, over the original weighting models, with a statistical significance (Wilcoxon matched-pairs signed-ranks test) for Disks 4&5. This conclusion is consistent

---

^{3} varying n between 3 and 6 gives similar results to the ones reported here.
^{4} c values are also recommended at:
  http://ir.dcs.gla.ac.uk/terrier/doc/dfr_description.html

Table 3. MAP scores using weighting models with default/optimised parameters and query expansion (QE); baseline = original weighting model; term quality = weighting model with term quality; * (**) = stat. significance at p<0.05 (p<0.01) with Wilcoxon matched-pairs signed-ranks test.

settings	model	WT10G		Disks 4&5	
		baseline	term quality	baseline	term quality
default	BM25	0.1874	0.1923 (+2.6%)	0.2363	0.2425 (+2.6%**)
	PL2	0.1753	0.1846 (+5.3%**)	0.2242	0.2348 (+4.7%**)
optimised	BM25	0.2096	0.2104 (+0.4%)	0.2499	0.2549 (+2.0%**)
	PL2	0.2093	0.2112 (+1.9%)	0.2530	0.2532 (+0.1%*)
optimised + QE	BM25	0.2362	0.2440 (+3.3%)	0.2933	0.2985 (+1.8%**)
	PL2	0.2241	0.2276 (+1.6%)	0.2966	0.2980 (+0.8%**)

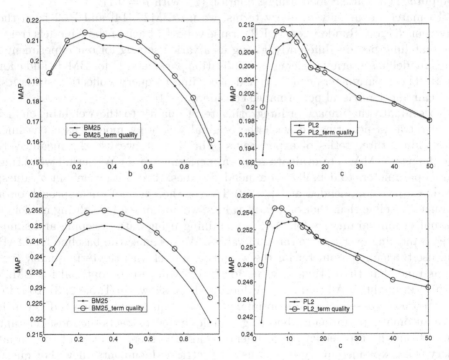

Fig. 2. WT10G in top row, Disks 4&5 in bottom row. The x axis is the weighting model parameter (b for BM25, c for PL2). The y axis is the Mean Average Precision.

for two different weighting models, with and without query expansion, for 350 topics and for two different collections, hence it is a solid indication that enriching term frequency with the proposed term quality can enhance retrieval. This finding is also supported by Figure 2, which shows the effect of varying the weighting model parameters without term quality (dot) and with term quality (circle), for the two collections. Two trends emerge in Figure 2: 1) the baseline

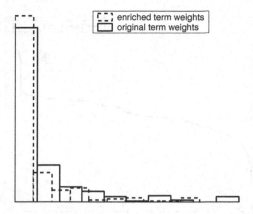

Fig. 3. BM25 term weights for all 350 queries for the top retrieved document, without *tqs* (original) and with *tqs* (enriched)

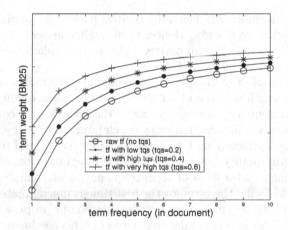

Fig. 4. *tf* component of BM25 term weight, without and with *tqs*

and term quality lines have similar shapes throughout, and 2) the term quality line (circle) moves generally at higher MAP values than the baseline line (dot). This means that our integration of term quality to the weighting model helps retrieval in a consistent way. This consistency can explain the fact that the optimal *tf* normalisation values are very similar and sometimes identical for the baseline and for using term quality.

Figure 3 shows an example of the effect of integrating term quality into BM25 as a histogram. We plot the weight of all 350 queries for the top retrieved document as a function, similarly to Figure 1. We compare the term weights computed by the baseline BM25 (solid line) and BM25 with term quality (dotted line)[5]. High term weights mean that a term is very relevant to the document, and vice

---

[5] PL2 behaves very similarly.

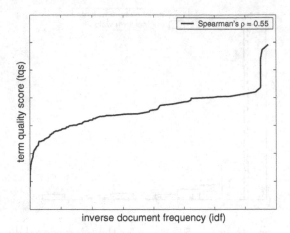

Fig. 5. *idf* versus *tqs* for all query terms used

versa. We see that term quality generally renders high term weights even higher and low term weights even lower. Hence, term quality appears to make the resulting term weights more discriminative, which could explain the improvement in retrieval performance shown in Table 3.

Figure 4 plots the effect of integrating term quality to term frequency, ignoring term frequency normalisation, and for three different term quality values. The x axis is the term frequency, and the y axis is the term weight computed using BM25. We plot the original term frequency (circle), term frequency with low term quality (dot), term frequency with high term quality (star), and term frequency with very high term quality (cross). We see that integrating term quality never breaks the non-linear saturation of term frequency. More simply, the gain (in informativeness) in seeing the term for the first time is much greater, than seeing that term subsequently, even after we have integrated our proposed term quality to term frequency. This is very similar to the effect of inverse document frequency upon term frequency [15] (this point is further discussed in the next paragraph). This shows that term quality is compatible to term frequency. Note that the effect of term quality on term frequency becomes more noticeable as term frequency decreases. More simply, a term with high tf will be boosted less by tqs, than a term with low tf. This is expected, given the difference in magnitude between term quality (between 0-1) and term frequency ($\gg 1$).

Finally, Figure 5 shows the relation between inverse document frequency (idf) and term quality (tqs), for the terms of all 350 queries used in these experiments. We see that, indeed, our proposed measure of term quality is correlated to inverse document frequency (Spearman's $\rho = 0.55$), as indicated previously (Figure 4). This correlation indicates that our proposed term quality is compatible to term frequency, and can explain why intergating tqs into the weighting model overall enhances retrieval performance.

5   Conclusion

We introduced a novel notion of term quality, which measures how informative a term generally is, regardless of a document/query. We derived this measure using part of speech (POS) information extracted from a corpus as POS n-grams. We tested this term quality in IR, by integrating it to the term frequency component of two weighting models. We reasoned that if this integration resulted in more accurate term weights, and retrieval performance improved, then we could consider term quality as useful evidence. Experimental results on two standard TREC collections, with default and optimal settings, and query expansion, showed that retrieval performance with term quality improved consistently, and with a statistical significance at all times for one collection, over strong baselines.

The main contribution of this work consists in posing the question: Can there be such a thing as an intrinsic notion of term quality? We showed that yes, there can, and also how to practically apply it to enhance retrieval performance. Future work includes exploring the integration of term quality into the retrieval process, as well as evaluating it in other tasks, such as text classification.

References

1. Amati, G.: Probabilistic Models for Information Retrieval based on Divergence from Randomness, PhD Thesis. University of Glasgow, UK (2003)
2. Brin, S., Page, L.: The Anatomy of a Large-Scale Hypertextual Web Search Engine. Computer Networks and ISDN Systems. 30, 107–117 (1998)
3. Brown, P.F., Pietra, V.J.D., deSouza, P.V., Lai, J.C., Mercer, R.L.: Class-Based N-Gram Models of Natural Language. Computational Linguistics 18(4), 467–479 (1992)
4. Craswell, N., Robertson, S., Zaragoza, H., Taylor, M.: Relevance Weighting for Query Independent Evidence. In: 28th Annual International ACM SIGIR Conference on Research and Development in Information Retrieval, pp. 416–423. ACM Press, New York (2005)
5. He, B., Ounis, I.: A Study of the Dirichlet Priors for Term Frequency Normalisation. In: He, B., Ounis, I. (eds.) 28th Annual International ACM SIGIR Conference on Research and Development in Information Retrieval, pp. 465–471. ACM Press, New York (2005)
6. Kraaij, W., Westerveld, T., Hiemstra, D.: The Importance of Prior Probabilities for Entry Page Search. In: SIGIR. 25th Annual International ACM SIGIR Conference on Research and Development in Information Retrieval, pp. 27–34. ACM Press, New York (2002)
7. Lioma, C., Ounis, I.: Light Syntactically-Based Index Pruning for Information Retrieval. In: ECIR 2007. LNCS, vol. 4425, pp. 88–100. Springer, Heidelberg (2007)
8. Macdonald, C., He, B., Plachouras, V., Ounis, I.: University of Glasgow at TREC2005: Experiments in Terabyte and Enterprise tracks with Terrier. In: TREC. 14^{nth} Text REtrieval Conference (2005)
9. Mikk, J.: Prior Knowledge of Text Content and Values of Text Characteristics. Journal of Quantitative Linguistics. 8(1), 67–80 (2001)

10. Ounis, I., Amati, G., Plachouras, V., He, B., Macdonald, C., Lioma, C.: Terrier: A High Performance and Scalable Information Retrieval Platform. In: Association for Computing Machinery (ACM) Conference on Research and Development in Information Retrieval (SIGIR) Workshop on Open Source Information Retrieval (OSIR), pp. 18–24 (2006)
11. Ozmutlu, S., Spink, A., Ozmutlu, H.C.: A Day in the Life of Web Searching: an Exploratory Study. Information Processing & Management 2, 319–345 (2004)
12. Peng, J., Macdonald, C., He, B., Ounis, I.: Combination of Document Priors in Web Information Retrieval. In: RIAO. 8_{th} Large-Scale Semantic Access to Content, pp. 28–39 (2007)
13. Plachouras, V., Ounis, I.: Multinomial Randomness Models for Retrieval with Document Fields. In: ECIR 2007. LNCS, vol. 4425, pp. 88–100. Springer, Heidelberg (2007)
14. Robertson, S., Walker, S.: Some Simple Approximations to the 2-Poisson Model for Probabilistic Weighted Retrieval. In: 17th Annual International ACM SIGIR Conference on Research and Development in Information Retrieval, pp. 232–241 (1994)
15. Robertson, S., Zaragoza, H., Taylor, M.: Simple BM25 Extension to Multiple Weighted Fields. In: CIKM. 13^{nth} ACM International Conference on Information and Knowledge Management, pp. 42–49. ACM Press, New York (2004)
16. Savicky, P., Hlavacova, J.: Measures of Word Commonness. Journal of Quantitative Linguistics 9(3), 215–231 (2002)
17. Schmid, H.: Probabilistic Part-of-Speech Tagging Using Decision Trees. In: International Conference on New Methods in Language Processing, pp. 44–49 (1994)
18. van Rijsbergen, C.J.: Information Retrieval. Butterworths, Butterworths, London, UK (1979)
19. Zubov, A.: Formalization of the Procedure of Singling Out of the Basic Text Contents. Journal of Quantitative Linguistics 11(1-2), 33–48 (2004)

Highly Frequent Terms and Sentence Retrieval

David E. Losada and Ronald T. Fernández

Departamento de Electrónica y Computación,
Universidad de Santiago de Compostela, Spain
dlosada@dec.usc.es, ronald.teijeira@rai.usc.es

Abstract. In this paper we propose a novel sentence retrieval method based on extracting highly frequent terms from top retrieved documents. We compare it against state of the art sentence retrieval techniques, including those based on pseudo-relevant feedback, showing that the approach is robust and competitive. Our results reinforce the idea that top retrieved data is a valuable source to enhance retrieval systems. This is especially true for short queries because there are usually few query-sentence matching terms. Moreover, the approach is particularly promising for weak queries. We demonstrate that this novel method is able to improve significantly the precision at top ranks when handling poorly specified information needs.

Keywords: Information Retrieval, Sentence Retrieval, Term Frequency.

1 Introduction

Retrieval of sentences that are relevant to a given information need is an important problem for which an effective solution is still to be found. Many Information Retrieval (IR) tasks rely on some form of sentence retrieval to support their processes [13]. For example, question answering systems tend to apply sentence extraction methods to focus the search for an answer on a well-selected set of sentences or paragraphs [5]. In query-biased summarization, there is a large body of work dedicated to building summaries using sentences extracted from the documents [12]. Topic detection and tracking (TDT) is another task where the availability of effective sentence retrieval techniques is crucial in order to isolate relevant material from a dynamic stream of texts (e.g. news) [19]. In web IR, a good ranking of sentences, in decreasing order of estimated relevance to the query, can also act as a solid tool to improve web information access [25]. Sentence retrieval is therefore a core problem in IR research and advances in this area could potentially trigger significant benefits across the field.

We adopt the sentence retrieval problem as defined in the TREC novelty tracks [8,18,17]. Given a textual information need, an initial ranking of documents is produced using some effective retrieval method and, next, the systems should process the retrieved documents to locate the sentences that are relevant

N. Ziviani and R. Baeza-Yates (Eds.): SPIRE 2007, LNCS 4726, pp. 217–228, 2007.

to the information need[1]. This is a realistic task. For instance, typical web search engines have to address similar problems when building query-biased summaries.

The problem of sentence retrieval is far from easy. Sentences are very short pieces of texts and, therefore, the sentence-query matching process has difficulties. Many strategies, techniques and models have been proposed to address this problem. Despite the variety of the approaches investigated, very simple variations of tf/idf measures can be labeled as state of the art sentence retrieval methods [2,11].

We are not interested here in proposing a new model for sentence retrieval, but we will extend a competitive sentence retrieval method to incorporate the influence from terms which are highly frequent in the retrieved set of documents. This is inspired by well-known research in query-biased summarization [20] that estimated the set of significant words of a document using the number of term occurrences within each document. We adapt this intuition to our current retrieval scenario. Rather than focusing on a single document to estimate which words are significant, we will estimate the significant terms from the top retrieved documents. The set of highly ranked documents for a given query is a very valuable source of information and, ideally, it provides a vocabulary focused on the query topics. Hence, we compute term statistics globally in the retrieved set of documents and adjust the tf/idf scores to take into account the contribution from the most significant terms. In this way, sentences that do not match any query term can still be retrieved provided that they contain some significant terms.

In the experiments we show that the approach is simple but very effective. It outperforms significantly a competitive baseline under three different benchmarks. The method is able to improve both precision at 10 sentences and the F measure. We also compare the relative merits of this method against pseudo-relevance feedback, showing that our approach is much more robust, especially when queries are poor.

The rest of the paper is organized as follows. Section 2 reviews some papers related to our research. In section 3 we explain the foundations of the method proposed and the evaluation conducted is reported in section 4. The paper ends with some conclusions and future lines of work.

2   Related Work

Sentence retrieval is an active research area where many researchers have proposed different alternatives to tackle the problem. Many studies applied query expansion either via pseudo-relevance feedback [6] or with the assistance of a terminological resource [27,9]. Nevertheless, the effect of pseudo-relevance feedback is very sensitive to the quality of the initial ranks and it is quite difficult to apply it effectively across different collections and types of queries [26]. More

---

[1] The TREC novelty tracks propose two different tasks: retrieval of relevant sentences and retrieval of relevant and novel sentences but we are only interested here in the first task.

evolved methods, such as selective feedback [1], are more stable but they usually require training data. On the other hand, to expand queries with synonyms or related terms from a lexical resource is problematic because noisy terms can be easily introduced into the new query [21]. Moreover, a large terminological resource, with good coverage, is not always available. As a matter of fact, lexical expansion was not significantly better than purely statistical expansion methods in the sentence retrieval tasks of the TREC novelty tracks [8,18,17]. Other expansion approaches based on co-occurrence data have been proposed. For instance, in [27] the authors expand the query with terms that co-occur highly with the query terms in the retrieved documents. Co-occurrence statistics from a background corpus have been applied in [15]. Nevertheless, there is not much evidence that these approaches can outperform the standard pseudo-feedback methods.

Rather than expanding queries with new terms, other studies have focused on improving the matching process by analyzing carefully the nature of the sentence components. In this line, in [11], patterns such as phrases, combinations of query terms and named entities were identified into sentences and the sentence retrieval process was driven by such artifacts. Although this technique was effective for detecting redundant sentences, it was not significantly better than a regular tf/idf baseline for finding relevant sentences. In [7], terms in sentences were categorized into four query-based categories, namely: highly relevant, scarcely relevant, non-relevant and highly non-relevant. Nevertheless, this classification was mainly guided by the topic subfield in which the term occurs (title, descriptive, narrative). Unlike this work, our approach uses the number of term occurrences in the retrieved set of documents as the main factor to estimate the significance of a term.

The work by Zhang and his colleagues [28] deserves special attention. They presented a high performing sentence retrieval system which combined query expansion and sentence expansion using Wordnet. Besides the linguistic-based expansion, they proposed two *impact factor* measures to estimate the significance of the query terms. The utility of these measures in removing some query terms (low impact factor words) was empirically evaluated but no benefits were observed. One of the impact factor measures was based on the frequency of the term in relevant documents. The intuition was that highly frequent terms are very significant and, therefore, they should not be removed from the query. We pursue a similar idea but with a different objective. Rather than polishing queries, our aim is to define a measure of how central a sentence is in the context of the retrieved documents. The highly frequent words in the retrieved set are the foundations of this measure.

In query-biased summarization, the notion of significant word (term with high frequency in the document) was applied for sentence scoring purposes [20]. This was combined with aspects such as term location (the presence of a term in the title of the document or in the leading paragraphs produced a positive weight) and some query-oriented weights to produce a final score for the sentences. Our method to adjust the basic tf/idf method is inspired by the notion of significant word applied in [20] but, rather than estimating the significant terms for a single document, we are interested in computing the significant words in a set of documents. As a matter of fact, the set of highly ranked documents for a given query is a good container of topic-related terms, as demonstrated by the success of different expansion methods for document retrieval [23]. Some papers have applied the results of [20] to other environments. For instance, in [24], the effects of query-biased summaries in web searching were studied. The summaries are constructed from sentences scored using title, location, relation to the query and text formatting information. However, the notion of significant words, as proposed in [20], has not received major attention since then. We believe that it is interesting to study the effect of such notion for sentence retrieval.

3   Highly Frequent Terms and Sentence Retrieval

The main component in our method consists of extracting a set of significant words from the set of documents retrieved for a given query. Ideally, these terms characterize well the query topics and, therefore, they can be used for improving sentence retrieval. Given a set of highly ranked documents, R_q, terms occurring more than a certain number of times in this set are collected into a *set of significant terms* (ST_q) as follows[2]:

$$ST_q = \{t \in V_q | tf_{t,R_q} > mno\} \tag{1}$$

where V_q is the set of unique terms appearing in R_q, tf_{t,R_q} is the term count of t in R_q $(tf_{t,R_q} = \sum_{D \in R_q} tf_{t,D})$ and mno is a parameter that determines the minimum number of occurrences required for a term to be considered as significant.

In order to select the sentences within the retrieved set of documents which are relevant to the query, a combination of a regular tf/idf score and a significant term score is applied. The tf/isf score (isf stands for inverse *sentence* frequency) is:

$$tf\_isf(s,q) = \sum_{t \in q} log(tf_{t,q} + 1) log(tf_{t,s} + 1) log(\frac{n+1}{0.5 + sf_t}) \tag{2}$$

where sf_t is the number of sentences in which t appears, n is the number of sentences in the collection and $tf_{t,q}$ $(tf_{t,s})$ is the number of occurrences of t in q (s). This formula was applied successfully in TREC novelty tracks for sentence retrieval purposes [2]. Along this paper this method will be referred to as TF/ISF.

Given a query q, the significant term score of a sentence s, $htf(s,q)$, is defined as:

$$htf(s,q) = \sqrt{|V_s \cap ST_q|} \tag{3}$$

---

[2] We include q as a subindex to stress the fact that this set is query-dependent. In our experiments, R_q was fixed to be the set of documents supplied by the track's organizers (set of highly ranked documents associated to each query).

where V_s is the set of unique terms appearing in s. The score depends on the number of highly frequent terms in the sentence and the square root was introduced to get the tf_isf and htf scores on a similar scale and, therefore, combine them adequately.

The sentence-query similarity is simply defined as the sum of the tf_isf and htf scores. Hereafter, this sentence retrieval method will be referred to as the HTF method. The combination method applied here is intentionally simplistic and we consider it as a first attempt to mix these factors. More evolved (and formal) methods will be studied in the near future.

Sentences with a poor overlap with the query can still be retrieved provided that they contain terms which are significant in the retrieved set. This method promotes the retrieval of sentences which are somehow *central* for the query topics. It is also a way to tackle the well-known vocabulary mismatching problem in retrieval engines. The notion of term significance is equated here to high term frequency in the retrieved set. Note that in our experiments stopwords were removed. This is important because, otherwise, the ST_q sets would be likely populated by many common words, which would introduce much noise. Of course, many other alternatives could have been proposed to estimate term significance (e.g. based on successful methods such as Local Context Analisis [26] or Divergence from Randomness [3]). Nevertheless, the present study is focused on a simple term frequency-based method which has proved successful in query-biased summarization [20]. A complete comparison of different alternatives to estimate term significance for sentence retrieval is out of the scope of this paper.

The parameter *mno* determines the size of the set ST_q and, consequently, influences directly the weight of the htf component in the similarity formula. Low *mno* values (e.g. *mno* < 5) do not make sense because the set ST_q would contain most of the terms in the retrieved set. This set would have many non-significant terms and the htf component in the formula above would be merely promoting long sentences. On the other hand, very high *mno* values are not advisable either because ST_q would be very small (empty in the extreme case) and, therefore, the method would be roughly equivalent to the basic TF/ISF technique. In this work, we test the effectiveness of our method with varying number of minimum occurrences and analyze the sensitiveness of the method with respect to the size of the set of significant terms. In the future, we will also study other alternatives, such as query-dependent *mno* values estimated from R_q.

The HTF method can be actually regarded as a form of pseudo-relevance feedback (at the document level) aimed at estimating the importance of the sentences within a retrieved set (with no query expansion). Term selection is done before sentence retrieval and the selected terms are not used for expansion (the terms are used for estimating the significance of a sentence). On the other hand, the standard query expansion via pseudo-relevance feedback or local co-occurrence is strongly sensitive to the quality of the original query [26]. If we use a few top ranked sentences to expand the query then it is likely that we end up introducing some noise in the new query. We expect that the highly frequent terms approach is more robust because, rather than doing first a sentence retrieval process to get some new terms, it analyzes globally the set of retrieved documents to locate important sentences. It seems reasonable to think that the behaviour of this method is more stable. One can rightly argue that an original poor query will also be harmful in our approach because the quality of the document ranking will be low. Nevertheless, once we have the rank of documents, the adjustment that we propose (htf score) does not involve the query text. On the contrary, pseudo-relevance feedback methods apply subsequently a sentence retrieval process from the query terms and, hence, they are much more dependent on the quality of the queries (the query text is used twice). Other evolved expansion methods, such as Local Context Analysis [26], which has proved to be more effective and robust than pseudo-relevance feedback for document retrieval, require also a second usage of the original query text for retrieving passages and, next, phrases are selected from the top passages to expand the query. The process is thus rather complicated whereas the method applied here is much simpler.

4   Experiments

The performance of the HTF method has been tested using three different collections of data. These datasets were provided in the context of the TREC-2002, TREC-2003 and TREC-2004 novelty tracks [8,18,17]. There are no newer TREC collections suitable for our experiments because we need relevance judgments at the sentence level. This sort of judgments is only available in the novelty track, whose last edition took place in 2004. The novelty track data was constructed as follows. Every year there were 50 topics available. In TREC-2002, the topics were taken from TRECs 6, 7 and 8 (the complete list of topics chosen for the novelty track can be found in [8]). In 2003 and 2004, the topics were created by assessors designated specifically for the task [18,17] (topics N1-N50 and N51-N100). For each topic, a rank of documents was obtained by NIST using an effective retrieval engine. In 2002 and 2003 the task aimed at finding relevant sentences in relevant documents and, therefore, the ranks included only relevant documents (i.e. given a topic the set of relevant documents to the topic were collected and ranked using a document retrieval engine). On the contrary, the TREC-2004 ranks contained also irrelevant documents (i.e. the initial search for documents was done against a regular document base, with relevant and irrelevant documents). Note that this means that the irrelevant documents are close matches to the relevant documents, and not random irrelevant documents [17]. In any case, the ranks of documents contained at most 25 relevant documents for each query.

The documents were segmented into sentences, the participants were given these ranks of sentence-tagged documents and they were asked to locate the relevant sentences. The relevance judgments in this task are complete because the assessors reviewed carefully the ranked documents and marked every sentence as relevant or non-relevant to the topic. In TREC-2002, very few sentences were judged as relevant (approximately 2% of the sentences in the documents). In TREC-2003 and TREC-2004 the documents were taken from the AQUAINT collection and the average percentage of relevant sentences was much higher than in 2002 (approximately 40% in 2003 and 20% in 2004). This therefore shapes an assorted evaluation design in which we can test the HTF method under different scenarios and conditions. We focus our interest on short queries, which are by far the most utilized ones, especially in environments such as the web [16]. In our experiments, short queries were constructed from the title tags of the TREC topics.

We used two different evaluation measures: precision at 10 sentences retrieved and the F-measure, which was the official measure in the TREC novelty experiments[3]. Regarding statistical significance tests, we applied two different tests, the t-test and the Wilcoxon test, and we only concluded that a given difference between two runs was significant when both tests agree (with a 95% confidence level).

To ensure that the baseline was competitive we ran some initial experiments with other popular retrieval methods. We experimented with Okapi BM25 [14] and a Language Modeling approach based on Kullback-Leibler Divergence (KLD) as described in [10] (with Dirichlet smoothing). The performance of BM25 is influenced by some parameters: $k1$ controls the term frequency effect, b controls a length-based correction and $k3$ is related to query term frequency. We tested exhaustively different parameter configurations ($k1$ between 0 and 2 in steps of 0.2, b between 0 and 1 in steps of 0.1 and different values of $k3$ between 1 and 1000). Similarly, we experimented with the KLD model for different values of the μ constant, which determines the amount of smoothing applied ($\mu = 10$, 100, 500, $1k$, $3k$, $5k$). Results are reported in Table 1. A run marked with an asterisk means that the difference in performance between the run and TF/ISF is statistically significant. In all collections, there was not statistically significant difference between the TF/ISF run and the best BM25 run. We also observed that BM25 was very sensitive to the parameter setting (many BM25 runs performed significantly worse than TF/ISF). On the other hand, KLD was inferior to both TF/ISF and BM25. These results reinforced previous findings about the robustness of the TF/ISF method [2,11] and demonstrated that this method is a very solid baseline.

We also tried out different combinations of the standard preprocessing strategies (stopwords vs no stopwords, stemming vs no stemming). Although there was no much overall difference, the runs with stopword processing and no stemming were slightly more consistent. Since the TF/ISF method takes the idf statistics from the sentences in the documents available for the task (which is a small set of sentences), we were wondering whether better performance may be obtained using idf data from a larger collection. To check this, we indexed a large collection

---

[3] The F-measure is the harmonic mean (evenly weighted) of sentence set recall and precision. In the TREC-2002 experiments, we computed the F-measure using the top 5% of the retrieved sentences and in the other collections we used the top 50% of the retrieved sentences. Similar thresholds were taken in TREC experiments [2].

Table 1. Comparing different baselines

	TF-ISF	TREC-2002 BM25	KLD
best run		$k1 = .4, b = 0, k3 = 1$	$\mu = 3000$
P@10	.19	.19	.16
F	.188	.190	.172*
	TF-ISF	TREC-2003 BM25	KLD
best run		$k1 = .6, b = 0, k3 = 1$	$\mu = 1000$
P@10	.74	.76	.73
F	.512	.512	.510*
	TF-ISF	TREC-2004 BM25	KLD
best run		$k1 = .2, b = 0, k3 = 1$	$\mu = 500$
P@10	.43	.44	.41
F	.370	.371	.369

Table 2. Evaluation results

		TREC-2002 HTF, mno=				PRF, # exp terms=			
	TF/ISF	7	10	15	20	5	10	20	50
P@10	.19	.23*	.22*	.22	.22	.19	.20	.17	.20
F	.188	.197	.197	.197	.192	.190	.181	.181	.178
		TREC-2003							
P@10	.74	.78*	.78	.79*	.79*	.78*	.79*	.78	.77
F	.512	.560*	.559*	.555*	.554*	.535*	.550*	.558*	.560*
		TREC-2004							
P@10	.43	.51*	.50*	.50*	.50*	.48*	.48*	.49*	.46
F	.370	.391*	.389*	.387*	.388*	.376	.382	.386*	.392*

of documents (the collection used in the TREC-8 adhoc experiments [22]) and ran some experiments where the idf statistics were taken from this index. The original TF/ISF method computed at the sentence level over the small document base was superior. It appears that the small index of sentences is good enough for sentence retrieval (at least for these short queries). We therefore set the baseline to be the original TF/ISF approach with stopword and no stemming[4].

In the HTF experiments we took the set of retrieved documents for each query and computed the set of highly frequent terms (ST_q). Several experiments were executed with varying minimum number of occurrences (mno) namely, 7, 10, 15, 20[5]. Results are shown in Table 2 (columns 3-6). The results are very encouraging. All the HTF runs produced better performance than the baseline's performance. The improvements are small in TREC-2003. This is not surprising because there is a high population of relevant sentences in this collection (note that the baseline performance is very high, e.g. P@10=74%) and, therefore, it is difficult to get further benefits. There are so many relevant sentences that there is

---

[4] We use short queries, while the groups participating in the TREC novelty tracks were allowed to use the whole topic. This means that the results presented here are not comparable to any of the results reported in the novelty tracks.

[5] In [20], terms occurring seven or more times in a document were regarded as significant for building a query-biased summary.

no need to apply an evolved method to get a reasonably good top 10. On the other hand, when the retrieved documents have less relevant sentences (2002 and 2004) the HTF retrieval approach produces very significant improvements. It should be noted that there was not a single HTF run yielding worse performance than the baseline. Actually, the method is quite insensitive to the minimum number of occurrences for a term to be considered highly frequent. The four values tested yielded very similar performance. This is a nice feature of the approach as it does not seem problematic to set a good value. Furthermore, the improvements are apparent in both performance ratios. This means that the HTF technique works effectively not only for supplying ten good sentences but also consistently provides relevant sentences across the whole rank.

We also wanted to study the relative merits of HTF against expansion via pseudo-relevance feedback (hereafter, PRF). Although HTF and PRF are intrinsically distinct and both alternatives could actually complement each other, it is still interesting to analyze how robust the HTF method is in comparison with PRF. We experimented with query expansion using an standard PRF technique [4] which consists of adding the most frequent terms from the top ranked documents. This technique, adapted to sentence retrieval (i.e. selecting expansion terms from the top ranked sentences), has proved to be successful to improve the performance of sentence retrieval [10]. There is no empirical evidence that any other advanced query expansion method (e.g. Local Context Analysis [26]) works better than PRF in sentence retrieval.

Since the characteristics of the collections are very different to one another, the comparison between HTF and PRF is general enough, with varying conditions of the type of data and the amount of relevant material. For instance, we expected modest improvements from PRF in 2002 data due to the scarcity of relevant material, and much better PRF results in 2003 and 2004 collections because there are more relevant sentences. Another interesting point to study is the effect of the number of expansion terms. This number affects performance in a critical way (many terms imply usually too much noise in the new query). Therefore we want to check this effect in sentence retrieval and compare it against the behaviour of the HTF approach.

A pool of pseudo-relevance feedback experiments was designed as follows. Given the TF/ISF rank, queries were expanded with the 5, 10, 20 or 50 highest frequent terms in the top 10 sentences. Next, the set of sentences was re-ranked using the new queries. Results are reported in Table 2 (columns 7-10). PRF is much less consistent than HTF. The average performance is clearly lower than the HTF's average performance. The number of PRF runs whose performance is significantly better than the baseline's performance is smaller than the corresponding number of HTF runs. Furthermore, some PRF runs performed worse than the baseline. Summing up, the HTF method works at least as well as PRF techniques and it is less sensitive to variations in its parameter. It is important to note that the HTF method is also convenient for efficiency reasons. It does not require an initial sentence retrieval process and the ST_q sets can be easily

Table 3. Poorest queries

	TREC-2002	
	TF/ISF	HTF	PRF
P@10	.03	.09	.03
		7(+) 0(-) 8(=)	1(+) 2(-) 12(=)
	TREC-2003	
	TF/ISF	HTF	PRF
P@10	.39	.52	.50
		9(+) 3(-) 3(=)	10(+) 2(-) 3(=)
	TREC-2004	
	TF/ISF	HTF	PRF
P@10	.13	.21	.15
		8(+) 2(-) 5(=)	4(+) 4(-) 7(=)

computed from a regular inverted file (summing the term counts in the retrieved documents).

Let us now pay attention to the behaviour of these sentence retrieval methods when handling poor queries. Such queries are very problematic for retrieval systems and it is important to analyze them in depth. For each collection, we analyzed the 15 queries that had yielded the lowest P@10 figures with the baseline method[6]. In Table 3 we report the P@10 values obtained with the TF/ISF run and the best HTF and PRF runs. For the HTF and PRF runs, we also show the number of queries whose P@10 is better than (+), worse than (-) or equal to (=) the P@10 obtained with the baseline run. The results are very conclusive. In TREC-2003 data, where the 15 worst topics retrieve a reasonably good number of relevant sentences in the top 10 (39% on average), both methods perform roughly the same. On the contrary, in TREC-2002 and TREC-2004, where the initial ranks are quite poor (3% and 13% of relevant sentences on average, respectively), the HTF method is much more robust and performs significantly better than PRF. The HTF method is consistent even in the presence of weak queries. With such queries, pseudo-relevance feedback suffers from poor performance (in some cases, it does not outperform the baseline), whilst the HTF runs still provide solid improvements. This is very valuable because it is usually easy to improve performance when the initial rank has a good population of relevant material but, on the other hand, it is quite hard to improve performance when the initial rank is not good enough. The ability of the HTF method to improve significantly the precision at top ranks when handling poor queries is an important property of this sentence retrieval approach. This suggests that this method is especially suitable for real applications with poorly specified information needs where users only want to go through a small number of sentences.

Having demonstrated that the HTF technique is competitive and outperforms solid sentence retrieval methods when initial ranks are poor, it is important to mention that it is actually compatible with pseudo-relevance feedback. Our method is capable of estimating the centrality of sentences within a retrieved set

---

[6] For reproductability purposes, these queries were: **T2002**: 305,312,314,315,330, 377,381,406,411,420,432,323,325,326,339. **T2003**: 48,12,14,25,19,20,1,45,24,28,29,30, 5,22,36. **T2004**: 57,77,61,71,86,93,94,97,56,62,65,70,78,80,84.

of documents. This effect could be combined with an expansion approach from a few top ranked sentences and, therefore, retrieval engines can incorporate both techniques. With weak queries the application of pseudo-relevance feedback is not advisable but stronger queries could attain benefits from the combined use of centrality and expansion.

5 Conclusions and Future Work

We have proposed a novel sentence retrieval mechanism based on extracting highly frequent terms from a retrieved set of documents. The experiments reported in this paper demonstrate that this approach outperforms clearly state-of-the-art sentence retrieval methods in the context of a query retrieval use case. We also showed that our method is more robust than standard pseudo-relevance feedback methods and it is simpler because it does not require an initial sentence retrieval process. The method proposed here is especially valuable in terms of the precision at top ranks when queries are weak. In the future we will explore more formal methods to combine the centrality and retrieval scores. We will also study alternative ways to estimate term significance.

Acknowledgements

We thank Tassos Tombros, David Elsweiler and Alvaro Barreiro for their very helpful discussions and recommendations. This work was partially supported by projects PGIDIT06PXIC206023PN and TIN2005-08521-C02-01. David E. Losada belongs to the "Ramón y Cajal" program, whose funds come from "Ministerio de Educación y Ciencia" and the FEDER program.

References

1. Abdul-Jaleel, N., Allan, J., Croft, B., Diaz, F., Larkey, L., Li, X., Smucker, M., Wade, C.: UMass at TREC 2004: Novelty and Hard. In: Proc. TREC-2004, the 13th Text Retrieval Conference (2004)
2. Allan, J., Wade, C., Bolivar, A.: Retrieval and novelty detection at the sentence level. In: Proc. SIGIR-2003, the 26th ACM Conference on Research and Development in Information Retrieval, Toronto, Canada, pp. 314–321. ACM Press, New York (2003)
3. Amati, G., van Rijsbergen, C.J.: Probabilistic models of information retrieval based on measuring the divergence from randomness. ACM Transactions on Information Systems (TOIS) 20(4), 357–389 (2002)
4. Buckley, C., Singhal, A., Mitra, M., Salton, G.: New retrieval approaches using SMART: TREC 4. In: Harman, D. (ed.) Proc. TREC-4, pp. 25–48 (1996)
5. Cardie, C., Ng, V., Pierce, D., Buckley, C.: Examining the role of statistical and linguistic knowledge sources in a general-knowledge question-answering system. In: Proc. ANLP-2000, the 6th Applied Natural Language Processing Conference, Seattle, Washington, pp. 180–187 (2000)
6. Collins-Thompson, K., Ogilvie, P., Zhang, Y., Callan, J.: Information filtering, novelty detection, and named-page finding. In: Proc. TREC-2002 (2002)

7. Dkaki, T., Mothe, J.: TREC novelty track at IRIT-SIG. In: Proc. TREC-2004, the 13th text retrieval conference (2004)
8. Harman, D.: Overview of the TREC 2002 novelty track. In: Proc. TREC-2002 (2002)
9. Kwok, K.L., Deng, P., Dinstl, N., Chan, M.: TREC 2002 web, novelty and filtering track experiments using PIRCS. In: Proc. TREC-2002, the 11th text retrieval conference (2002)
10. Larkey, L., Allan, J., Connell, M., Bolivar, A., Wade, C.: UMass at TREC 2002:cross language and novelty tracks. In: Proc. TREC-2002 (2002)
11. Li, X., Croft, B.: Novelty detection based on sentence level patterns. In: Proc. CIKM-2005. ACM Conf. on Information and Knowledge Management, ACM Press, New York (2005)
12. Mani, I., Maybury, M.T.: Advances in Automatic Text Summarization. MIT Press, Cambridge (1999)
13. Murdock, V.: Aspects of sentence retrieval. PhD thesis, Univ. Massachussetts (2006)
14. Robertson, S.E., Walker, S., Jones, S., HancockBeaulieu, M.M., Gatford, M.: Okapi at TREC-3. In: Harman, D. (ed.) Proc. TREC-3, the 3rd Text Retrieval Conference, pp. 109–127. NIST (1995)
15. Schiffman, B.: Experiments in novelty detection at columbia university. In: Proc. TREC-2002, the 11th text retrieval conference (2002)
16. Silverstein, C., Henzinger, M., Marais, H., Moricz, M.: Analysis of a very large web search engine query log. ACM SIGIR Forum 33(1), 6–12 (1999)
17. Soboroff, I.: Overview of the TREC 2004 novelty track. In: Proc. TREC-2004, the 13th text retrieval conference (2004)
18. Soboroff, I., Harman, D.: Overview of the TREC 2003 novelty track. In: Proc. TREC-2003, the 12th text retrieval conference (2003)
19. Stokes, N., Carthy, J.: First story detection using a composite document representation. In: Proc. of HTL-01, the Human Language Technology Conference, San Diego, USA (March 2001)
20. Tombros, A., Sanderson, M.: Advantages of query biased summaries in information retrieval. In: Proc. SIGIR-1998, the 21st ACM Int. Conf. on Research and Development in Information Retrieval, August 1998, pp. 2–10. ACM Press, New York (1998)
21. Voorhees, E.: Using Wordnet to disambiguate word senses for text retrieval. In: Proc. SIGIR-1993, Pittsburgh, PA, pp. 171–180 (1993)
22. Voorhees, E., Harman, D.: Overview of the eight text retrieval conference. In: Proc. TREC-8, the 8th text retrieval conference (1999)
23. Voorhees, E., Harman, D. (eds.): The TREC AdHoc Experiments, chapter The TREC AdHoc Experiments, pp. 79–97. MIT Press, Cambridge (2005)
24. White, R., Jose, J., Ruthven, I.: A task-oriented study on the influencing effects of query-biased summarisation in web searching. Information Processing and Management 39, 707–733 (2003)
25. White, R., Jose, J., Ruthven, I.: Using top-ranking sentences to facilitate effective information access. Journal of the American Society for Information Science and Technology (JASIST) 56(10), 1113–1125 (2005)
26. Xu, J., Croft, B.: Query expansion using local and global document analysis. In: Proc. SIGIR-1996, Zurich, Switzerland, July 1996, pp. 4–11 (1996)
27. Zhang, H.P., Xu, H.B., Bai, S., Wang, B., Cheng, X.Q.: Experiments in TREC 2004 novelty track at CAS-ICT. In: Proc. TREC-2004 (2004)
28. Zhang, M., Song, R., Lin, C., Ma, S., Jiang, Z., Jin, Y., Liu, Y., Zhao, L.: THU TREC 2002: Novelty track experiments. In: Proc. TREC-2002, the 11th text retrieval conference (2002)

Implicit Compression Boosting with Applications to Self-indexing

Veli Mäkinen[1,*] and Gonzalo Navarro[2,**]

[1] Department of Computer Science, University of Helsinki, Finland
vmakinen@cs.helsinki.fi
[2] Department of Computer Science, University of Chile
gnavarro@dcc.uchile.cl

Abstract. *Compression boosting* (Ferragina & Manzini, SODA 2004) is a new technique to enhance zeroth order entropy compressors' performance to k-th order entropy. It works by constructing the Burrows-Wheeler transform of the input text, finding optimal partitioning of the transform, and then compressing each piece using an arbitrary zeroth order compressor. The optimal partitioning has the property that the achieved compression is boosted to k-th order entropy, for any k. The technique has an application to text indexing: Essentially, building a *wavelet tree* (Grossi et al., SODA 2003) for each piece in the partitioning yields a k-th order compressed full-text *self-index* providing efficient substring searches on the indexed text (Ferragina et al., SPIRE 2004). In this paper, we show that using explicit compression boosting with wavelet trees is not necessary; our new analysis reveals that the size of the wavelet tree built for the complete Burrows-Wheeler transformed text is, in essence, the sum of those built for the pieces in the optimal partitioning. Hence, the technique provides a way to do compression boosting implicitly, with a trivial linear time algorithm, but fixed to a specific zeroth order compressor (Raman et al., SODA 2002). In addition to having these consequences on compression and static full-text self-indexes, the analysis shows that a recent *dynamic* zeroth order compressed self-index (Mäkinen & Navarro, CPM 2006) occupies in fact space proportional to k-th order entropy.

1 Introduction

The indexed string matching problem is that of, given a long text $T[1,n]$ over an alphabet Σ of size σ, building a data structure called *full-text index* on it, to solve two types of queries: (a) Given a short pattern $P[1,m]$ over Σ, *count* the occurrences of P in T; (b) *locate* those *occ* positions in T. There are several classical full-text indexes requiring $O(n \log n)$ bits of space which can answer counting queries in $O(m \log \sigma)$ time (like suffix trees [1]) or $O(m + \log n)$ time (like suffix arrays [18]). Both locate each occurrence in constant time once the

---

* Funded by the Academy of Finland under grant 108219.
** Partially Funded by Fondecyt Grant 1-050493, Chile.

N. Ziviani and R. Baeza-Yates (Eds.): SPIRE 2007, LNCS 4726, pp. 229–241, 2007.
© Springer-Verlag Berlin Heidelberg 2007

counting is done. Similar complexities are obtained with modern compressed data structures [6,11,9], requiring space $nH_k + o(n \log \sigma)$ bits (for some small k), where $H_k \leq \log \sigma$ is the k-th order empirical entropy of T. These indexes are often called *compressed self-indexes* refering to their space requirement and to their ability to work without the text and even fully replace it, by delivering any text substring without accessing T.

The main building blocks in compressed self-indexes are the Burrows-Wheeler transform T^{bwt} [3] and function $rank_c(T^{bwt}, i)$ that counts how many times symbol c appears in $T^{bwt}[1, i]$. Function $rank_c$ can be efficiently provided by building the *wavelet tree* [11] on T^{bwt}; this reduces the problem to $rank$ queries on binary sequences, which are already studied by Jacobson [14] in his seminal work on compressed data structures. Using a more recent binary $rank$ solution [23] inside wavelet trees, one almost automatically achieves a compressed self-index taking $nH_0 + o(n \log \sigma)$ bits of space [11,9,16]. Let us call this index *Succinct Suffix Array* (SSA) following [16].

What has remained unnoticed so far is that SSA actually takes only $nH_k + o(n \log \sigma)$ bits of space. This result makes some of the more complicated techniques to achieve the same result obsolete. However, our analysis builds on the existence of the compression-boosted version of SSA [9], as we show that the internal parts of the structures in the boosted version are compressed exactly the same way in the basic SSA. This shows a remarkable property of wavelet trees when used together with the encoding of Raman, Raman, and Rao [23].

In the following, we first define the entropy concepts more formally, then explain the encoding in [23], wavelet trees [11], Burrows-Wheeler transform [3], and compression boosting [7] in order to give our new analysis in a self-contained manner. We conclude with the application to space-efficient construction of (dynamic) full-text self-indexes.

2   Definitions

We assume our text $T = t_1 \ldots t_n$ to be drawn from an alphabet $\{0, 1, \ldots \sigma - 1\}$. Let n_c denote the number of occurrences of symbol c in T, i.e., $n_c = |\{i \mid t_i = c\}|$. Then the zero-order *empirical entropy* is defined as $H_0(T) = \sum_{0 \leq c < \sigma} \frac{n_c}{n} \log \frac{n}{n_c}$. This is the lower bound for the average code word length of any compressor that fixes the code words to the symbols independently of the context they appear in.

A tighter lower bound for texts is the k-th order empirical entropy $H_k(T)$, where the compressor can fix the code word based on the k-symbol context following the symbol to be coded.[1] Formally, it can be defined as $H_k(T) = \sum_{w \in \Sigma^k} \frac{n_w}{n} H_0(T|w)$, where n_w denotes the number of occurrences of substring

---

[1] It is more logical (and hence customary) to define the context as the k symbols preceding a symbol, but we use the reverse definition for technical convenience. If this is an issue, the texts can be handled reversed to obtain results on the more standard definition. It is anyway known that both definitions do not differ by much [8].

w in T and $T|w$ denotes the concatenation of the symbols appearing immediately before those n_w occurrences [19]. Substring $w = T[i + 1, i + k]$ is called the k-*context* of symbol t_i. We take T here as a *cyclic string*, such that t_n precedes t_1, and thus the amount of k-contexts is exactly n.

3   Previous Results

3.1   Entropy-Bound Structures for Bit Vectors

Raman et al. [23] proposed a data structure to solve *rank* and *select* (inverse of *rank*) queries in constant time over a static bit vector $A = a_1 \ldots a_n$ with binary zero-order entropy H_0. The structure requires $nH_0 + o(n)$ bits.

The idea is to split A into *superblocks* $S_1 \ldots S_{n/s}$ of $s = \log^2 n$ bits. Each superblock S_i is in turn divided into $2 \log n$ blocks $B_i(j)$, of $b = (\log n)/2$ bits each, thus $1 \le j \le s/b$. Each such block $B_i(j)$ is said to belong to *class c* if it has exactly c bits set, for $0 \le c \le b$. For each class c, a universal table G_c of $\binom{b}{c}$ entries is precomputed. Each entry corresponds to a possible block belonging to class c, and it stores all the local *rank* answers for that block. Overall all the G_c tables add up $2^b = \sqrt{n}$ entries, and $O(\sqrt{n} \operatorname{polylog}(n))$ bits.

Each block $B_i(j)$ of the sequence is represented by a pair $D_i(j) = (c, o)$, where c is its class and o is the index of its corresponding entry in table G_c. A block of class c thus requires $\log(c + 1) + \log \binom{b}{c}$ bits. The first term is $O(\log \log n)$, whereas all the second terms add up $nH_0 + O(n/\log n)$ bits. To see this, note that $\log \binom{b}{c_1} + \log \binom{b}{c_2} \le \log \binom{2b}{c_1+c_2}$, and that $nH_0 \ge \log \binom{b(n/b)}{c_1+\ldots+c_{n/b}}$. The pairs $D_i(j)$ are of variable length and are all concatenated into a single sequence.

Each superblock S_i stores a pointer P_i to its first block description in the sequence (that is, the first bit of $D_i(1)$) and the *rank* value at the beginning of the superblock, $R_i = rank(A, (i - 1)s)$. P and R add up $O(n/\log n)$ bits. In addition, S_i contains s/b numbers $L_i(j)$, giving the initial position of each of its blocks in the sequence, relative to the beginning of the superblock. That is, $L_i(j)$ is the position of $D_i(j)$ minus P_i. Similarly, S_i stores s/b numbers $Q_i(j)$ giving the *rank* value at the beginning of each of its blocks, relative to the beginning of the superblock. That is, $Q_i(j) = rank(A, (i - 1)s + (j - 1)b) - R_i$. As those relative values are $O(\log n)$, sequences L and Q require $O(n \log \log n / \log n)$ bits.

To solve $rank(A, p)$, we compute the corresponding superblock $i = 1 + \lfloor p/s \rfloor$ and block $j = 1 + \lfloor (p - (i - 1)s)/b \rfloor$. Then we add the *rank* value of the corresponding superblock, R_i, the relative *rank* value of the corresponding block, $Q_i(j)$, and complete the computation by fetching the description (c, o) of the block where p belongs (from bit position $P_i + L_i(j)$) and performing a (precomputed) local *rank* query in the universal table, $rank(G_c(o), p - (i - 1)s - (j - 1)b)$.

The overall space requirement is $nH_0 + O(n \log \log n / \log n)$ bits, and *rank* is solved in constant time. We do not cover *select* because it is not necessary to follow this paper.

The scheme extends to sequences over small alphabets as well [9]. Let $B = a_1 \ldots a_b$ be the symbols in a block, and call n_a the number of occurences of symbol $a \in [1, q]$ in B. We call $(n_1, \ldots, n_q)$ the *class* of B. Thus, in our (c, o) pairs, c will be a number identifying the class of B and o an index within the class. A simple upper bound to the number of classes is $(b + 1)^q$ (as a class is a tuple of q numbers in $[0, b]$, although they have to add up b). Thus $O(q \log \log n)$ bits suffice for c (a second bound on the number of classes is q^b as there cannot be more classes than different sequences). Just as in the binary case, the sum of the sizes of all o fields adds up $nH_0(A) + O(n/\log_q n)$ [9].

3.2   Wavelet Trees and Entropy-Bound Structures for Sequences

We now extend the result of the previous section to larger alphabets. The idea is to build a wavelet tree [11] over sequences represented using *rank* structures for small alphabets.

A binary wavelet tree is a balanced binary tree whose leaves represent the symbols in the alphabet. The root is associated with the whole sequence $A = a_1 \cdots a_n$, its left child with the subsequence of A obtained by concatenating all positions i having $a_i < \sigma/2$, and its right child with the complementary subsequence (symbols $a_i \geq \sigma/2$). This subdivision is continued recursively, until each leaf contains a repeat of one symbol. The sequence at each node is represented by a bit vector that tells which positions (those marked with 0) go to the left child, and which (marked with 1) go to the right child. It is easy to see that the bit vectors alone are enough to determine the original sequence: To recover a_i, start at the root and go left or right depending on the bit vector value B_i at the root. When going to the left child, replace $i \leftarrow rank_0(B, i)$, and similarly $i \leftarrow rank_1(B, i)$ when going right. When arriving at the leaf of character c it must hold that the original a_i is c. This requires $O(\log \sigma)$ *rank* operations over bit vectors.

It also turns out that operations *rank* and *select* on the original sequence can be carried out via $O(\log \sigma)$ operations of the same type on the bit vectors of the wavelet tree [11]. For example, to solve $rank_c(A, i)$, start at the root and go to the left child if $c < \sigma/2$ and to the right child otherwise. When going down, update i as in the previous paragraph. When arriving at the leaf of c, the current i value is the answer.

A multiary wavelet tree, of arity q, is used in [9]. In this case the sequence of each wavelet tree node ranges over alphabet $[1, q]$, and symbol rank/select queries are needed over those sequences. One needs $\log_q \sigma$ operations on those sequences to perform the corresponding operation on the original sequence.

Either for binary or general wavelet trees, it can be shown that the H_0 entropies in the representations of the sequences at each level add up to $nH_0(A)$ bits [11,9]. However, as we have $O(\sigma)$ bit vectors, the sublinear terms sum up to $o(\sigma n)$. The space occupancy of the sublinear structures can be improved to $o(n \log \sigma)$ by concatenating all the bit vectors of the same level into a single sequence, and handling only $O(\log \sigma)$ such sequences[2]. It is straightforward to do *rank*, as well as obtaining symbol a_i, without any extra information [9].

---

[2] Note that $o(n \log \sigma)$ is sublinear in the size of A measured in bits.

If we now represent the concatenated bit vectors of the binary wavelet tree by the *rank* structures explained in the previous section, we obtain a structure requiring $nH_0(A) + O(n \log \log n / \log_\sigma n) = nH_0(A) + o(n \log \sigma)$ bits, solving *rank* in $O(\log \sigma)$ time. Within the same bounds one can solve *select* as well [23,11].

Theorem 1 ([11]). *Let L be a string and B_v the corresponding binary sequence for each node v of the wavelet tree of L. Then $\sum_v |B_v| H_0(B_v) = |L| H_0(L)$.*

One can also use multiary wavelet trees and represent the sequences with alphabet size q using the techniques for small alphabets (see the end of previous section). With a suitable value for q, one obtains a structure requiring the same $nH_0(A) + o(n \log \sigma)$ bits of space, but answering *rank* and *select* in constant time when $\sigma = O(\text{polylog}(n))$, and $O(\lceil \log \sigma / \log \log n \rceil)$ time in general [9].

3.3   Full-Text Self-indexes

Many full-text self-indexes are based on representing the Burrows-Wheeler transform [3] of a text using wavelet trees to support efficient substring searches. We follow the description given in [16].

The Burrows-Wheeler Transform. The *Burrows-Wheeler transform (BWT)* [3] of a text T produces a permutation of T, denoted by T^{bwt}. We assume that T is terminated by an endmarker "\$" $\in \Sigma$, smaller than other symbols. String T^{bwt} is the result of the following transformation: (1) Form a *conceptual* matrix $\mathcal{M}$ whose rows are the cyclic shifts of the string T, call F its first column and L its last column; (2) sort the rows of $\mathcal{M}$ in lexicographic order; (3) the transformed text is $T^{bwt} = L$.

The BWT is reversible, that is, given T^{bwt} we can obtain T. Note the following properties [3]:

a. Given the i-th row of $\mathcal{M}$, its last character $L[i]$ precedes its first character $F[i]$ in the original text T, that is, $T = \ldots L[i]F[i]\ldots$.
b. Let $L[i] = c$ and let r_i be the number of occurrences of c in $L[1,i]$. Let $\mathcal{M}[j]$ be the r_i-th row of $\mathcal{M}$ starting with c. Then the character corresponding to $L[i]$ in the first column F is located at $F[j]$ (this is called the *LF mapping*: $LF(i) = j$). This is because the occurrences of character c are sorted both in F and L using the same criterion: by the text following the occurrences.

The BWT can then be reversed as follows:

1. Compute the array $C[1, \sigma]$ storing in $C[c]$ the number of occurrences of characters $\{\$, 1, \ldots, c-1\}$ in the text T. Notice that $C[c] + 1$ is the position of the first occurrence of c in F (if any).
2. Define the *LF mapping* as follows: $LF(i) = C[L[i]] + rank_{L[i]}(L, i)$.
3. Reconstruct T backwards as follows: set $s = 1$ (since $\mathcal{M}[1] = \$t_1 t_2 \ldots t_{n-1}$) and, for each $i \in n - 1, \ldots, 1$ do $T[i] \leftarrow L[s]$ and $s \leftarrow LF[s]$. Finally put the endmarker $T[n] = \$$.

The BWT transform by itself does not compress T, it just permutes its characters. However, this permutation is more compressible than the original T. Actually, it is not hard to compress L to $O(nH_k + \sigma^{k+1} \log n)$ bits, for any $k \geq 0$ [19]. The idea is as follows (we will reuse it in our new analysis later): Partition L into minimum number of pieces $L^1 L^2 \cdots L^\ell$ such that the symbols inside each piece $L^p = L[i_p, j_p]$ have the same k-context. Note that the k-context of a symbol $L[i]$ is $\mathcal{M}[i][1, k]$. By the definition of k-th order entropy, it follows that $|L^1| H_0(L^1) + |L^2| H_0(L^2) + \cdots + |L^\ell| H_0(L^\ell) = nH_k$. That is, if one is able to compress each piece up to its zero-order entropy, then the end result is k-th order entropy.

Theorem 2 ([19]). *Let $L = L^1 L^2 \ldots L^\ell$ be a partition of L, the BWT of T, according to contexts of length k in $\mathcal{M}$. Then $\sum_{1 \leq i \leq \ell} |L^j| H_0(L^j) = nH_k(T)$.*

Using, say, arithmetic coding on each piece, one achieves $nH_k + \sigma^{k+1} \log n$ bits encoding of T for a *fixed* k. The latter term comes from the encoding of the symbol frequencies in each piece separately. This fact is the base of *compression boosting* [7]; they give a linear time algorithm to find, for a given zero order compressor, the *optimal partitioning* of L such that when each piece is compressed using the given zero order compressor, the compression result is the best over all possible partitions. Notice that the partitions fixed by the k-contexts are a subset of all partitions, and hence the resulting compression can be bounded by k-th order entropy for *any* k.

Suffix Arrays. The *suffix array* $\mathcal{A}[1, n]$ of text T is an array of pointers to all the suffixes of T in lexicographic order. Since T is terminated by the endmarker "$\$$", all lexicographic comparisons are well defined. The i-th entry of $\mathcal{A}$ points to text suffix $T[\mathcal{A}[i], n] = t_{\mathcal{A}[i]} t_{\mathcal{A}[i]+1} \ldots t_n$, and it holds $T[\mathcal{A}[i], n] < T[\mathcal{A}[i+1], n]$ in lexicographic order.

Given the suffix array and T, the occurrences of the pattern $P = p_1 p_2 \ldots p_m$ can be counted in $O(m \log n)$ time. The occurrences form an interval $\mathcal{A}[sp, ep]$ such that suffixes $t_{\mathcal{A}[i]} t_{\mathcal{A}[i]+1} \ldots t_n$, for all $sp \leq i \leq ep$, contain the pattern P as a prefix. This interval can be searched for using two binary searches in time $O(m \log n)$. Once the interval is obtained, a locating query is solved simply by listing all its pointers in constant time each.

We note that the suffix array $\mathcal{A}$ is essentially the matrix $\mathcal{M}$ of the BWT (Sect. 3.3), as sorting the cyclic shifts of T is the same as sorting its suffixes given the endmarker "$\$$": $\mathcal{A}[i] = j$ if and only if the i-th row of $\mathcal{M}$ contains the string $t_j t_{j+1} \ldots t_{n-1} \$ t_1 \ldots t_{j-1}$.

Backward Search. The FM-index [6] is a self-index based on the Burrows-Wheeler transform. It solves counting queries by finding the interval of $\mathcal{A}$ that contains the occurrences of pattern P. The FM-index uses the array C and function $rank_c(L, i)$ of the LF mapping to perform backward search of the pattern. Fig. 1 shows the counting algorithm. Using the properties of the BWT, it is easy to see that the algorithm maintains the following invariant [6]: At the i-th phase, variables sp and ep point, respectively, to the first and last row of $\mathcal{M}$ prefixed

Algorithm. FMCount($P[1, m]$,$L[1, n]$)
(1) $i \leftarrow m$;
(2) $sp \leftarrow 1$; $ep \leftarrow n$;
(3)   **while** ($sp \leq ep$) **and** ($i \geq 1$) **do**
(4) $c \leftarrow P[i]$;
(5) $sp \leftarrow C[c] + rank_c(L, sp - 1)+1$;
(6) $ep \leftarrow C[c] + rank_c(L, ep)$;
(7) $i \leftarrow i - 1$;
(8)   **if** ($ep < sp$) **then return** 0 **else return** $ep - sp + 1$;

Fig. 1. FM-index algorithm for counting the number of occurrences of $P[1, m]$ in $T[1, n]$

by $P[i, m]$. The correctness of the algorithm follows from this observation. Note that P is processed backwards, from p_m to p_1.

Note that array C can be explicitly stored in little space, and to implement $rank_c(L, i)$ in little space we can directly use the wavelet tree as explained in Sect. 3.2. The space usage is $nH_0 + o(n \log \sigma)$ bits and the m steps of backward search take overall $O(m \log \sigma)$ time [16].

4   Implicit Compression Boosting

In Sect. 3.3 we showed that if L is partitioned into ℓ pieces $L^1 L^2 \cdots L^\ell$ according to the k-contexts, then it is enough to achieve zero-order entropy within each partition to obtain k-th order total entropy. We now prove that the simple solution of Sect. 3.3 supporting backward search requires only nH_k bits of space. We start with an important lemma.

Lemma 1. *Let* $L = L^1 L^2 \cdots L^\ell$ *be any partition of* L, *the BWT of* T. *The number of bits used by a partition* L^j *in the wavelet tree of* L *is upper bounded by* $|L^j| H_0(L^j) + O(|L^j| \log \sigma \log \log n / \log n + \sigma \log n)$.

Proof. The bits corresponding to L^j form a substring of the bit vectors at each node of the wavelet tree, as their positions are mapped to the left and right child using $rank_0$ or $rank_1$, thus order is preserved. Let us consider a particular node of the wavelet tree and call B its bit sequence. Let us also call B^j the substring of B corresponding to partition L^j, and assume B^j has l^j bits set. Consider the blocks of b bits that compose B, according to the partitioning of [23] (Section 3.1). Let $B_{blk}^j = B_1^j B_2^j \ldots B_t^j$ be the concatenation of those bit blocks that are *fully contained* in B^j, so that B_{blk}^j is a substring of B^j of length $b \cdot t$. Assume B_i^j has l_i^j bits set, so that B_{blk}^j has $l_1^j + \ldots + l_t^j \leq l^j$ bits set. The space the o fields of the (c, o) representations of blocks B_j^i take in the compressed B_{blk}^j is

$$\sum_{i=1}^{t} \left\lceil \log \binom{b}{l_i^j} \right\rceil \leq \log \binom{b \cdot t}{l_1^j + \ldots + l_t^j} + t \leq \log \binom{|B^j|}{l^j} + t \leq |B^j| H_0(B^j) + t$$

where all the inequalities hold by simple combinatorial arguments [21] and have
been reviewed in Section 3.1.

Note that those B^j bit vectors are precisely those that would result if we
built the wavelet tree just for L^j. According to Theorem 1, adding up those
$|B^j|H_0(B^j)$ over all the $O(\sigma)$ wavelet tree nodes gives $|L^j|H_0(L^j)$. To this we
must add three space overheads. The first is the extra t bits above, which add
up $O(|L^j|\log \sigma / \log n)$ over the whole wavelet tree because $b \cdot t \leq |B^j|$ and the
$|B^j|$ lengths add up $|L^j|$ at each wavelet tree level. The second overhead is the
space of the blocks that overlap with B^j and thus were not counted: As B^j is
a substring of B, there can be at most 2 such blocks per wavelet tree node.
At worst they can take $O(\log n)$ bits each, adding up $O(\sigma \log n)$ bits over the
whole wavelet tree. The third overhead is that of the c fields, which add up
$O(|L^j|\log \sigma \log \log n / \log n)$.

The above lemma lets us split the wavelet tree "horizontally" into pieces. Let us
add up all the zero-order entropies for the pieces. If we partition L according to
contexts of length k in $\mathcal{M}$, and add up all the space due to all partitions in the
wavelet tree, we get $\sum_{1 \leq j \leq \ell} |L^j|H_0(L^j) = nH_k(T)$ (Theorem 2). To this we must
add (i) $O(|L^j|\log \sigma / \log n)$, which sums up to $O(n \log \sigma / \log n) = o(n \log \sigma)$ bits
over all the partitions; (ii) $O(\sigma \log n)$ bits per partition, which gives $O(\ell \sigma \log n)$;
and (iii) $O(|L^j|\log \sigma \log \log n / \log n)$, which sums up to $O(n \log \sigma \log \log n / \log n)$
$= o(n \log \sigma)$. In the partitioning we have chosen we have $\ell \leq \sigma^k$, thus the upper
bound $nH_k + o(n \log \sigma) + O(\sigma^{k+1} \log n)$ holds for the total number of bits spent
in the wavelet tree. The next theorem immediately follows.

Theorem 3. *The space required by the wavelet tree of L, the BWT of T, if the
bitmaps are compressed using [23], is $nH_k(T) + o(n \log \sigma) + O(\sigma^{k+1} \log n)$ bits
for any $k \geq 0$. This is $nH_k(T) + o(n \log \sigma)$ bits for any $k \leq \alpha \log_\sigma n - 1$ and any
constant $0 < \alpha < 1$. Here n is the length of T and σ its alphabet size.*

Note that this holds *automatically and simultaneously for any* k, and we do not
even have to care about k in the index. Fig. 2 illustrates. Our result obviously
applies as well to the BWT alone, without wavelet tree on top, if we use a
suitable local zero-order encoder [9].

5   Discussion

We have shown that the space produced by *any* splitting of L into pieces is
achieved in the simple arrangement having just one wavelet tree. In [7] they
introduce an algorithm to find the optimal partitioning. We have just used their
analysis to show that it is not necessary to apply such a partitioning algorithm to
achieve the boosted result. Their technique, on the other hand, has more general
applications unrelated to wavelet trees.

Several full-text self-indexes in the literature build on the wavelet tree of the
BWT of the text [16,9], and engage in different additional arrangements to reach
k-th order compression. In [16], they first run-length compress the BWT in order

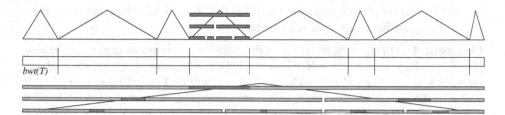

Fig. 2. Illustration of the argument used in the theorem. On top of $bwt(T) = T^{bwt}$, the wavelet trees induced by the optimal partitioning. One of those is highlighted so that we show the bitmaps stored in the wavelet tree. Below $bwt(T)$, a single wavelet tree built for the whole sequence. The bitmaps of this large wavelet tree are also shown, and they contain the bitmaps of the small highlighted wavelet tree on top.

to reduce its length to $O(nH_k)$ and then apply the BWT. In [9] they explicitly cut the BWT into pieces L^j so that the sum of nH_0 sizes of the pieces adds up nH_k. In both cases, the simpler version they build on (just the wavelet tree of the BWT) would have been sufficient. Thus, we have achieved a significant simplification in the design of full-text indexes.

Also the paper where the wavelet tree is originally proposed [11] as an internal tool to design one of the most space-efficient compressed full-text indexes, would benefit from our simplification. They cut L into a table of *lists* (columns) and *contexts* (rows). All the entries across a row correspond to a contiguous piece of L, that is, some context L^j. A wavelet tree is built over each table row so as to ensure, again, that the sum of zero-order entropies over the rows adds up to global k-th order entropy. Our finding implies that all rows could have been concatenated into a single wavelet tree and the same space would have been achieved. This would greatly simplify the original arrangement. Interestingly, in [12] they find out that, if they use gap encoding over the successive values along a *column*, and they then concatenate all the columns, the total space is $O(nH_k)$ without any table partitioning as well. Both results stem from the same fact: the cell entropies can be added in any order to get nH_k.

Finally, it is interesting to point out that, in a recent paper [5], the possibility of achieving k-th order compression when applying wavelet trees over the BWT is explored (among many other results), yet they resort to run-length compression to achieve this. Once more, our finding is that this is not really necessary to achieve k-th order compression if the levels of the wavelet tree are represented using the technique of block identifier encoding [22].

6   Application to Space-Efficient Construction of (Dynamic) Self-indexes

Another consequence of our result is that we obtain an $O(n \log n \log \sigma)$ time construction algorithm for a compressed self-index requiring $nH_k + o(n \log \sigma)$

bits *working space* during construction: This is obtained by enhancing our recent result on dynamic compressed self-indexes:

Theorem 4 ([17]). *There is a data structure maintaining a collection $\{T_1, T_2, \ldots T_m\}$ of texts in $nH_0(\mathcal{C}) + o(n \log \sigma)$ bits supporting counting of occurrences of a pattern P in $O(|P| \log n \log \sigma)$ time, and inserting and deleting a text T in $O(|T| \log n \log \sigma)$ time. After counting, any occurrence can be located in time $O(\log^2 n \log \log n)$. Any substring of length ℓ from any T in the collection can be displayed in time $O(\log n(\ell \log \sigma + \log n \log \log n))$. Here n is the length of the concatenation $\mathcal{C} = 0\, T_1 0\, T_2 \cdots 0\, T_m$, and we assume $\sigma = o(n)$.*

The dynamic index above uses wavelet tree with the static encoding [23] (see Sect. 3.1) replaced with a dynamic version of the same encoding: The dynamic bit vector representation [17] achieves the same $nH_0 + o(n)$ space as the static, but supports *rank* and *select*, and in addition insertions and deletions of bits, in $O(\log n)$ time. This can then be used to improve the dynamic index of Chan et al. [4] to obtain the above result.

Exactly the same analysis as in Sect. 4 applies to this dynamic variant, and Theorem 4 is boosted into the following.

Corollary 1. *There is a data structure maintaining a collection $\{T_1, T_2, \ldots T_m\}$ of texts in $nH_k(\mathcal{C}) + o(n \log \sigma)$ bits, for any $k \leq \alpha \log_\sigma n - 1$ and constant $0 < \alpha < 1$, supporting the same operations of Theorem 4 within the same complexities.*

Now, just inserting text T into an empty collection, yields the promised space-efficient construction algorithm for compressed self-index. This index can be easily converted into a more efficient static self-index, where a static wavelet tree requires the same space and reduces the $O(\log n \log \sigma)$ time complexities to $O(\lceil \log \sigma / \log \log n \rceil)$ [9].

Therefore, we have obtained the *first* compressed self-index with space essentially equal to the k-th order empirical entropy of the text collection, which in addition can be built within this working space. Alternative dynamic indexes or constructions of self-indexes [6,13,2] achieve at best $O(nH_k)$ bits of space (with constants larger than 4), and in many cases worse time complexities.

Note also that, from the dynamic index just built, it is very easy to obtain the BWT of T. It is a matter of finding the characters of L one by one. This takes $O(n \log n \log \sigma)$ time, just as the construction, and gives an algorithm to build the BWT of a text within entropy bounds. The best result in terms of space complexity takes $O(n)$ bits working space, $O(n \log^2 n)$ time in the worst case, and $O(n \log n)$ time in the expected case [15]. Using $O(n \log \sigma)$ working space, there is a faster algorithm achieving $O(n \log \log \sigma)$ time requirement [13]. Finally, one can achieve the optimal $O(n)$ time with the price of $O(n \log^\epsilon n \log \sigma)$ bits of space, for some $0 < \epsilon < 1$ [20].

7   Final Practical Considerations

Our main finding is that all the sophistications [16,9,11] built over the simple "wavelet tree on top of the BWT" scheme in order to boost its zero-order to high-order compression are unnecessary; the basic arrangement already achieves high-order entropy if combined with a local zero-order encoder [23].

Still, the sophisticated techniques have practical value. In the actual implementation of such methods (*Pizza&Chili* site, *http://pizzachili.dcc.uchile.cl*), zero-order entropy is achieved by using *uncompressed* bit streams over a *Huffman-tree shaped* wavelet tree, instead of *compressed* bit streams over a *balanced* wavelet tree. In this case the locality property does not hold, and high-order entropy would not be achieved if just the simple wavelet tree of the BWT was used.

Huffman-shaped trees were chosen to reach zero-order compression because of the considerable difficulty in implementing Raman et al.'s scheme [23]. As both alternatives were believed to yield similar compression ratios, the Huffman-shaped option seemed to be far more attractive from a practical point of view.

The situation is rather different now that we know that Raman et al.'s scheme yields high-order by itself, thus avoiding the need of any further complication to achieve high-order compression such as run-length compression (Run-Length FM-index, RLFM [16]) or compression boosting plus managing multiple wavelet trees (Alphabet-Friendly FM-index, AFFM [9]). Those complications not only make the implementation effort comparable to that of using Raman et al.'s scheme, but also involve a considerable space overhead for extra structures.

A prototype combining Raman et al.'s compression with balanced wavelet trees has already been implemented by Francisco Claude, a student of the second author. Unlike the existing implementations, this one offers a space-time tradeoff, related to the partial sums sampling rate. A preliminary comparison with the implementations of the SSA (bare Huffman-shaped wavelet tree over the BWT), RLFM, and AFFM, shows that our technique is 2–3 times slower for counting when using the same amount of space, which confirms the original predictions about implementation overheads. In exchange, it can still operate with reasonable efficiency using less than 75% of the space needed by the alternatives. This makes it a relevant choice when space reduction is the main concern.

It is interesting that our technique can achieve such a low space even when it has to pay for space overheads like the c components in the (c, o) pairs. This opens the door to the study of other alternatives that, even when they do not surpass the "$nH_k + o(n \log \sigma)$ for $k \le \alpha \log_\sigma n$" theoretical barrier, do behave better in practice. We point out that this barrier is not as good as it may seem when one puts numbers to the condition on k and realizes that the achievable k values are rather low. Worse than that, it is unlikely that this theoretical limit can be sensibly improved [10]. Yet, those limits are worst-case, and different methods may not have to pay the $\Theta(\sigma^{k+1} \log n)$ space overhead in practice. For example, in our case, this overhead comes from the fact that we are unable to analyze better the compression of blocks that are split among contexts, and thus we assume the worst about them. On the other hand, the c components are real space overhead in our scheme ($2n$ bits!), and that perhaps could be improved.

References

1. Apostolico, A.: The myriad virtues of subword trees. In: Combinatorial Algorithms on Words, NATO ISI Series, pp. 85–96. Springer, Heidelberg (1985)
2. Arroyuelo, D., Navarro, G.: Space-efficient construction of LZ-index. In: Deng, X., Du, D.-Z. (eds.) ISAAC 2005. LNCS, vol. 3827, pp. 1143–1152. Springer, Heidelberg (2005)
3. Burrows, M., Wheeler, D.: A block sorting lossless data compression algorithm. Technical Report Technical Report 124, Digital Equipment Corporation (1994)
4. Chan, H.-L., Hon, W.-K., Lam, T.-W.: Compressed index for a dynamic collection of texts. In: Sahinalp, S.C., Muthukrishnan, S.M., Dogrusoz, U. (eds.) CPM 2004. LNCS, vol. 3109, pp. 445–456. Springer, Heidelberg (2004)
5. Ferragina, P., Giancarlo, R., Manzini, G.: The myriad virtues of wavelet trees. In: Bugliesi, M., Preneel, B., Sassone, V., Wegener, I. (eds.) ICALP 2006. LNCS, vol. 4051, pp. 560–571. Springer, Heidelberg (2006)
6. Ferragina, P., Manzini, G.: Opportunistic data structures with applications. In: Proc. FOCS 2000, pp. 390–398 (2000)
7. Ferragina, P., Manzini, G.: Compression boosting in optimal linear time using the Burrows-Wheeler transform. In: Proc. SODA 2004, pp. 655–663 (2004)
8. Ferragina, P., Manzini, G.: Indexing compressed texts. J. of the ACM 52(4), 552–581 (2005)
9. Ferragina, P., Manzini, G., Mäkinen, V., Navarro, G.: Compressed representations of sequences and full-text indexes. ACM TALG, 3(2):article 20 (2007)
10. Gagie, T.: Large alphabets and incompressibility. IPL 99(6), 246–251 (2006)
11. Grossi, R., Gupta, A., Vitter, J.: High-order entropy-compressed text indexes. In: Proc. SODA 2003, pp. 841–850 (2003)
12. Grossi, R., Gupta, A., Vitter, J.: When indexing equals compression: Experiments with compressing suffix arrays and applications. In: Proc. SODA 2004, pp. 636–645 (2004)
13. Hon, W.-K., Sadakane, K., Sung, W.-K.: Breaking a time-and-space barrier in constructing full-text indexes. In: Proc. FOCS 2003, pp. 251–260 (2003)
14. Jacobson, G.: Space-efficient static trees and graphs. In: Proc. FOCS 1989, pp. 549–554 (1989)
15. Kärkkäinen, J.: Fast BWT in small space by blockwise suffix sorting. In: Proc. DIMACS Working Group on the Burrows-Wheeler Transform (2004)
16. Mäkinen, V., Navarro, G.: Succinct suffix arrays based on run-length encoding. Nordic J. of Computing 12(1), 40–66 (2005)
17. Mäkinen, V., Navarro, G.: Dynamic entropy-compressed sequences and full-text indexes. In: Lewenstein, M., Valiente, G. (eds.) CPM 2006. LNCS, vol. 4009, pp. 307–318. Springer, Heidelberg (2006)
18. Manber, U., Myers, G.: Suffix arrays: a new method for on-line string searches. SIAM J. on Computing, 935–948 (1993)
19. Manzini, G.: An analysis of the Burrows-Wheeler transform. J. of the ACM 48(3), 407–430 (2001)
20. Na, J.C.: Linear-time construction of compressed suffix arrays using $o(n \log n)$-bit working space for large alphabets. In: Apostolico, A., Crochemore, M., Park, K. (eds.) CPM 2005. LNCS, vol. 3537, pp. 57–67. Springer, Heidelberg (2005)
21. Pagh, R.: Low redundancy in dictionaries with $O(1)$ worst case lookup time. In: Wiedermann, J., van Emde Boas, P., Nielsen, M. (eds.) ICALP 1999. LNCS, vol. 1644, pp. 595–604. Springer, Heidelberg (1999)

22. Raman, R., Raman, V., Srinivasa, S.: Succinct dynamic data structures. In: Dehne, F., Sack, J.-R., Tamassia, R. (eds.) WADS 2001. LNCS, vol. 2125, pp. 426–437. Springer, Heidelberg (2001)
23. Raman, R., Raman, V., Srinivasa Roa, S.: Succinct indexable dictionaries with applications to encoding k-ary trees and multisets. In: Proc. SODA 2002, pp. 233–242 (2002)

A Web-Page Usage Prediction Scheme Using Weighted Suffix Trees

Christos Makris, Yannis Panagis, Evangelos Theodoridis,
and Athanasios Tsakalidis

Computer Engineering and Informatics Department, University of Patras,
Rio 26504, Greece
{makri,panagis,theodori,tsak}@ceid.upatras.gr

Abstract. In this paper we consider the problem of web page usage prediction in a web site by modeling users' navigation history with weighted suffix trees. This user's navigation prediction can be exploited either in an on-line recommendation system in a website or in a web-page cache system. The method proposed has the advantage that it demands a constant amount of computational effort per user action and consumes a relatively small amount of extra memory space. These features make the method ideal for an on-line working environment. Finally, we have performed an evaluation of the proposed scheme with experiments on various website logfiles and we have found that its prediction quality is fairly good, in many cases outperforming existing solutions.

Keywords: World Wide Web, web mining, online web page recommendation, weighted sequences.

1 Introduction

The vast size of the World Wide Web (WWW) nowadays makes it the largest database ever existed. Back to the beginning of this decade it was estimated to contain over 350 million pages [3] while recently it has been estimated that only the indexed part of WWW by a web search engine consists of at least 11.3 billion pages [13]. Every attempt to shape this huge volume of data that follows a very loose schema is quite difficult and extreme challenging. According to [9] the application of data mining techniques in order to extract useful information woven among web data is an essential task. Web data may be either web data pages or data describing the activity of users. Actual web data consists of web pages, web page structure, linkage structure between the web pages, surfing navigational behavior of the users and user profiles including demographic and registration information about the users [9,29].

Web data mining can be divided into three general categories: *web content mining*, *web structure mining* and finally *web usage mining* [36]. In this paper we focus to the last area that tries to exploit the navigational traces of users in order to extract knowledge about their preferences and their behavior. The task of modeling and predicting a user's navigational behavior on a website

N. Ziviani and R. Baeza-Yates (Eds.): SPIRE 2007, LNCS 4726, pp. 242–253, 2007.

or a web-domain can be useful in a handful of web applications such as web caching [22,31], web page recommendation [8,4], web search engines [2,24] and personalization [10]. Refer to [12] for several other applications of web usage mining.

According to [12] most of the web usage mining techniques are based on association rules, sequential patterns and clustering. The approaches based on association rules build a set of rules of the form: {*page1.html, page2.html*} → *page3.html* which means that users having visited page1 and page2, have also visited page3. Methods that use association rules can be found in [19,20]. Sequential patterns maintain navigational patterns as sequences and try to discover frequent subsequences that describe better the data. Sequences are either used to produce association rules or to produce tree structures or Markov chains to represent navigation patterns. Methods proposed in [19,23,22] fall into this general category. Lately, in [37,25,14] the authors proposed the use of web access motifs and string matching techniques. The idea behind the use of these methods is that string algorithmics seem to adapt quite well to the analysis of hypertext navigation. Finally, clustering techniques have been used to group similar user sessions according to a distance function, like longest common subsequence [1] or sequence alignment [17]. Also more complex methods have been proposed in [27,35,26] in order to enhance the results of the clustering method.

In this paper we propose an efficient method for modeling user navigation history. Our method is based upon several ideas from the existing literature and upon a string processing structure previously used in computational biology problems; the weighted suffix tree [18]. The provided estimation of user's navigational intention can be exploited either in an on-line recommendation system in a website or in a web-page cache system. The method proposed here has the advantage that it demands a constant amount of computational effort per one user's action and consumes a relatively small amount of extra memory. These features make our method ideal for an on-line working environment. The structure of the paper is as follows. In section 2 we present some basic definitions and background information. Section 3 presents the approach proposed by this paper. In section 4 we describe the methodology followed to evaluate our method and present experimental results, while in section 5 we conclude and discuss future directions.

2   Definitions and Background

Let Σ be a finite alphabet which consists of a set of characters (or symbols). The cardinality of an alphabet, denoted by $|\Sigma|$, expresses the number of distinct characters in the alphabet. A string or word is a sequence of zero or more characters drawn from an alphabet. The set of all words over the alphabet Σ is denoted by Σ^+. A word w of length n is represented by $w[1..n] = w[1]w[2]\ldots w[n]$, where $w[i] \in \Sigma$ for $1 \le i \le n$, and $n = |w|$ is the length of w.

A subword u of length p is said to occur at position i in the word w if $u = w[i..i + p - 1]$. In other words u is a substring of length p occurring at position

i in word w. In the case that for a given position of a word w we consider the presence of a set of characters each with a given probability of appearance, we have the concept of a weighted sequence X.

Definition 1. *A weighted sequence/word $X = X[1]X[2]\ldots X[n]$ is a sequence of positions, where each position $X[i]$ consists of a set of ordered pairs. Each pair has the form $(\sigma, \pi(\sigma))$, where $\pi(\sigma)$ is the probability of having the character σ at position i. For every position $X[i], 1 \leq i \leq n, \sum \pi(\sigma) = 1$.*

The suffix tree is a fundamental data structure supporting a wide variety of efficient string processing algorithms. In particular, the suffix tree is well known to allow efficient and simple solutions to many problems concerning the identification and location either of a set of patterns or repeated substrings (contiguous or not) in a given sequence. The reader can find an extended literature on such applications in [30].

The weighted suffix tree can be considered as a generalization of the ordinary suffix tree in order to handle weighted sequences by incorporating the notion of probability of appearance for every suffix stored in a leaf. The construction of this structure has been proposed in [33], together with a set of several applications.

Definition 2. *Let X be a weighted sequence. For every suffix starting at position i we define a list of possible weighted subwords so that the probability of appearance for each one of them is greater than $1/k$. We denote each of them as $X_{i,j}$, where j is the subword rank in arbitrary numbering. We define $WST(X)$ the weighted suffix tree of a weighted sequence X, as the compressed trie of a portion of all the weighted subwords starting within each suffix X_i of $X\$$, $\$$ not in Σ, having a probability of appearance greater than $1/k$. Let $L(v)$ denote the path-label of node v in $WST(X)$, which results by concatenating the edge labels along the path from the root to v. Leaf v of $WST(X)$ is labeled with index i if $\exists j > 0$ such that $L(v) = X_{i,j}[i..n]$ and $\pi(X_{i,j}[i\ldots n]) \geq 1/k$, where $j > 0$ denotes the j-th weighted subword starting at position i.*

We suppose that the navigation history in a web site is maintained as web access sequences (WAS). Web access sequences are sequences of web pages that express each session (clicking sequence) of the users and can be generated by the logfiles of the web site; if we represent each web page as a symbol in an appropriate alphabet then the WAS is essentially a string.

3   Prediction/Recommendation Model

In our model we assume that each navigation session is described by a WAS_i and stored for offline preprocess. S is the set of all existing web access sequences that have taken place in the web site. A brief outline of our method follows: In an offline manner, running either in an idle time period or in the background, the system processes the set S of the navigation sessions in order to group them into clusters. Each of the clusters contains sequences very similar to each other.

Then each of the clusters C_i is represented by a weighted sequence WS_i and, finally, using these sequences a generalized weighted suffix tree gWST($i : WAS_i$) is constructed. This structure is further maintained and used as a web page prediction/recommendation tool. Each one of the pre-processing steps is going to be described in a more detailed way in the following sections.

3.1 WAS Maintenance

Each of the user navigation sessions is implicitly maintained in the logfiles of the websites server. Either we program properly the web server to store each WAS in separate repository or we can program an extraction process from the logfiles that is executed at the beginning of the preprocessing procedure. Each webpage of the site has been labeled with a unique symbol. All these symbols form the alphabet of the web access sequences. Assume for the sake of description that there are N sequences that form a set $S = \{WAS_1, WAS_2, \ldots, WAS_N\}$.

3.2 WAS Clustering

At this step of the preprocessing procedure we construct the similarity matrix of S. This $N \times N$ matrix expresses the similarity of each pair (i, j) of sequences. As a metric of the distance/similarity between WAS_i and WAS_j we chose to use a hybrid metric taking into account the global alignment and the local alignment of the two sequences. More formally:

$$D(i,j) = (1 - p) * LA(WAS_i, WAS_j) + p * GA(WAS_i, WAS_j) \quad (1)$$

where $LA(WAS_i, WAS_j)$ is the score of the local alignment of the sequences WAS_i and WAS_j, $GA(WAS_i, WAS_j)$ is the score of the global alignment for these sequences, and p is a parameter that expresses the importance that we give to the scores of the two different alignments.

In order to define the value of p there are several possibilities. One choice is to use p equal to 0.5 giving the same influence to both alignments. The second and, more proper, choice is to define the value of p to be relative to the ratio of the lengths of the two sequences. More formally, assuming without loss of generality that $|WAS_j| \geq |WAS_i|$, we define:

$$p = \frac{|WAS_i|}{|WAS_j|} \quad (2)$$

The intuition behind this definition is that when the two sequences, and thus navigational behaviors, have almost the same length we should take more into account the global alignment than the local. When the lengths of the two sequences are very different p is close to zero and local alignment has more influence than the global one. Our claim is that common navigational preferences of the users that are depicted by the two sequences are not captured only by aligning the sequences in their full length; smaller but closely aligned subsections of the sequences can capture common behavior, too.

The score function for the alignment has to be appropriate in order to express navigational properties of the sessions. A straightforward approach would assume that the match of two characters/pages in the alignments should be awarded a positive constant score, while a mismatch or a match with a space with a negative constant score. A specific page of a web site may play dissimilar roles to different users so a simple character/page match in may mean nothing in the task of similarity testing of two sequences. However, a web page may have no importance to a user; we call these web pages *Unimportant*, such as web site start-pages or a page visited by a wrong click and immediately a back-step is performed. Usually, a user stays to unimportant pages a very short period of time, a few seconds at most, and then visits another page. Also a page is unimportant for a user when it is recorded in logfile for inordinately large period of time, usually above 25-30 minutes. This usually means that the user has changed web page by typing a new address or by clicking a banner. Another role that can play a page is the one of index, we call these pages *Hub*. Users try in these pages to locate the link to the desired page by reading roughly the content and staying in those pages not more than 5-6 minutes. Finally, there are the *Content* pages with the desired for the user content. Users spend in Content pages approximately 10 to 20 minutes. Consequently, there is need to store in each WAS the corresponding period of time along with each character/page. Techniques for session extraction from log files and for extracting additional information can be found in [33,28,6]. As mentioned before, from all the pages we form initially an alphabet Σ from the web-pages, which is augmented by at most three times by characterizing each of the pre-existing symbols as Unimportant, Hub and Content.

For calculating the global and the local alignments the classical approach of dynamic programming [21] has been chosen using recursive formulas for the alignments of the prefixes WAS_i and WAS_j. Both in global and local alignment of two sequences the alignments of symbols with spaces should not be concentrated in consecutive positions. Consecutive alignments with spaces are defined as gaps. Very long gaps mean that in the one WAS, the user has followed a very different path (maybe lost) in comparison with the other so we would not like to favor such gaps. In order to tackle this phenomenon, we apply affine gap penalties in the alignments. The objective is then to find an alignment that maximizes the quantity:

$$W_m * (\#matches) - W_{ms} * (\#mismatches) - W_g * (\#gaps) - W_s * (\#spaces) \quad (3)$$

where $W_m = 1$ denotes the weight of a match, $W_{ms} = -1$ denotes the weight of a mismatch W_g denotes the weight of starting a new gap and W_s denotes the weight of an alignment with a space. We would like to favor small gaps so we have chosen $W_g = -1$ while $W_s =$-5. The gap opening penalty is, intentionally, kept smaller than the gap penalty in order to favor frequent gap openings with small length instead of longer and more infrequent gaps. The classic recursive equations can easily adapt to incorporate affine gaps, too [15].

When similarity matrix D has been constructed we use it in order to group the sequences into clusters. The distance matrix can be viewed as N points in N-dimensional space. Running a clustering algorithm upon these N points will produce clusters consisting of very similar web access sequences. D is given as an input to the k-windows clustering algorithm [30]; a well known unsupervised partitional clustering algorithm. Our decision to use k-windows as a clustering method was followed for a variety of reasons, such as the enhanced quality of the produced clusters and its inherent parallel nature (more information can be found in [30]).

One of the drawbacks of the traditional implementation of the k-windows algorithm is that it uses as an underlying data structure a d-dimensional range tree [32] which is very inefficient in high-dimensions. In order to overcome this drawback we propose a variation of k-windows algorithm that utilizes R-trees [16] as an underlying data structure for algorithm operations. As an alternative to k-windows, the k-means algorithm can also be used.

3.3 WAS Cluster Representation

When the WAS clustering procedure is over each one of the clusters is expressed as a weighted sequence. The produced weighted sequences implicitly capture most of the navigational behavior that is observed in the corresponding cluster. Assuming that we examine a cluster with f web access sequences, in order to produce the weighted sequences the web access sequences are combined using their multiple sequence alignment. Although the calculation of a multiple sequence alignment is extremely time and space consuming using dynamic programming approaches, in some cases it may be affordable because the length of the sessions is usually short (5 to 10 pages is a very frequently observed session length) due to limited human recall [11] or to bounded web site height [5].

As an alternative, someone could possibly use the approach of progressive or iterative pairwise alignment in order to produce the multiple sequence alignment [15]. Progressive of iterative algorithms are based on the idea that the solution can be computed by modifying an already suboptimal solution. The simplest approach is to initially align the pair of WAS with the maximum similarity. Then successively merge in, the WAS with smallest distance from any of the strings already in the multiple alignment. This method can be viewed as finding the multiple alignments consistent with a maximum spanning tree formed from similarity data (matrix D). A more realistic variant of the above approach might choose the sequence with the maximum average similarity in order to incorporate it in the multiple sequence alignment aligning that sequence with the most alike. In our case we have used the *Star-alignment heuristic* presented in [15].

Finally, when the multiple sequence alignment of the cluster's sequences has been calculated, we construct the weighted sequence that represents the cluster. At each position and for each character/page we maintain the percentage of their appearance, without taking the space into account.

3.4   WST Utilization - Recommendation/Prediction Method

As soon as all weighted sequences have been produced, their corresponding generalized weighted suffix tree is constructed. For this construction the algorithm of [18] is used tuned by a parameter $1/k$. The computational time and the space utilized by this structure are linear to the sum of the lengths of all weighted sequences. The generalized weighted suffix tree will represent all possible weighted subwords of all of the weighted sequences such that the probability of appearance for each one of them is greater than $1/k$. The experimental evaluation showed up that a proper value for parameter $1/k$ is between 0.01 and 0.1.

The weighted suffix tree can implicitly capture the *most important* navigational experience and at a small computational and memory price. It works like a prediction model without storing explicitly strings and probabilities of occurrence. Also inherits all the virtues of suffix trees (which is a well studied data structure): the linear space, the linear construction time and many well designed practical implementations.

The recommendation/prediction algorithm works as follows: when a new user arrives to the system, he is assigned to the root of the generalized weighted suffix tree (gWST). As the user navigates through the web site selecting web pages by clicking links, his position in the gWST is advanced, too. Assume that he currently is at an internal node u and he is ready to perform his next step. The system proposes to him the web pages indicated by the outgoing edges of u. Each of the outgoing edges is labeled by a substring and starts with a different character. All those different characters/web pages are proposed to the user. The proposition of all descendants is not realistic since the cardinality of them can grow more than 100 (we refer to this approach as *unbounded* in the evaluation section just for the sake of comparisons). If the number of the proposed web pages is relatively large, above 20, a selection of at most 20 webpages is presented (or prefetched) to the user. Those pages are either randomly selected (*bounded-random*) or we pick the "heaviest" ones(*bounded-weight*). The weight of each one of the choices is defined as the number of times that this node has been visited by a user before and it is implemented by a simple counter and a priority queue upon each node.

In the extreme case that under an internal node there is no edge starting with the character that corresponds to the user's choice, then the user is assigned to the root again and the tree navigation starts again according to the following actions of the user.

4   Evaluation

For the evaluation of the proposed system various experiments have been performed. Through this process we would like to study and estimate: the demands in memory of the proposed system, the preprocessing effort to construct the structure and the success of the prediction/recommendation that the system creates.

We have performed several experiments, choosing different values for the parameters of the system in each one. All experiments have been performed on a AMD64 dual core@ 2Ghz with 1 Gbyte of main memory under Windows XP. The programs were coded in Perl and C. The experiments simulate the user behavior by using a web-server log file. The 3-10% of the information of the logfile has been used in order to construct the data structure (gWST) of the system while the rest has been used to evaluate the recommendations.

For the purposes of this experiment we have utilized the web server's logfile of our department.[1] This logfile corresponds to a two month period and has approximately 260 Mbytes size. We have also used for this set of experiments the well-known data sets in many studies in this area; the log file of NASA web server over the months of July and August 1995 [2]. For extracting the web access sequences from the logfiles we have extended the Follow web-log analysis tool [38], in order to have the properties described in section 3.2. The initial tool and the extension were both written in Perl. With the preprocessing of the logfiles a set of web access sequences has been produced. A percentage of these (3-10%), randomly selected, has been utilized for capturing the user behavior of the website by clustering them and constructing the weighted suffix tree while the rest has been used as an evaluation set. For each WAS, we perform the movements of the user described by the sequence and the system behavior as described in section 3.4. In order to evaluate the performance of our method we have used two known metrics as in [26,7]:

- *Hit-Ratio*: For each WAS we maintain an evaluation score. For each character of the sequence/step of the user in the site we increase the score by one if the next step (page of the web site) has been proposed by the system while in the case that the user performs a step not in the set recommended by the WST (section 3.4) we keep the score unchanged. Finally for each WAS, we calculate a weighted score defined by the ratio of the previous score over the length of the sequence (which is the best score).
- *Click-Soon-Ratio*: For each WAS we maintain an evaluation score. For each character of the sequence/step of the user in the site we increase the score by one if at least one of the recommendations/predictions is requested by the user until the end of the sequence (thus while the user session is active). If none of the recommendations is used while the session is active we keep the score unchanged. Finally for each WAS, we calculate a weighted score defined by the ratio of the previous score over the length of the sequence (which is the best score too).

Figure 1. recapitulates the settings and the results of the experiments on the NASA logfile. A small amount (2%) of the data set has been used to train our model in both experiments. In the first one, the $1/k$ building parameter of the weighted suffix tree is fine-tuned while in the second one the number of initial clusters is varied. At a first glance we observe that the weighted scheme for

---

[1] http://www.ceid.upatras.gr
[2] http://ita.ee.lbl.gov/html/contrib/NASA-HTTP.html

Training Set	# clusters	1/k prob. Occurrenc	#recomm endation	Hit- Ratio			Click-Soon			#nodes
				unbounded	bounded random	bounded weighted	unbounded	bounded random	bounded weighted	
2%	10	0,5	20	51,90%	29,20%	48,70%	74,10%	37,60%	67,70%	226
2%	10	0,1	20	73,90%	48,40%	65,50%	91,20%	59,90%	87,60%	726
2%	10	0,05	20	82,90%	49,80%	74,00%	97,20%	59,87%	95,02%	8069
2%	10	0,01	20	93,20%	54,30%	82,50%	97,60%	63,69%	95,80%	13365
2%	5	0,1	20	65,70%	40,50%	58,90%	97,00%	50,60%	79,80%	499
2%	10	0,1	20	71,50%	44,60%	64,05%	97,02%	53,90%	85,60%	1316
2%	15	0,1	20	77,50%	51,01%	68,85%	97,06%	61,80%	89,60%	4838
2%	20	0,1	20	78,60%	53,02%	69,70%	96,78%	64,30%	92,20%	2492
2%	20	0,01	20	94,41%	50,30%	83,20%	97,30%	59,31%	96,23%	24442
3%	20	0,01	20	94,82%	51,40%	83,86%	97,68%	60,99%	96,12%	50187
6%	20	0,01	20	94,70%	51,35%	83,63%	97,34%	61,11%	96,45%	42532
10%	20	0,01	20	95,50%	50,12%	84,12%	98,50%	59,90%	97,29%	46197

Fig. 1. Experiments on NASA logfile

the selection of the recommendations, in both metrics, is always much better from the random selection but slightly worse from the impractical unbounded proposing. Decreasing $1/k$ implies increasing the effectiveness of the model in both metrics but with larger memory usage. Suitable selection for $1/k$ seems to be values below 0.1. The number of utilized clusters seems to affect the capability of the model leading to less memory consuption but in a more indirect way than the $1/k$ parameter. Finally, in the third experiment we have varied the size of the training data set from 2% to 10%. It seems that with larger training sets and small values of $1/k$ the model can achieve even more better prediction with the weighted scheme in both metrics. (84.12% and 97.29% resp.). With such high hit-ratio scores this model is capable to perform as an online recommendation system and with those click-soon ratio scores as a prefetching web-page system. Comparing our experimental results with those of [14], which is one of the latest techniques competitive to ours and the latest comparison work in the topic, we can claim that our method outperforms most of them. More specifically:

- our model seems to have similar performance and sometimes slightly better in the click-soon metric (the best scores in [14] vary from 85 − 94%)
- in hit-ratio metrics our method is achieving clearly better performance (the best scores in [14] vary from 50 − 60%)

in both cases by utilizing a much smaller training set which indicates that our model has a stronger predictive capability.

The results of experiments on the other data set, the logfile of www.ceid. upatras.gr, are quite similar to the above ones and slightly better again using the same portion of information for training purposes. This comes from the fact that this website is a smaller one with much less navigational patterns that are captured better by our model. This data set consists of 27805 web access sequences while from NASA logfile they have been produced over 55000 sequences. Due to lack of space the data table of these results is not presented.

Concerning, memory utilization, the consumption of memory of the main data structure is quite reasonable and does not exceed the size of few MBytes

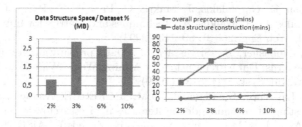

Fig. 2. Data structure space. Overall preprocessing time, data structure construction time.

while the preprocessing procedure is more demanding due to the computation of distance matrix D, whose size is quadratic to the size of the training set (see fig 2).

5 Conclusions and Open Issues

In this paper we have proposed a technique for predicting web page usage patterns by modeling users' navigation history using string processing techniques, and we have validated experimentally the superiority of our proposed technique. Future work includes different ways of modeling web user access patterns, choice of different clustering techniques, investigation of different metrics, exhaustive experimental comparisons with similar techniques and exploitation of the trade-off between the degree of the weighted suffix tree (is directly affected from the threshold), and the performance of our scheme.

Acknmowledgments. This work is partially supported by the Western Greece Operational Programme, Research and Development Cooperations in National Priority sectors, under grant no.: PEP_DEL11.

References

1. Banerjee, A., Ghosh, J.: Clickstream clustering using weighted longest common subsequences. In: Proc. of the Web Mining Workshop at the 1st SIAM Conference on Data Mining (2001)
2. Brin, S., Page, L.: The anatomy of a large-scale hypertextual Web search engine. In: Proceedings of the Seventh int. Conf. on World Wide Web, 7th edn., pp. 107 117 (1998)
3. Chakrabarti, S., van der Berg, M., Dom, B.: Focused crawling: a new approach to topic-specific Web resource discovery. In: Proceedings of 8th Int. World Wide Web Conf (1999)
4. Chen, M., LaPaugh, A.S., Singh, J.P.: Predicting category accesses for a user in a structured information space. In: Proc. of the 25th Annual int. ACM SIGIR Conf. on Research and Development in information Retrieval, ACM Press, New York (2002)

5. Christopoulou, E., Garofalakis, J., Makris, C., Panagis, Y., Psaras-Chatzigeorgiou, A., Sakkopoulos, E., Tsakalidis, A.: Techniques and Metrics for Improving Website Structure. J. Web Eng. 2(1-2), 90–114 (2003)
6. Cooley, R., Mobasher, B., Srivastava, J.: Data preparation for mining WWW browsing patterns. Data Preparation for Mining World Wide Web Browsing Patterns. Knowl. Inf. Syst. 1(1), 5–32 (1999)
7. Cosley, D., Lawrence, S., Pennock, D.M.: REFEREE: An open framework for practical testing of recommender systems using ResearchIndex. In: Bressan, S., Chaudhri, A.B., Lee, M.L., Yu, J.X., Lacroix, Z. (eds.) CAiSE 2002 and VLDB 2002. LNCS, vol. 2590, pp. 35–46. Springer, Heidelberg (2003)
8. Dean, J., Henzinger, M.R.: Finding related pages in the World Wide Web. In: Enslow, P.H. (ed.) Proceeding of the Eighth international Conference on World Wide Web (Toronto, Canada), pp. 1467–1479. Elsevier North-Holland, New York, NY (1999)
9. Dunham, M.: Data Mining Introductory and Advanced Topics. Prentice Hall, Englewood Cliffs (2003)
10. Eirinaki, M., Vazirgianis, M.: Web Mining for Web Personalization. ACM Transactions on Internet Tehnology (TOIT) 3(1) (2003)
11. Eysenck, M., Keane, M.: Cognitive Psychology. 1990. Lawrence Erlbaum Associates, Mahwah (1990)
12. Facca, F.M., Lanzi, P.L.: Mining interesting knowledge from weblogs: a survey. Data Knowl. Eng. 53(3), 225–241 (2005)
13. Gulli, A., Signorini, A.: The indexable web is more than 11.5 billion pages. In: Special interest Tracks and Posters of the 14th international Conference on World Wide Web, Chiba, Japan, May 10 - 14, pp. 10–14 (2005)
14. Gunduz, U., Ozsu, M.T.: A Web page prediction model based on click-stream tree representation of user behavior. In: Proceedings of the Ninth ACM SIGKDD international Conference on Knowledge Discovery and Data Mining, ACM Press, New York (2003)
15. Gusfield, D.: Algorithms on Strings, Trees, and Sequences - Computer Science and Computational Biology. Cambridge University Press, Cambridge (1997)
16. Guttman, A.: R-trees: a dynamic index structure for spatial searching. In: Proceedings of the 1984 ACM SIGMOD international Conference on Management of Data, ACM Press, New York (1984)
17. Hay, B., Wets, G., Vanhoof, K.: Clustering navigation patterns on a website using a sequence alignment method. In: Intelligent Techniques for Web Personalization: IJCAI 2001. 17th Int. Joint Conf. on Artificial Intelligence, Seattle, WA, USA, August 4, 2001, pp. 1–6 (2001)
18. Iliopoulos, C.S., Makris, C., Panagis, Y., Perdikuri, K., Theodoridis, E., Tsakalidis, A.: The weighted suffix tree: an efficient data structure for handling molecular weighted sequences and its applications. Fundamenta Informaticae 71(2-3), 259–277 (2006)
19. Joshi, K.P., Joshi, A., Yesha, Y.: On using a warehouse to analyze web logs. Distributed and Parallel Databases 13(2), 161–180 (2003)
20. Nanopoulos, A., Katsaros, D., Manolopoulos, Y.: Exploiting web log mining for web cache enhancement. In: Kohavi, R., Masand, B., Spiliopoulou, M., Srivastava, J. (eds.) WEBKDD 2001 - Mining Web Log Data Across All Customers Touch Points. LNCS (LNAI), vol. 2356, Springer, Heidelberg (2002)
21. Needleman, S., Wunsch, C.: A general method applicable to the search for similarities in the amino acid sequence of two proteins. J. Mol. Biol. 48(3), 443–453

22. Padmanabhan, V.N., Mogul, J.C.: Using predictive prefetching to improve World Wide Web latency. SIGCOMM Comput. Commun. Rev. 26, 3 (1996)
23. Pei, J., Han, J., Mortazavi-asl, B., Zhu, H.: Mining access patterns efficiently from web logs. In: Pacific-Asia Conference on Knowledge Discovery and Data Mining, pp. 396–407 (2000)
24. Qiu, F., Cho, J.: Automatic identification of user interest for personalized search. In: WWW 2006. Proceedings of the 15th Int. Conf. on World Wide Web, pp. 727–736 (2006)
25. Ruddle, R.A.: Using String-matching to Analyze Hypertext Navigation. In: HT 2006. Proceedings of the 17th ACM Conference on Hypertext and Hypermedia, ACM Press, New York (2006)
26. Rangarajan, S.K., Phoha, V.V., Balagani, K.S., Selmic, R.R., Iyengar, S.S.: Adaptive Neural Network Clustering of Web Users. Computer 37(4), 34–40 (2004)
27. Shahabi, C., Chen, Y.-S.: Improving user profiles for e-commerce by genetic algorithms. E-Commerce and Intelligent Methods Studies in Fuzziness and Soft Computing 105(8) (2002)
28. Smith, D., Pricer, J.: Sessioninzing clickstream data. Teradataview , 8–9 (2000)
29. Spiliopoulou, M.: Web usage mining for Web site evaluation. Commun. ACM 43(8), 127–134 (2000)
30. Tasoulis, D.K., Vrahatis, M.N.: Novel Approaches to Unsupervised Clustering Through the k-Windows Algorithm. Knowledge Mining: Series Studies in Fuzziness and Soft Computing Series: Studies in Fuzziness and Soft Computing 185, 51–78 (2005)
31. Schechter, S., Krishnan, M., Smith, M.D.: Using path profiles to predict HTTP requests. In: Enslow, P.H., Ellis, A. (eds.) Proceedings of the Seventh international Conference on World Wide Web (Brisbane, Australia), pp. 457–467. Elsevier Science Publishers, B. V., Amsterdam, The Netherlands (1998)
32. Willard, D.: New data structures for orthogonal range queries. SIAM J. on Computing 14, 232–253 (1985)
33. Xiao, Y., Dunham, M.: Efficient mining of traversal patterns. Dunham. Efficient mining of traversal patterns. Data and Knowledge Engineering 39(2), 191–214 (2001)
34. Xie, Y., Phoha, V.V.: Web user clustering from access log using belief function. In: Proc. of the First Inter. Conf. on Knowledge Capture (K-CAP 2001), pp. 202–208. ACM Press, New York (2001)
35. Ypma, A., Heskes, T.: Clustering web surfers with mixtures of hidden markov models. In: Proceedings of the 14th Belgian-Dutch Conference on AI (2002)
36. Zaiane, O. R.: Resource and Knowledge Discovery from the Internet and Multimedia Repositories. Phd Thesis (1999)
37. Zhao, Q., Bhowmick, S.S., Gruenwald, L.: WAM-Miner: in the search of web access motifs from historical web log data. In: Proc. of the 14th ACM CIKM '05, pp. 421–428. ACM Press, New York (2005)
38. Follow, version 1.5. http://www.mnot.net/follow/

Enhancing Educational-Material Retrieval Using Authored-Lesson Metadata

Olivier Motelet[1], Benjamin Piwowarski[2], Georges Dupret[2], Jose A. Pino[1], and Nelson Baloian[1]

[1] DCC - Universidad de Chile
[2] Yahoo! Research Latin America

Abstract. Many authors believe that in order to achieve coherence and flexibility at the same time in multimedia-based learning units, it is highly recommendable to structure the different components as a graph. In a lesson graph, educational resources are encapsulated into learning objects (LO) along with their respective metadata and are interconnected through different kind of rhetorical and semantical relationships. The LOs of these graphs are stored within repositories, where their metadata are used to ease their retrieval. In this paper we propose to integrate the processes of searching LOs and editing the lesson graph. This new framework extends traditional keyword and metadata search to take advantage of the information stored implicitly in the lesson graph structure, making LOs retrieval more effective and the expression of queries more intuitive. The retrieval of the learning material consists of two processes: (1) The user first defines the topological location of a required piece of educational material within the lesson graph, this is, its relationships with other pieces. (2) Then, the user issues a traditional keyword query, which is processed by an IR system modified to take the graph structure into account. Experiments show the advantages of this approach.

1 Introduction

The last ten years have witnessed the emergence of various repositories dedicated to store and share educational resources under the form of learning objects, or LO in short. Although LOs have different definitions in the literature, this article considers a LO as a piece of educational material (a slide, a web page, a simulation, etc.) associated with some metadata. The main goal of LOs is to enable the re-usability of learning/teaching material. International organizations have worked towards supporting that purpose by defining characterization and interoperability standards for LOs [1].

One of the main ongoing efforts in this direction is the specification of the Learning Object Metadata (LOM) standard for the metadata characterizing a LO [2]. Unlike most standards that mainly describe physical attributes of the digital resources, LOM offers a large set of educational attributes designed to help teachers and learners to retrieve existing educational material and activities.

In practice, however, most users issue simple keyword queries and neglect the use of the metadata in spite of the difficulties they face while retrieving the

N. Ziviani and R. Baeza-Yates (Eds.): SPIRE 2007, LNCS 4726, pp. 254–263, 2007.

material they expect to obtain [3]. In this paper, we present a user-oriented way of taking advantage of the metadata: The user tells the system the location in the lesson graph where to find the LO he needs and what type of relations this object should hold with its neighbors. We then make use of this information to improve retrieval.

The graph representation is an intuitive way of visualizing the course structure and is amenable to last minute modifications during the lecture delivery itself. Many course authoring tools designed to create adaptive and flexible lesson units use a graph for the course structure [4]. Typically, a teacher starts the authoring of a course by building a graph of LOs connected by semantic and/or rhetoric relationships and fills the nodes with teaching material. The new search paradigm for learning material we propose comes into play naturally when the teacher introduces a new, empty node, i.e., not yet referring concrete material, into the graph and links it to other nodes. These nodes, the links and the associated metadata provide a *context* to the search that we use to improve retrieval. The main challenge we face is the transformation of the information contained implicitly in the graph into a form that can be used to search the LO database.

The following section introduces the notion of lesson graph based on LOs. Next querying a LO repository from within a lesson graph is presented. Section 4 describes our method for taking advantage of the lesson graph associated with the query in order to enhance LO retrieval. The results of experiments in a small-scale but realistic setting are reported and support our approach: We show in the experiment section how our method improves over simple keyword queries. Finally, related work is reviewed and conclusions are obtained.

2   Authoring a Course as a Graph

Metadata for Learning Objects (LOM) has two purposes, to describe the LOs and to interconnect LOs. The LOM specification concerns about 60 metadata attributes describing mostly technical, educational and general aspects of educational resources. Attributes are identified by a series of names separated by slashes, e.g., general/title, where "general" is the category and "title" the attribute name. Attributes can be classified in three sets: (1) Predefined vocabulary values (e.g., *easy* and *difficult* are vocabulary values for the educational/difficulty attribute). The specification proposes values for the vocabulary but this is generally tailored towards the needs of particular teaching communities. (2) Free text that can be associated with a given language. Several texts corresponding to several languages can be proposed. (3) Primitive types, such as identifier, date, time, or integer. Most attributes have a value defined on a set, e.g., a set of strings for the general/keywords attribute.

Links between LOs are defined by a special kind of attribute called a relation. Links are typed, such as *introducesTo* or *exemplifiedBy*. The set of links defines the structure of the graph. Figure 1 illustrates such a LO graph where six LOs, labeled from L1 to L6, describe a part of a programming course for an object

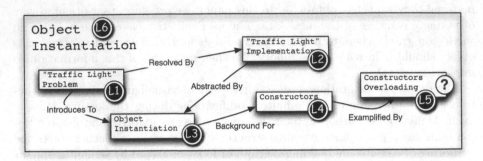

Fig. 1. Start of a lesson graph about "object instantiation"

oriented language. L1 (problem definition) and L2 (Java code) introduce the problem of how to organize a crossroad with two traffic lights. The topic of this exercise is object instantiation in a program. L3 and L4 refer to slides defining respectively object instantiation and the concept of constructors. L5 is an example of a query within a LO graph and will be described in the next section. L6 is a LO of coarser granularity and acts as a container for L1 to L5. Practically, lesson graphs such as the one presented in Figure 1 can be authored with LessonMapper2, a graph authoring tool of LOs characterized with LOM [5].

Several authors argue that the LOM relation types specification is insufficient for lesson authoring and needs to be extended [6,7]. We focus on the work of Trigg [8], which defines an extensive set of relations for supporting narration. According to our institution teachers' needs, we empirically selected a subset of these relations, emphasizing the semantical, rhetorical and organizational aspects of course authoring: *introducesTo, assessedBy, supportedBy, abstractedBy, exemplifiedBy, comparableWith, backgroundFor, summarizedBy, explainedBy, resolvedBy, refutedBy, isPartOf.* Each of these relations has an opposite: a relation from a LO *a* to another LO *b* implies an inverse relation from *b* to *a*. It is important to note that the suggested relations do not apply to all teaching contexts, but are easily complemented to suit other situations.

3   Querying a Repository from Inside the Lesson Graph

Standard querying is done by searching matching metadata and keywords. To describe L5, we could for example choose "constructors overloading" and retrieve the LOs with metadata related to it. This is the approach taken by the Lucene search engine [9] that we use in our experiments (Section 5).

Querying a LO repository can be done using a purely graphical approach. For instance, consider node L5 of Figure 1. This node is not associated with an existing document, instead it is a **query node**: A query reflecting the need for a learning object with certain characteristics is thus expressed as a position in the lesson graph. In Figure 1, the LOs satisfying the query node L5 are examples of the concepts introduced by L4.

Since authoring a lesson consists of adding nodes to the lesson graph, it is natural to integrate this task with retrieval and to use the implicit information associated with the new node position. This can be seen as associating a context extracted from the graph to the traditional term or metadata search. In this work, we will modify the ranking of retrieved objects as produced by a Lucene search process with the results of classifiers designed to take the neighbor graph into account.

4   Using the Lesson Graph to Evaluate Potential Results

Motelet et al. [10] propose a system that takes advantage of the semantics of a lesson graph in order to infer information about the LOM semantics of the LOs of a lesson graph. Unlike other models (e.g., [11]), this system suggests metadata where they are missing using an exhaustive-diffusion process of the node characteristics along the graph edges. This characteristic is particularly interesting in the context of the lesson graph authoring process where metadata are typically incomplete. Two types of metadata information are generated: value suggestions and value restrictions. We propose to generate this additional information for the query nodes and to use it when querying a LO repository from within a lesson graph.

This section first describes the use of value restrictions and value suggestions as classifiers. Then the combination of the generated classifiers with a machine learning algorithm is discussed.

4.1   Using Value Restrictions as Classifiers

Value restrictions are deduced from graph consistency analysis. In Figure 1 for instance, we expect L1 to be simpler than L2 because it introduces it. While the difficulty of a LO is a relative notion that may be hard to standardize, it is possible to compare difficulty levels of two LOs inside the same lesson graph. In terms of value restrictions, it means that the value of the LOM attribute educational/difficulty of L1 should be as low or lower than the educational/difficulty of L3. If L1 introduces more than one resource, its level of educational/difficulty should be compatible with each element it introduces. Rules about graph consistency may be defined for any LOM attribute and any relation as far as it is meaningful. In [10], the authors propose that such an assumption about graph consistency be tailored to suit other teacher requirements if necessary.

When searching a repository, the LOs that comply with the value restrictions associated with the query should be promoted. To implement this idea, a score is computed for each LO as $\frac{\#CompliedRestrictions}{\#Restrictions}$. If there are no generated restrictions for a certain LOM attribute, the scores are set to 1.

4.2   Using Value Suggestions as Classifiers

Value suggestions are generated based on a set of assumptions about attribute similarities between related LOs. In our example, since L1 *introduces* L3 we expect that both share a similar value for the attribute general/keyword. We evaluate this similarity based on probabilities observed in the repository over the same attributes. For instance, analyzing the repository on which we conducted the experiments of Section 5, we observe that a keyword of the attribute general/keyword of a LO has 54% chance to appear in the same attribute field of the LOs it *introduces*.

Formally, if we define V_{att} as the set of possible values for an attribute *att*, a value suggestion for a LOM attribute of a given LO is a set of weighted values $\{(v, w(v)) : v \in V_{att}\}$ where $w(v)$ is the probability that the v value suits the LO. We can compute suggestions for elements in the repository in the same way as we compute them for queries based on the neighboring nodes. To estimate the similarity between a query node q and a node e from the repository, we need to measure how similar the nodes' suggestions are. We propose to adapt the traditional cosine measure for this task:

$$sim_{att}(q, e) = \frac{\sum_v w_q(v) \times w_e(v)}{\sqrt{\sum_v w_q(v)^2} \times \sqrt{\sum_v w_e(v)^2}}$$

where v scans the values of *att* and $w_q(v)$ and $w_e(v)$ are the weights associated with v in the suggestions of q and e, respectively. This measure quantifies the intuition that the value suggestions represent the relation of the node q and e with their neighbors in their respective graphs. In other words, these values summarize the context of the nodes.

Note that the original attribute values from node e are discarded and only the suggestions extracted from its neighbors are considered for evaluating the cosine measure. There are two reasons for this. First, because we are attempting to match the lesson graph and the repository graph, it is natural to simulate what would be the state of e if it were the currently introduced node. Setting e as a new node in its environment with the same kind of information as the teacher provides for q has the effect of making the two associated LOs representations more homogeneous, and improves retrieval as shown by our numerical experiments. The second reason for discarding the original attribute values of the node e is more pragmatic and is related to the keyword search: Lucene indexes the LOs in the repository according to the original attributes and thus there is an information redundancy if we re-introduce these values in the classifiers. Instead, using only the value suggestions originating from the node neighbors, the context is better taken into account. To distinguish the case where we discard the original attribute values from the case where they are preserved, we refer to the first as **context-only diffusion** and the second as **full diffusion**. Section 5.2 shows the benefit of context-only diffusion over full diffusion.

4.3 Combining Classifiers

As explained above, graph consistency and context similarity can be used to evaluate the relevance of repository LOs for each metadata attribute. In our implementation, we suppose the attributes are independent and have a total of 17 classifiers based on value restrictions and 17 classifiers based on the value suggestions. These classifiers are called **graph-based classifiers**. Combination of these classifiers is done by RankBoost [12].

RankBoost is a machine learning algorithm that searches for an optimal combination of several weak or uncertain classifiers. In preliminary experiments not reported in this paper, we first evaluated separately each graph-based classifier. As the Lucene and graph-based classifiers operate on almost distinct variable sets, we opted for a series of new classifiers whose score is the score of one graph-based classifier multiplied by the Lucene score. We call them **mixed classifiers**. Each of these classifiers orders repository LOs: The corresponding ranks are the input to the RankBoost algorithm.

Thanks to the fact that the relevance can be considered binary, we used the simplest version of the RankBoost algorithm: At each iteration of the algorithm, a base classifier is chosen, along with a threshold rank and a weight α. The result of learning is a set of step functions f_i, one for each base classifier. The final score of a LO L for a query q is given by $\sum_i f_i(r_i(q, \mathsf{L}))$ where $r_i(q, \mathsf{L})$ is the rank of L according to the ith classifier and f_i is the function learned by RankBoost for this classifier. We furthermore required that each function f_i is decreasing with respect to the rank r_i, in order to avoid over-fitting (as suggested in [12]).

5   Experiments

5.1   Description

A learning object repository was implemented within our institution and populated with 170 learning objects about a single topic: An introductory Java course. This repository contains fine grained LOs, each corresponding to teaching/learning material for about 5 minutes. In contrast with the available repositories, relation semantics linking repository LOs are based on the proposal made in Section 2 (see [13] for getting access to a repository snapshot). Eleven teachers of Java Programming, not involved in the project presented in this article, were asked to complete a lesson graph about *object instantiation and method call*. We informed them that the lesson should tackle the following topics: *constructor, new, method call, '.' , constructors with arguments, method overloading, delegation, object design, separation of concern*. They were also informed that topic ordering was flexible. Topics were purposely defined for various granularities and without apparent coherency so that each teacher felt free to create a new course.

The lesson was presented as a graph built in LessonMapper2. The tool was previously introduced to the teachers, along with examples of lesson graphs and available relation types. Each teacher was confronted with 4 different situations: (1) A graph with one LO, (2) a graph with 5 LOs (including the first graph), (3) a graph with 9 LOs (including the second), and (4) a graph with 13 resources (including the third). The content of the presented graphs were not previously known to teachers: The presented LOs were original and not similar to the teachers' own courses. The repository did not contain any LOs used in the presented graphs.

The proposed graphs were purposely incomplete in order to motivate the teacher to complete them. For each situation, teachers were asked to complete the corresponding lesson graph with 2 new LOs of their choice. The teachers had to thoroughly describe the required LOs so that the interviewer could identify which repository LO matched the teacher's intent. The matching LOs were not communicated to the teachers. Instead of that, they were asked to search for them in the repository using common **keyword queries** and locating the expected material inside the graphs, i.e., by defining **query nodes**. Teachers knew that keyword queries were used to search the metadata of the available LOs of the repository and not their content. Eventually, query terms referring to some special vocabulary values were defined in natural language and then replaced by the interviewer with the proper vocabulary value.

The four situations gave rise to respectively 23, 21, 22 and 22 test cases, including keyword query and position in lesson graph along with the relevant results (see [13] for getting access to the test cases). In the first situation, one teacher formulated 3 queries instead of two. In the second situation, one query had no answer in the repository and was subsequently ignored.

5.2 Result Analysis

Using the queries (composed of a graph with a query node and of a set of keywords), we evaluated three different systems: First, we used only keywords and the Lucene IR system. Lucene indexed the metadata of the LOs contained in the repository. During the preprocessing, metadata values were stemmed and tags were removed. Only metadata were indexed since most of the educational

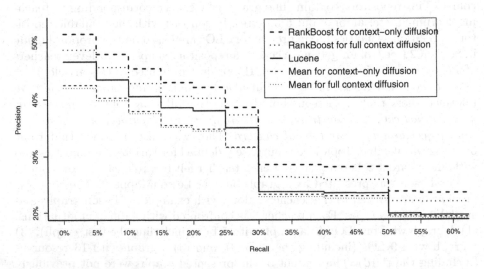

Fig. 2. Precision recall results of mixed classifiers combined with RankBoost and Lucene. We also plotted the average precision of the mixed classifiers for both full and context-only diffusions.

resources of the used repository were multimedia documents with proprietary format where relevant information is difficult to access. Similarly, keyword queries are stemmed.

In the next step, we used the mixed classifiers combined with RankBoost and each of the two diffusion mechanisms in turn (full diffusion and context-only diffusion) described in Section 4. We trained and tested RankBoost using a 4-fold cross-validation. Each fold corresponded to one of the different situations of the experiment. When testing one fold, the data of the three others were used for training.

Figure 2 shows the standard precision-recall [14] curves for the data of the experience described above. Lucene outperforms the mean of the mixed classifiers taken individually. Nevertheless, the RankBoost combination of the mixed classifiers outperforms significantly Lucene alone, especially when graph-based classifiers use the context-only diffusion algorithm. Table 1 summarizes tests confirming the statistical significance of this difference.

Table 1. T-tests on precision differences between RankBoost (context-only diffusion) and Lucene

Recall interval	Mean	95% confidence interval	p-value
0% − 25%	0.05226301	0.01948787 − 0.08503815	0.002109
0% − 50%	0.05020075	0.02076609 − 0.07963540	0.001053
0% − 75%	0.04613934	0.01865014 − 0.07362853	0.001251
0% − 100%	0.04106811	0.01429855 − 0.06783768	0.003039

To gain insight on these results, we studied the mean proportional differences between the RankBoost-based evaluation and Lucene alone using 100 samples taken randomly with repetition into the 88 test cases (Bootstrap sampling). The Rankboost combination of the mixed classifiers using context-only diffusion presents a constantly positive gain of over 10% over Lucene alone.

Analyzing the way the various classifiers are combined, it is possible to identify which LOM attributes contribute to enhance retrieval: We observe that the `interactivityLevel`, `interactivityType` and `semanticDensity` classifiers have been singled out by Rankboost as operating on the most helpful attributes.

We could not experiment with other publicly available repositories because they lack semantically rich relations between LOs. Therefore, future work will consist in developing a bigger and more heterogeneous repository in order to prove that our approach is scalable.

6   Discussion

This paper described how a graph structure can be used as an additional source of evidence when searching in LO repositories during lesson authoring. We show that a query can be defined topologically, i.e., according to the location of the node within the lesson graph and semantically by a set of keywords. The keyword search is processed by a standard IR system (in our case Lucene) while the topological information is used to compare the graph surrounding the query node with the surrounding graphs of each LO in the repository. Information from both sources is then combined in order to enhance the retrieval of LOs to be reused by lesson authors. The idea of defining a query as a graph is also present in approaches related to concept maps [15,16] but those works ignore the semantics of the links between concept nodes and characteristics other than node titles. Experiments show that our system makes successful use of this information even though we focused on a reduced teacher community and a repository about a very specific topic.

Most existing works attempting to enhance LO retrieval are based on user profiles [17,18]. In contrast, our approach requires the teacher to author her lesson as a graph of LOs but does not need any personal or historical data. Nevertheless, generating a lesson as a graph imposes on the teacher the tedious task of generating the metadata [19]. Consequently, various research efforts focus on automatic and semi-automatic generation of such metadata ([20,21,22]). Such support systems are necessary to make the authoring of a lesson as a graph of LOs a reasonable requirement in the context of a teacher community trying to systematically share teaching/learning material among its members.

The difficult use of metadata makes the extension of a query with contextual information extracted from the graph –like we propose in this work– a promising approach. It integrates nicely and intuitively into the lesson authoring process. One can even argue that the necessity of locating the LOs being edited or searched helps the teacher to clarify her aims. Our interaction with the teachers that participated in the experiments tends to confirm this hypothesis. Moreover, while for now metadata are exclusively used when the LO is retrieved, our approach directly takes advantage of them when they are created because they help retrieval, generating a positive insentive for lesson authors to produce high quality metadata.

References

1. IMS: IMS global learning consortium (2007), http://www.imsglobal.org
2. LOM: IEEE LTSC p1484.12.1 learning object metadata specification final draft (2002), http://ieeeltsc.org/wg12LOM/
3. Najjar, J., Klerkx, J., Vuoikari, R., Duval, E.: Finding appropriate learning objects: An empirical evaluation. In: World Conference on Educational Multimedia, Hypermedia and Telecommunications ED-MEDIA 2005, Montreal, Canada, pp. 1407–1414 (2005)
4. McCalla, G.: The search for adaptability, flexibility, and individualization: Approaches to curriculum in intelligent tutoring systems. In: Foundations and Frontiers of Adaptive Learning Environments, pp. 91–122. Springer, Heidelberg (1992)
5. LessonMapper2: Web site (2007), http://www.dcc.uchile.cl/~omotelet/LessonMapper2.xhtml

6. Engelhardt, M., Hildebrand, A., Lange, D., Schmidt, T.C.: Reasoning about elearning multimedia objects. In: SWAMM. Proc. of WWW 2006, Intern. Workshop on Semantic Web Annotations for Multimedia (2006)
7. Fischer, S.: Course and exercise sequencing using metadata in adaptive hypermedia learning systems. J. Educ. Resour. Comput. 1, 5 (2001)
8. Trigg, R.: A Network-Based Approach to Text Handling for the Online Scientific Community. PhD thesis, University of Maryland (1983)
9. Lucene: Full-featured text search engine library in java (2007), http://lucene.apache.org
10. Motelet, O., Baloian, N., Piwowarsky, B., Pino, J.A.: Taking advantage of the semantics of a lesson graph based on learning objects. In: The 13th International Conference on Artificial Intelligence in Education (AIED 2007), IOS Press, Amsterdam (2007) (to be published)
11. Hatala, M., Richards, G.: Value-added metatagging: Ontology and rule based methods for smarter metadata. In: Schroeder, M., Wagner, G. (eds.) RuleML 2003. LNCS, vol. 2876, pp. 65–80. Springer, Heidelberg (2003)
12. Freund, Y., Iyer, R.D., Schapire, R.E., Singer, Y.: An efficient boosting algorithm for combining preferences. Journal of Machine Learning Research 4, 933–969 (2003)
13. Motelet, O.: Experiment data (repository snapshot and test cases) (2007), http://reflex.dcc.uchile.cl/lm/lessonMapper2/IRTests.zip
14. Baeza-Yates, R., Ribeiro-Neto, B.: Modern Information Retrieval. Addison-Wesley, Reading (1999)
15. Carvalho, M., Hewett, R., Canas, A.: Enhancing web searches from concept map-based knowledge models. In: Proceedings of the SCI Conference, AAAI Press, Stanford, California, USA (2001)
16. Leake, D., Maguitman, A., Reichherzer, T., Canas, A.J., Carvalho, M., Arguedas, M., Eskridge, T.: "googling" from a concept map: Towards automatic concept-map-based query formation. In: Canas, A.J., Novak, J.D., Gonzalez, F.M. (eds.) Concept Maps: Theory, Methodology, Technology Proc. of the First Int. Conference on Concept Mapping (2004)
17. Murray, T.: Authoring intelligent tutoring systems: an analysis of state of the art. International Journal of Artificial Intelligence in Education 10, 98–129 (1999)
18. Rafaeli, S., Dan-Gur, Y., Barak, M.: Social recommender systems: Recommendations in support of e-learning. J. of Dist. Educ. Tech. 3, 29–45 (2005)
19. Recker, M.: Perspectives on teachers as digital library users. D-Lib Magazine 12 (2006)
20. Ochoa, X., Cardinaels, K., Meire, M., Duval, E.: Frameworks for the automatic indexation of learning management systems content into learning object repositories. In: World Conference on Educational Multimedia, Hypermedia and Telecommunications ED- MEDIA, pp. 1407–1414 (2005)
21. Saini, P.S., Ronchetti, M., Sona, D.: Automatic generation of metadata for learning objects. [23] 275–279
22. Motelet, O., Baloian, N.A.: Hybrid system for generating learning object metadata. [23] 563–567
23. Kinshuk, Koper, R. (eds.): ICALT 2006. Proceedings of the 6th IEEE International Conference on Advanced Learning Technologies, Kerkrade, The Netherlands, 5-7 July 2006. IEEE Computer Society Press, Los Alamitos (2006)

Approximate String Matching
with Lempel-Ziv Compressed Indexes

Luís M.S. Russo[1,*], Gonzalo Navarro[2,**], and Arlindo L. Oliveira[1]

[1] INESC-ID, R. Alves Redol 9, 1000 Lisboa, Portugal
lsr@algos.inesc-id.pt, aml@inesc-id.pt
[2] Dept. of Computer Science, University of Chile
gnavarro@dcc.uchile.cl

Abstract. A compressed full-text self-index for a text T is a data structure requiring reduced space and able of searching for patterns P in T. Furthermore, the structure can reproduce any substring of T, thus it actually replaces T. Despite the explosion of interest on self-indexes in recent years, there has not been much progress on search functionalities beyond the basic exact search. In this paper we focus on indexed approximate string matching (ASM), which is of great interest, say, in computational biology applications. We present an ASM algorithm that works on top of a Lempel-Ziv self-index. We consider the so-called hybrid indexes, which are the best in practice for this problem. We show that a Lempel-Ziv index can be seen as an extension of the classical q-samples index. We give new insights on this type of index, which can be of independent interest, and then apply them to the Lempel-Ziv index. We show experimentally that our algorithm has a competitive performance and provides a useful space-time tradeoff compared to classical indexes.

1 Introduction and Related Work

Approximate string matching (ASM) is an important problem that arises in applications related to text searching, pattern recognition, signal processing, and computational biology, to name a few. It consists in locating all the occurrences of a given pattern string $P[0, m-1]$ in a larger text string $T[0, u-1]$, letting the occurrences be at distance $ed()$ at most k from P. In this paper we focus on edit distance, that is, the minimum number of character insertions, deletions, and substitutions of single characters to convert one string into the other.

The classical sequential search solution runs in $O(um)$ worst-case time (see [1]). An optimal average-case algorithm requires time $O(u(k+\log_\sigma m)/m)$ [2,3], where σ is the size of the alphabet Σ. Those good average-case algorithms are called *filtration* algorithms: they traverse the text fast while checking for a simple necessary condition, and only when this holds they verify the text area using a

---

* Supported by the Portuguese Science and Technology Foundation by grant SFRH/BD/12101/2003 and Project DBYEAST POSC/EIA/57398/2004.
** Supported in part by Fondecyt Grant 1-050493 (Chile).

N. Ziviani and R. Baeza-Yates (Eds.): SPIRE 2007, LNCS 4726, pp. 264–275, 2007.
© Springer-Verlag Berlin Heidelberg 2007

classical ASM algorithm. For long texts, however, sequential searching might be impractical because it must scan all the text. To avoid this we use an *index* [4].

There exist indexes specifically devoted to ASM, e.g. [5,6,7,8], but these are oriented to worst-case performance. There seems to exist an unbreakable space-time barrier with indexed ASM: Either one obtains exponential times (on m or k), or one obtains exponential index space (e.g. $O(u \log^k u)$). Another trend is to reuse an index designed for exact searching, all of which are linear-space, and try to do ASM over it. Indexes such as suffix trees [9], suffix arrays [10], or based on so-called q-grams or q-samples, have been used. There exist several algorithms, based on suffix trees or arrays, which focus on worst-case performance [11,12,13]. Given the mentioned time-space barrier, they achieve a search time independent of u but exponential on m or k. Essentially, they simulate the sequential search over all the possible text suffixes, taking advantage of the fact that similar substrings are factored out in suffix trees or arrays.

Indexes based on q-grams (indexing all text substrings of length q) or q-samples (indexing non-overlapping text substrings of length q) are appealing because they require less space than suffix trees or arrays. The algorithms on those indexes do not offer worst-case guarantees, but perform well on average when the *error level* $\alpha = k/m$ is low enough, say $O(1/\log_\sigma u)$. Those indexes basically simulate an on-line filtration algorithm, such that the "necessary condition" checked involves exact matching of pattern substrings, and as such can be verified with any exact-searching index. Such filtration indexes, e.g. [14,15], cease to be useful for moderate k values, which are still of interest in applications.

The most successful approach, in practice, is in between the two techniques described above, and is called "hybrid" indexing. The index determines the text positions to verify using an approximate-matching condition instead of an exact one. This requires a search of the first kind (whose time is exponential on the length of the string or the number of errors). Yet, these searches are done over short strings and allowing few errors, so that the exponential cost is controlled. Indexes of this kind offer average-case guarantees of the form $O(mn^\lambda)$ for some $0 < \lambda < 1$, and work well for higher error levels. They have been implemented over q-gram indexes [16], suffix arrays [17], and q-sample indexes [18].

Yet, many of those linear-space indexes are very large anyway. For example, suffix arrays require 4 times the text size and suffix trees require at the very least 10 times [19]. In recent years a new and extremely successful class of indexes has emerged. *Compressed full-text indexes* use data compression techniques to produce less space-demanding data structures [20,21,22,23,24]. It turns out that data compression algorithms exploit the internal structure of a string much in the same way indexes do, and therefore it is possible to build a compressed index that takes space proportional to that of the compressed text, gives indexed searching, and replaces the text as it can reproduce any text substring (in which case they are called *self-indexes*). The size of those indexes is measured in terms of the empirical text entropy, H_k [25], which gives a lower bound on the number of bits per symbol achievable by a k-th order compressor. In this work we are interested in indexes based on Lempel-Ziv compression [21,22,26,27,28].

Despite the great success of self-indexes, they have been mainly used for exact searching. Only very recently some indexes taking $O(u)$ or $O(u\sqrt{\log u})$ *bits* have appeared [29,30,7]. Yet, those are again of the worst-case type, and thus all their times are exponential on k. In this paper we present a practical algorithm that runs on a compressed self-index and belongs to the most successful class of hybrid algorithms.

2 Our Contribution in Context

One can easily use *any* compressed self-index to implement a filtration ASM method that relies on looking for exact occurrences of pattern substrings, as this is what all self-indexes provide. Indeed, this has been already attempted [31] using the FM-index [21] and a Lempel-Ziv index [22]. The Lempel-Ziv index worked better because it is faster to extract the text to verify (recall that in self-indexes the text is not directly available). The specific structure of the Lempel-Ziv index used allowed several interesting optimizations (such as factoring out the work of several text extractions) that we will not discuss further here.

Lempel-Ziv indexes split the text into a sequence of so-called *phrases* of varying length. They are efficient to find the (exact) occurrences that lie within phrases, but those that span two or more phrases are costlier.

Our goal in this paper is to provide efficient approximate searching over a small and practical self-index. Based on the described previous experiences, (1) we want an algorithm of the hybrid type, which implies that the self-index should do approximate search for pattern pieces; (2) we want a Lempel-Ziv-based index, so that the extraction of text to verify is fast; (3) we wish to avoid the problems derived from pieces spanning several Lempel-Ziv phrases. We will focus on an index [28] whose suffix-tree-like structure is useful for this approximate searching.

Mimicking q-sample indexes is particularly useful for our goals. Consider that the text is partitioned into contiguous q-samples. Any occurrence O of P is of length at least $m - k$. Wherever an occurrence lies, it must contain at least $j = \lfloor (m - k - q + 1)/q \rfloor$ complete q-samples. The following lemma, simplified from [4], gives the connection to use approximate searching for pattern substrings with a q-samples index [18].

Lemma 1. *Let A and B be strings such that $ed(A, B) \le k$. Let $A = A_1 A_2 \ldots A_j$, for strings A_i and for any $j \ge 1$. Then there is a substring B' of B and an i such that $ed(B', A_i) \le \lfloor k/j \rfloor$.*

Therefore, if we assume that $B = P$ and A is contained in O, we index all the different text q-samples into, say, a trie data structure. Then the trie is traversed to find q-samples that match within P with at most $\lfloor k/j \rfloor$ errors. All the contexts around all occurrences of the matching q-samples are examined for full occurrences of P. Note in passing that we could also take $A = P$ and B contained in O, in which case we choose how to partition P but we must be able to find any text substring with the index (exactly [15] or approximately [16,17], depending on j). Thus we must use a suffix tree or array [17], or even a q-gram index if we never use pieces of P longer than q [15,16].

A Lempel-Ziv parsing can be regarded as an irregular sampling of the text, and therefore our goal in principle is to adapt the techniques of [18] to an irregular parsing (thus we must stick to the interpretation $B = P$). As desired, we would not need to consider occurrences spanning more than one phrase. Moreover, the trie of phrases stored by all Lempel-Ziv self-indexes is the exact analogous of the trie of q-samples, thus we could search without requiring further structures.

The irregular parsing poses several challenges. There is no way to ensure that there will be a minimum number j of phrases contained in an occurrence. Occurrences could even be fully contained in a phrase!

We develop several tools to face those challenges. (1) We give a new variant of Lemma 1 that distributes the errors in a convenient way when the samples are of varying length. (2) We introduce a new filtration technique where the samples that overlap the occurrence (not only those contained in the occurrence) can be considered. This is of interest even for classical q-sample indexes. (3) We search for q-samples within long phrases to detect occurrences even if they are within a phrase. This technique also includes novel insights.

We implement our scheme and compare it with the best technique in practice over classical indexes [17], and with the previous developments over compressed self-indexes [31]. The experiments show that our technique provides a relevant space-time tradeoff for indexed ASM.

3   An Improved q-Samples Index

In this section we extend classical q-sample indexes by allowing samples to overlap the pattern occurrences. This is of interest by itself, and will be used for an irregular sampling index later. Remind that a q-samples index stores the locations, in T, of all the substrings $T[qi..qi + q - 1]$.

3.1   Varying the Error Distribution

We will need to consider parts of samples in the sequel, as well as samples of different lengths. Lemma 1 gives the same number of errors to all the samples, which is disadvantageous when pieces are of different lengths. The next lemma generalizes Lemma 1 to allow different numbers of errors in each piece (all proofs are in the Appendix for lack of space).

Lemma 2. *Let A and B be strings, let $A = A_1 A_2 \ldots A_j$, for strings A_i and some $j \geq 1$. Let $k_i \in \mathbb{R}$ such that $\sum_{i=1}^{j} k_i > ed(A, B)$. Then there is a substring B' of B and an i such that $ed(A_i, B') < k_i$.*

Lemma 1 is a particular case of Lemma 2: set $k_i = k/j + \epsilon$ for sufficiently small $\epsilon > 0$. Our lemma reminds Lemma 2 of [4], and they can be proved to be equivalent. The current formulation is more advantageous for us because one does not need to know j. It can be used to adapt the error levels to the length of the pieces. For example, to try to maintain a constant error level, take $k_i = (1 + \epsilon) \, k \cdot |A_i|/|A|$ with $\epsilon > 0$.

3.2 Partial q-Sample Matching

Contrary to all previous work, let us assume that A in Lemma 2 is not only that part of an approximate occurrence O formed by full q-samples, but instead that $A = O$, so that A_1 is the suffix of a sample and A_j is the prefix of a sample. An advantage of this is that now the number of involved q-samples is at least $j = \lceil (m - k)/q \rceil$, and therefore we can permit fewer errors per piece (e.g. $\lfloor k/j \rfloor$ using Lemma 1). On the other hand, we would like to allow fewer errors for the pieces A_1 and A_j. Yet, notice that any text q-sample can participate as A_1, A_j, or as a fully contained q-sample in different occurrences at different text positions. Lemma 2 tells us that we could allow $k_i = (1+\epsilon)\, k \cdot |A_i|/|A|$ errors for A_i, for any $\epsilon > 0$. Conservatively, this is $k_i = (1+\epsilon)\, k \cdot q/(m - k)$ for $1 < i < j$, and less for the extremes.

In order to adapt the trie searching technique to those partial q-samples, we should not only search all the text q-samples with $(1+\epsilon)\, k \cdot q/(m-k)$, but also all their prefixes and suffixes with fewer errors. This includes, for example, verifying all the q-samples whose first or last character appears in P (cases $|A_1| = 1$ and $|A_j| = 1$). This is unaffordable. Our approach will be to redistribute the errors across A using Lemma 2 in a different way to ensure that only sufficiently long q-sample prefixes and suffixes are considered.

Let v be a non-negative integer parameter. We associate to every letter of A a weight: the first and last v letters have weight 0 and the remaining letters have weight $(1+\epsilon)/(|A| - 2v)$. We define $|A_i|_v$ as the sum of the weights of the letters of A_i. For example if A_i is within the first v letters of A then $|A_i|_v = 0$; if it does not contain any of the first or last v letters then $|A_i|_v = (1 + \epsilon)\, |A_i|/(|A| - 2v)$.

We can now apply Lemma 2 with $k_i = k \cdot |A_i|_v$ provided that $k > 0$. Note that $\sum_{i=1}^{j} k_i = (1 + \epsilon)\, k > k$. In this case, if $|A_1| \le v$ we have that $k_1 = 0$ and therefore A_1 can never be found with *strictly less* than zero errors. The same holds for A_j. This effectively relieves us from searching for any q-sample prefix or suffix of length at most v.

Parameter v is thus doing the job of discarding q-samples that have very little overlap with the occurrence $O = A$, and maintaining the rest. It balances between two exponential costs: one due to verifying all the occurrences of too short prefixes/suffixes, and another due to permitting too many errors when searching for the pieces in the trie. In practice tuning this parameter will have a very significant impact on performance.

3.3 A Hybrid q-Samples Index

We have explained all the ideas necessary to describe a hybrid q-samples index. The algorithm works in two steps. First we determine all the q-samples O_i for which $ed(O_i, P') < k \cdot |O_i|_v$ for some substring P' of P. In this phase we also determine the q-samples that contain a suffix O_1 for which $ed(O_1, P') < k \cdot |O_1|_v$ for some prefix P' of P (note that we do not need to consider substrings of P, just prefixes). Likewise we also determine the q-samples that contain a prefix O_j' for which $ed(O_j, P') < k \cdot |O_j|_v$ for some suffix P' of P (similar observation). The q-samples that classify are potentially contained inside an approximate occurrence of P, i.e. O_i may be a substring of a string O such that $ed(O, P) \leq k$. In order to verify whether this is the case, in the second phase we scan the text context around O_i with a sequential algorithm.

As the reader might have noticed, the problem of verifying conditions such as $ed(O_i, P') < k \cdot |O_i|_v$ is that we cannot know a priori which i does a given text q-sample correspond to. Different occurrences of the q-sample in the text could participate in different positions of an O, and even a single occurrence in T could appear in several different O's. We do not know either the size $|O|$, as it may range from $m - k$ to $m + k$.

A simple solution is as follows. Conservatively assume $|O| = m - k$. Then, search P for each different text q-sample in three roles: (1) as a q-sample contained in O, so that $|O_i| = q$, assuming pessimistically $|O_i|_v = (1+\epsilon) \min(q/(m - k - 2v), 1)$; (2) as an O_1, matching a prefix of P for each of the q-sample suffixes of lengths $v < \ell < q$, assuming $|O_1| = \ell$ and thus $|O_1|_v = (1 + \epsilon) \min((\ell - v)/(m - k - 2v), 1)$; (3) as an O_j, matching a suffix of P for each of the q-sample prefixes, similarly to case (2) (that is, $|O_j|_v = |O_1|_v$). We assume that $q < m - k$ and therefore the case of O contained inside a q-sample does not occur.

In practice, one does not search for each q-sample in isolation, but rather factors out the work due to common q-gram prefixes by backtracking over the trie and incrementally computing the dynamic programming matrix between every different q-sample and any substring of P (see [4]). We note that the trie of q-samples is appropriate for role (3), but not particularly efficient for roles (1) and (2) (finding q-samples with some specific suffix). In our application to a Lempel-Ziv index this will not be a problem because we will have also a trie of the reversed phrases (that will replace the q-grams).

4   Using a Lempel-Ziv Self-index

We now adapt our technique to the irregular parsing of phrases produced by a Lempel-Ziv-based index. Among the several alternatives [21,22,26,27,28], we will focus on the ILZI [28], yet the results can be carried over to similar indexes.

The ILZI partitions the text into phrases such that every suffix of a phrase is also a phrase (similarly to LZ78 compressors [32], where every prefix of a phrase is also a phrase). It uses two tries, one storing the phrases and another storing the reverse phrases. In addition, it stores a mapping that permits moving from one trie to the other, and it stores the compressed text as a sequence of phrase identifiers. This index [28] has been shown to require $O(uH_k)$ bits of space, and to be efficient in practice. We do not need more details for this paper.

4.1   Handling Different Lengths

As explained, the main idea is to use the phrases instead of q-samples. For this sake Lemma 2 solves the problem of distributing the errors homogeneously across phrases. However, other problems arise especially for long phrases. For example, an occurrence could be completely inside a phrase. In general, backtracking over long phrases is too costly.

We resort again to q-samples, this time within phrases. We choose two non-negative integer parameters q and $s < q$. We will look for any q-gram of P that appears with less than s errors within any phrase. All phrases spotted along this process must be verified. Still, some phrases not containing any pattern q-gram with $< s$ errors can participate in an occurrence of P (e.g. if $\lfloor (m-k-q+1)/q \rfloor \cdot s \leq k$ or if the phrase is shorter than q). Next we show that those remaining phrases have certain structure that makes them easy to find.

Lemma 3. *Let A and B be strings and q and s be integers such that $0 \leq s < q \leq |A|$ and for any substrings B' of B and A' of A with $|A'| = q$ we have that $ed(A', B') \geq s$. Then for every prefix A' of A there is a substring B' of B such that $ed(A', B') \leq ed(A, B) - s\lfloor (|A| - |A'|)/q \rfloor$.*

The lemma implies that, if a phrase is close to a substring of P, but none of its q-grams are sufficiently close to any substring of P, then the errors must be distributed uniformly along the phrase. Therefore we can check the phrase progressively (for increasing prefixes), so that the number of errors permitted grows slowly. This severely limits the necessary backtracking to find those phrases that escape from the q-gram-based search.

Parameter s permits us balancing between two search costs. If we set it low, then the q-gram-based search will be stricter and faster, but the search for the escaping phrases will be costlier. If we set it high, most of the cost will be absorbed by the q-gram search.

4.2 A Hybrid Lempel-Ziv Index

The following lemma describes the way we combine previous results to search using a Lempel-Ziv index.

Lemma 4. *Let A and B be strings such that $0 < ed(A, B) \leq k$. Let $A = A_1 A_2 \ldots A_j$, for strings A_i and some $j \geq 1$. Let q, s and v be integers such that $0 \leq s < q \leq |A|$ and $0 \leq v < |A|/2$. Then there is a substring B' of B and an i such that either:*

1. *there is a substring A' of A_i with $|A'| = q$ and $ed(A', B') < s$, or*
2. *$ed(A_i, B') < k \cdot |A_i|_v$ in which case for any prefix A' of A_i there exists a substring B'' of B' such that $ed(A', B'') < k \cdot |A_i|_v - s\lfloor (|A_i| - |A'|)/q \rfloor$.*

As before the search runs in two phases. In the first phase we find the phrases whose text context must be verified. In the second phase we verify those text contexts for an approximate occurrence of P. Lemma 4 gives the key to the first phase. We find the relevant phrases via two searches:

(1) We look for any q-gram contained in a phrase which matches within P with less than s errors. We backtrack in the trie of phrases for every $P[y_1..]$, descending in the trie and advancing y_2 in $P[y_1, y_2]$ while computing the dynamic programming matrix between the current trie node and $P[y_1, y_2]$. We look for all trie nodes at depth q that match some $P[y_1, y_2]$ with less than s errors. Since every suffix of a phrase is a phrase in the ILZI, every q-gram within any phrase can be found starting from the root of the trie of phrases. All the phrases Z that descend from each q-gram trie node found must be verified (those are the phrases that start with that q-gram). We must also spot the phrases suffixed by each such Z. Hence we map each phrase Z to the trie of reverse phrases and also verify all the descent of the reverse trie nodes. This covers case 1 of Lemma 4.

(2) We look for any phrase A_i matching a portion of P with less than $k \cdot |A_i|_v$ errors. This is done over the trie of phrases. Yet, as we go down in the trie (thus considering longer phrases), we can enforce that the number of errors found up to depth d must be less than $k \cdot |A_i|_v - s\lfloor(|A_i| - d)/q\rfloor$. This covers case 2 in Lemma 4, where the equations vary according to the roles described in Section 3.3 (that is, depending on i):

(2.1) $1 < i < j$, in which case we are considering a phrase contained inside O that is not a prefix nor a suffix. The $k \cdot |A_i|_v$ formula (both for the matching condition and the backtracking limit) can be bounded by $(1+\epsilon)$ $k \cdot \min(|A_i|/(m - k - 2v), 1)$, which depends on $|A_i|$. Since A_i may correspond to any trie node that descends from the current one, we determine a priori which $|A_i| \leq m - k$ maximizes the backtracking limit. We apply the backtracking for each $P[y_1..]$.

(2.2) $i = j$, in which case we are considering a phrase that starts by a suffix of O. Now $k \cdot |A_i|_v$ can be bounded by $(1 + \epsilon)$ $k \cdot \min((d - v)/(m - k - 2v), 1)$, yet still the limit depends on $|A_i|$ and must be maximized a priori. This time we are only interested in suffixes of P, that is, we can perform m searches with $y_2 = m$ and different y_1. If a node verifies the condition we must consider also those that descend from it, to get all the phrases that start with the same suffix.

(2.3) $i = 1$, in which case we are considering a phrase that ends in a prefix of O. This search is as case $i = j$, with similar formulas. We are only interested in prefixes of P, that is $y_1 = 0$. As the phrases are suffix-closed, we can conduct a single search for $P[0..]$ from the trie root, finding all phrase suffixes that match each prefix of P. Each such suffix node must be mapped to the reverse trie and the descent there must be included. The case $i = j = 1$ is different, as it includes the case where O is contained inside a phrase. In this case we do not require the matching trie nodes to be suffixes, but also prefixes of suffixes. That is, we include the descent of the trie nodes and map each node in that descent to the reverse trie, just as in case 1.

5   Practical Issues and Testing

We implemented a prototype to test our algorithm on the ILZI compressed index [28]. As a baseline we used efficient sequential bit-parallel algorithms (namely BPM, the bit-parallel dynamic programming matrix of Myers [33], and EXP, the exact pattern partitioning by Navarro and Baeza-Yates [34]).

For the real prototype we used a stricter backtracking than as explained in previous sections. For each pattern substring $P[y_1, y_2]$ to be matched, we computed the maximum number of errors that could occur when matching it in the text, also depending on the position $O[x_1, x_2]$ where it would be matched, and maximizing over the possible areas of O where the search would be necessary. For

Table 1. Memory peaks, in Megabytes, for the different approaches when $k = 6$

	ILZI	Hybrid	LZI	DLZI	FMIndex
English	55	257	145	178	131
DNA	45	252	125	158	127
Proteins	105	366	217	228	165

example, the extremes of P can be matched with fewer errors than the middle. This process involves precomputing tables that depend on m and k. We omit the details for lack of space.

We also included in the comparison an implementation of a filtration index using the simple approach of Lemma 1 with $A = P$ and $B = O$, as briefly described in the beginning of Section 2 [31]. The indexes used in that implementation are the LZ-index [22] (LZI) and Navarro's implementation of the FM-index [21]. We also compare an improved variant over the LZ-index (DLZI [31]). Note that the FM-Index does not divide the text into blocks, however it takes longer to locate occurrences.

The machine was a Pentium 4, 3.2 GHz, 1 MB L2 cache, 1GB RAM, running Fedora Core 3, and compiling with gcc-3.4 -O9. We used the texts from the Pizza&Chili corpus (http://pizzachili.dcc.uchile.cl), with 50 MB of English and DNA and 64 MB of proteins. The pattern strings were sampled randomly from the text and each character was distorted with 10% of probability. All the patterns had length $m = 30$. Every configuration was tested during at least 60 seconds using at least 5 repetitions. Hence the numbers of repetitions varied between 5 and 130,000. To parametrize the hybrid index we tested all the j values from 1 to $k + 1$ and reported the best time. To parametrize we choose $q = \lfloor m/h \rfloor$ and $s = \lfloor k/h \rfloor + 1$ for some convenient h, since we can prove that this is the best approach and it was corroborated by our experiments. To determine the value of h and v we also tested the viable configurations and reported the best results. In our examples choosing v and h such that $2v$ is slightly smaller than q yielded the best configuration.

The average query time, in seconds, is shown in Fig. 1 and the respective memory heap peaks for indexed approaches are shown in Table 1. The hybrid index provides the fastest approach to the problem, however it also requires the most space. Aside from the hybrid index our approach is always either the fastest or within reasonable distance from the fastest approach. For low error level, $k = 1$ or $k = 2$, our approach is significantly faster, up to an order of magnitude better. This is very important since the compressed approaches seem to saturate at a given performance for low error levels: in English $k = 1$ to 3, in DNA $k = 1$ to 2, and in proteins $k = 1$ to 5. This is particularly troublesome since indexed approaches are the best alternative only for low error levels. In fact the sequential approaches outperform the compressed indexed approaches for higher error levels. In DNA this occurs at $k = 4$ and in English at $k = 5$.

Our index performed particularly well on proteins, as did the hybrid index. This could owe to the fact that proteins behave closer to random text, and this

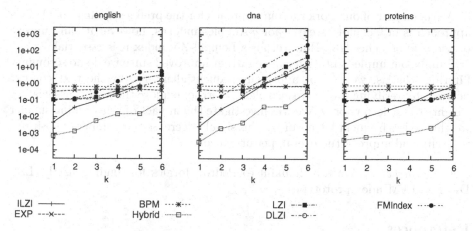

Fig. 1. Average user time for finding the occurrences of patterns of size 30 with k errors. The y axis units are in seconds and common to the three plots.

means that the parametrization of ours and the hybrid index indeed balances between exponential worst cases.

In terms of space the ILZI is also very competitive, as it occupies almost the same space as the plain text, except for proteins that are not very compressible. We presented the space that the algorithms need to operate and not just the index size, since the other approaches need intermediate data structures to operate.

6 Conclusions and Future Work

In this paper we presented an adaptation of the hybrid index for Lempel-Ziv compressed indexes. We started by addressing the problem of approximate matching with q-samples indexes, where we described a new approach to this problem. We then adapted our algorithm to the irregular parsing produced by Lempel-Ziv indexes. Our approach was flexible enough to be used as a hybrid index instead of an exact-searching-based filtration index. We implemented our algorithm and compared it with the simple filtration approach built over different compressed indexes, with sequential algorithms, and with a good uncompressed index.

Our results show that our index provides a good space/time tradeoff, using a small amount of space (at best 0.9 times the text size, which is 5.6 times less than a classical index) in exchange for searching from 6.2 to 33 times slower than a classical index, for $k = 1$ to 3. This is better than the other compressed approaches for low error levels. This is significant since indexed approaches are most valuable, if compared to sequential approaches, when the error level is low. Therefore our work significantly improves the usability of compressed indexes for approximate matching.

A crucial part of our work was our approach for the prefixes/suffixes of O. This approach is in fact not essential for q-samples indexes, however it can improve previous approaches [18]. However for a Lempel-Ziv index it is essential.

Finally, our implementation can be further improved since we do no secondary filtering, that is, we do not apply any sequential filter over the text context before fully verifying them. We also plan to further explore the idea of associating weights to the letters of O. We will investigate the impact of as assigning smaller weights to less frequent letters of O. This should decrease the number of positions to verify and improve the overall performance.

Acknowledgments. We are thankful to Pedro Morales for lending us the LZI, DLZI and FMIndex prototypes.

References

1. Navarro, G.: A guided tour to approximate string matching. ACM Comput. Surv. 33(1), 31–88 (2001)
2. Chang, W.I., Marr, T.G.: Approximate string matching and local similarity. In: Crochemore, M., Gusfield, D. (eds.) CPM 1994. LNCS, vol. 807, pp. 259–273. Springer, Heidelberg (1994)
3. Fredriksson, K., Navarro, G.: Average-optimal single and multiple approximate string matching. ACM Journal of Experimental Algorithmics 9(1.4) (2004)
4. Navarro, G., Baeza-Yates, R., Sutinen, E., Tarhio, J.: Indexing methods for approximate string matching. IEEE Data Engineering Bulletin 24(4), 19–27 (2001)
5. Cole, R., Gottlieb, L.A., Lewenstein, M.: Dictionary matching and indexing with errors and don't cares. In: STOC, pp. 91–100 (2004)
6. Maaß, M., Nowak, J.: Text indexing with errors. In: CPM, pp. 21–32 (2005)
7. Chan, H.L., Lam, T.W., Sung, W.K., Tam, S.L., Wong, S.S.: A linear size index for approximate pattern matching. In: Lewenstein, M., Valiente, G. (eds.) CPM 2006. LNCS, vol. 4009, pp. 49–59. Springer, Heidelberg (2006)
8. Coelho, L., Oliveira, A.: Dotted suffix trees: a structure for approximate text indexing. In: Crestani, F., Ferragina, P., Sanderson, M. (eds.) SPIRE 2006. LNCS, vol. 4209, pp. 329–336. Springer, Heidelberg (2006)
9. Weiner, P.: Linear pattern matching algorithms. In: IEEE 14th Annual Symposium on Switching and Automata Theory, pp. 1–11. IEEE Computer Society Press, Los Alamitos (1973)
10. Manber, U., Myers, E.: Suffix arrays: a new method for on-line string searches. SIAM Journal on Computing, 935–948 (1993)
11. Gonnet, G.: A tutorial introduction to Computational Biochemistry using Darwin. Technical report, Informatik E.T.H., Zuerich, Switzerland (1992)
12. Ukkonen, E.: Approximate string matching over suffix trees. In: Apostolico, A., Crochemore, M., Galil, Z., Manber, U. (eds.) Combinatorial Pattern Matching. LNCS, vol. 684, pp. 228–242. Springer, Heidelberg (1993)
13. Cobbs, A.: Fast approximate matching using suffix trees. In: Galil, Z., Ukkonen, E. (eds.) Combinatorial Pattern Matching. LNCS, vol. 937, pp. 41–54. Springer, Heidelberg (1995)
14. Sutinen, E., Tarhio, J.: Filtration with q-samples in approximate string matching. In: Hirschberg, D.S., Meyers, G. (eds.) CPM 1996. LNCS, vol. 1075, pp. 50–63. Springer, Heidelberg (1996)

15. Navarro, G., Baeza-Yates, R.: A practical q-gram index for text retrieval allowing errors. CLEI Electronic Journal 1(2) (1998)
16. Myers, E.W.: A sublinear algorithm for approximate keyword searching. Algorithmica 12(4/5), 345–374 (1994)
17. Navarro, G., Baeza-Yates, R.: A hybrid indexing method for approximate string matching. Journal of Discrete Algorithms 1(1), 205–239 (2000)
18. Navarro, G., Sutinen, E., Tarhio, J.: Indexing text with approximate q-grams. J. Discrete Algorithms 3(2-4), 157–175 (2005)
19. Kurtz, S.: Reducing the space requirement of suffix trees. Pract. Exper. 29(13), 1149–1171 (1999)
20. Sadakane, K.: New text indexing functionalities of the compressed suffix arrays. J. Algorithms 48(2), 294–313 (2003)
21. Ferragina, P., Manzini, G.: Indexing compressed text. Journal of the ACM 52(4), 552–581 (2005)
22. Navarro, G.: Indexing text using the Ziv-Lempel trie. J. Discrete Algorithms 2(1), 87–114 (2004)
23. Grossi, R., Vitter, J.S.: Compressed suffix arrays and suffix trees with applications to text indexing and string matching. SIAM J. Comput. 35(2), 378–407 (2005)
24. Navarro, G., Mäkinen, V.: Compressed full-text indexes. ACM Computing Surveys 39(1) article 2 (2007)
25. Manzini, G.: An analysis of the Burrows-Wheeler transform. Journal of the ACM 48(3), 407–430 (2001)
26. Kärkkäinen, J., Ukkonen, E.: Lempel-Ziv parsing and sublinear-size index structures for string matching. In: South American Workshop on String Processing, pp. 141–155. Carleton University Press (1996)
27. Arroyuelo, D., Navarro, G., Sadakane, K.: Reducing the space requirement of LZ-Index. In: Lewenstein, M., Valiente, G. (eds.) CPM 2006. LNCS, vol. 4009, pp. 318–329. Springer, Heidelberg (2006)
28. Russo, L.M.S., Oliveira, A.L.: A compressed self-index using a Ziv-Lempel dictionary. In: Crestani, F., Ferragina, P., Sanderson, M. (eds.) SPIRE 2006. LNCS, vol. 4209, pp. 163–180. Springer, Heidelberg (2006)
29. Huynh, T., Hon, W., Lam, T., Sung, W.: Approximate string matching using compressed suffix arrays. In: Sahinalp, S.C., Muthukrishnan, S.M., Dogrusoz, U. (eds.) CPM 2004. LNCS, vol. 3109, pp. 434–444. Springer, Heidelberg (2004)
30. Lam, T., Sung, W., Wong, S.: Improved approximate string matching using compressed suffix data structures. In: Deng, X., Du, D.-Z. (eds.) ISAAC 2005. LNCS, vol. 3827, pp. 339–348. Springer, Heidelberg (2005)
31. Morales, P.: Solución de consultas complejas sobre un índice de texto comprimido (solving complex queries over a compressed text index). Undergraduate thesis, Dept. of Computer Science, University of Chile, G. Navarro, advisor (2005)
32. Ziv, J., Lempel, A.: Compression of individual sequences via variable length coding. IEEE Transactions on Information Theory 24(5), 530–536 (1978)
33. Myers, G.: A fast bit-vector algorithm for approximate string matching based on dynamic programming. Journal of the ACM 46(3), 395–415 (1999)
34. Navarro, G., Baeza-Yates, R.: Very fast and simple approximate string matching. Information Processing Letters 72, 65–70 (1999)

Algorithms for Weighted Matching

Leena Salmela and Jorma Tarhio*

Helsinki University of Technology
{lsalmela,tarhio}@cs.hut.fi

Abstract. We consider the matching of weighted patterns against an
unweighted text. We adapt the shift-add algorithm for this problem. We
also present an algorithm that enumerates all strings that produce a score
higher than a given score threshold when aligned against a weighted pattern and then searches for all these strings using a standard exact multipattern algorithm. We show that both of these approaches are faster
than previous algorithms on patterns of moderate length and high significance levels while the good performance of the shift-add algorithm
continues with lower significance levels.

1   Introduction

In a weighted matching problem the text or the pattern is a weighted sequence
where in each position a weight is assigned to each character of the alphabet.
In this paper we consider the case where the pattern is weighted and the text
unweighted and we are interested in finding alignments where the score, which
is the sum of the weights in the pattern corresponding to the aligned characters
in the text, is larger than some given score threshold.

Weighted patterns arise for example in the modeling of transcription factor
binding sites in bioinformatics. In bioinformatics weighted patterns are called position weight matrices, position specific scoring matrices or profiles. The weight
of a nucleotide in a given position describes the log probability of that nucleotide
appearing in that position in a transcription factor binding site. Therefore the
score of an alignment is the log probability of that alignment being a transcription factor binding site. Many methods in bioinformatics rely on the large scale
scanning of these weighted patterns against a genome and there are large public
databases, like TRANSFAC [5] containing such patterns.

In this paper we adapt some standard string matching techniques to the
weighted matching problem and compare the performance of these algorithms
against the algorithm by Liefooghe et al. [4]. In Section 4, we adapt the shift-add
[1] algorithm to handle weighted patterns and in Section 5 we consider the enumeration of all strings matching a given weighted pattern and searching for these
strings by a standard multipattern algorithm. We compare our new approaches
to the previous algorithm by Liefooghe et al. [4] in Section 6. The preliminary
experimental results show that for high significance levels the enumeration approach is the fastest for pattern lengths 7 to 19 while the shift-add algorithm is

---

* Work by Jorma Tarhio was supported by Academy of Finland.

N. Ziviani and R. Baeza-Yates (Eds.): SPIRE 2007, LNCS 4726, pp. 276–286, 2007.

i	1	2	3	4	5	6	7	8	9	10	11	12
a	7	−6	−5	−10	−8	−10	4	−10	−10	−2	−10	−10
c	−5	−8	−10	14	−10	−8	−10	−10	−10	11	−10	−10
t	6	13	−10	−8	−10	12	−10	−10	−10	−3	−10	9
g	−5	−6	13	−10	14	−1	11	14	14	−10	14	6

Fig. 1. An example weighted pattern corresponding to the EGR-1 family extracted from TRANSFAC

the fastest for shorter and longer patterns. For the longest patterns either the algorithm by Liefooghe et al. or the shift-add algorithm is the fastest. For lower significance levels the shift-add algorithm is the fastest.

After submitting this paper we learned that Pizzi et al. [7] have also developed an algorithm based on the enumeration approach. However, they use a different multipattern algorithm to search for the enumerated strings while we use an algorithm tuned for very large pattern sets and low expected number of hits.

2 Definitions

We consider the matching of weighted patterns against an unweighted text. The text is a sequence of characters from an alphabet Σ of size σ. The weighted pattern assigns weights to all characters of the alphabet for each position of the pattern.

Definition 1. *A weighted pattern of length m is an $m \times \sigma$ matrix p of integer coefficients $p[i, c]$ which give the weight of the character $c \in \Sigma$ at position i where $1 \leq i \leq m$.*

Figure 1 shows an example of a weighted pattern. Here we will only consider weighted patterns with integer weights. Weighted patterns are obtained from entropy or log odd matrices that have real coefficients but in practice these are rounded to integer matrices to allow for more efficient computation.

Given a weighted pattern and a string of characters from the alphabet Σ the score of this string is defined as follows:

Definition 2. *Given a weighted pattern p of length m and a string t of length m from the alphabet Σ, the score of the pattern aligned with the string is defined as:*

$$\text{score}(p, t) = \sum_{i=1}^{m} p[i, t_i]$$

In the weighted matching problem we are interested in finding all those alignments of a text with the pattern that yield a large enough score:

Definition 3. *Given a weighted pattern p of length m, a score threshold α and an unweighted text $t_{1...n}$, find all such alignments i of the pattern with the text that $\text{score}(p, t_{i...i+m-1}) \geq \alpha$.*

Given a weighted matching problem, p-value [2,10] is a measure that can be used to estimate the statistical significance of the returned alignments. The p-value is defined as follows:

Definition 4. *Given a weighted matching problem with pattern p and score threshold α, p-value(p, α) is the probability that a given background model of the sequence produces a score equal to or greater than the score threshold α.*

In this paper we assume that the background model is the standard random string model where each character of the sequence is chosen independently and uniformly. In this case the p-value can be computed with the following recursion:

$$p\text{-value}(p[1...0], \alpha) = \begin{cases} 1 \text{ if } \alpha \leq 0 \\ 0 \text{ otherwise} \end{cases}$$

$$p\text{-value}(p[1...i], \alpha) = \frac{1}{\sigma} \sum_{c \in \Sigma} p\text{-value}(p[1...i-1], \alpha - p[i, c])$$

3   Previous Work

The brute force algorithm for the weighted matching problem calculates the score for each alignment of the pattern with the text and reports those alignments that yield a score higher than the score threshold. Lately various techniques have been proposed to speed up this scheme. Here we will review those techniques that are relevant to our work. See [8] for a survey on previous work.

Several algorithms use the lookahead technique [11] which provides a way to prune the calculation in a single alignment. For all suffixes of the pattern, there is a maximum score that they can contribute to the overall score. If after matching the prefix of the pattern, the score is not at least the score threshold minus maximum score of the suffix, there cannot be a match at this alignment. By calculating the maximum score for each pattern suffix, the overall computation time can be significantly reduced.

In Section 6 we will compare our algorithms to the algorithm by Liefooghe et al. [4]. Their algorithm uses the lookahead technique and in addition it divides the pattern into submatrices and precalculates for all possible strings the score yielded by each submatrix. For example, if we had a pattern of length 12, we could divide it to three submatrices of length four and then precalculate the scores of each submatrix for all the σ^4 possible strings. At matching time we can then just lookup the scores of each submatrix in a table.

4   Shift-Add for Weighted Matching

In this section we will adapt the shift-add algorithm [1] to weighted matching. Originally the shift-add algorithm was designed for the k-mismatch problem where the task is to find all substrings of the text that match the pattern with at most k mismatches. The algorithm works as follows.

For each pattern position i from 1 to m the algorithm has a variable s_i indicating with how many mismatches the suffix of length i of the text read so far matches the pattern prefix of length i. If the variables s_i can be represented in b bits, we can concatenate all these variables into a single vector $s = s_m s_{m-1} \cdots s_1$ of length mb. In the preprocessing phase we initialize for each symbol c in the alphabet a vector $T[c]$ where the bits in the position of s_i are 0^b if c equals p_i and $0^{b-1}1$ otherwise. The vector s (and hence also the variables s_i) can then in the matching phase be all updated at the same time when the next character c from the text is read:

$$s = (s \ll b) + T[c]$$

The algorithm has found a match if $s_m \leq k$.

If the variables s_i count mismatches, the maximum value that they can reach is m. However, in the k-mismatch problem it is enough to be able to represent values in the range $[0, k+1]$ yielding $b = \lceil \log(k+1) \rceil$. However, we need an additional bit so that the possible carry bits do not interfere with the next variable. With this modification the update operation of the algorithm becomes:

$$s = (s \ll b) + T[c]$$
$$of = (of \ll b) \mid (s \mathbin{\&} (10^{b-1})^m)$$
$$s = s \mathbin{\&} (01^{b-1})^m$$

Here the first line updates the variables s_i, the second one keeps track of those variables s_i that have overflowed and the last one clears the carry bits. When checking for a match, we now also need to check that the variable s_m has not overflowed which can be seen from the of vector. The shift-add algorithm for the k-mismatch problem has time complexity $O(n \lceil \frac{mb}{w} \rceil)$ where $b = \lceil \log(k+1) \rceil + 1$ and w is the size of the computer word in bits.

We will now present the shift-add algorithm for weighted matching with positive restricted weights. Then we will show how a general weighted pattern matching problem can be transformed into such a restricted problem. The weights of the weighted matching problem with positive restricted weights have the following properties:

1. $\forall i, 1 \leq i \leq m, \forall c \in \Sigma, \; 0 \leq p[i, c] \leq \alpha$
2. $\forall i, 1 \leq i \leq m \; \exists c \in \Sigma$ such that $p[i, c] = 0$

where p is the weighted pattern of length m and α is the score threshold. Property 1 is needed for the correct operation of the shift-add algorithm while Property 2 merely serves as a way to lower the score threshold and thus lower the number of bits needed for the variables s_i as will be seen later.

The adaptation of the shift-add algorithm to weighted matching with positive restricted weights is quite straightforward. Now instead of counting mismatches, we will be calculating scores so the variables s_i contain the score of the suffix of length i of the text read so far as compared to the prefix of length i of the pattern. For the update operation the bits corresponding to s_i in the preprocessed vectors $T[c]$ now contain the weight of the character c at position i. The update operation is exactly as in the shift-add algorithm for the k-mismatch problem. If after the update operation the score $s_m \geq \alpha$ or the variable s_m has overflowed, a match is reported.

Property 1 of the weighted matching problem with positive restricted weights states that all weights are non-negative and thus

$$\text{score}(p_{1\ldots i}, t_{j\ldots j+i+1}) \leq \text{score}(p_{1\ldots i+1}, t_{j\ldots j+i+2}) .$$

Because the score can only increase when reading a new character, we can truncate the score values to α. Property 1 further states that all weights are at most α. Thus, if we truncate the score values to α, after the update operation the variables $s_i \leq 2\alpha$ so 1 carry bit is enough. Therefore we need to reserve $b = \lceil \log \alpha \rceil + 1$ bits for each variable s_i and the time complexity of the algorithm is $O(n \lceil \frac{m(\lceil \log \alpha \rceil + 1)}{w} \rceil)$.

In the weighted matching problem the weights can be, and in practice often are, negative. The following observation points us to a way to transform any weighted matching problem to a weighted matching problem with positive restricted weights. Let p be a weighted pattern of length m and let p' be a weighted pattern such that for some i, $1 \leq i \leq m$, $p'[i,c] = p[i,c] + h$ for all $c \in \Sigma$ and some constant h, and for all $j \neq i$, $1 \leq j \leq m$, and all $c \in \Sigma$, $p'[j,c] = p[j,c]$. Then the following holds for the scores of p and p' aligned with any string t of length m:

$$\text{score}(p', t) = \text{score}(p, t) + h$$

Therefore the weighted pattern matching problem for a text t, pattern p and score threshold α returns exactly the same alignments as the weighted pattern matching problem for a text t, pattern p' and score threshold $\alpha' = \alpha + h$.

Now given a weighted pattern matching problem with a score threshold α and a pattern p containing any integer weights we can transform the problem into an equivalent problem with a score threshold α' and a pattern p' containing only non-negative weights.

To reduce the score threshold (and thus also the number of bits needed for the variables s_i) we further transform the pattern so that in each position at least one of the weights equals zero by adding an appropriate negative constant h to all weights in that position and by adjusting the score threshold also by h. Furthermore, if now any weight is larger than the score threshold, it can be truncated to the score threshold without affecting the returned alignments because the score of an alignment cannot get smaller as more characters are read. The scores of those alignments will however be lower. As a result we have transformed a weighted matching problem into a weighted matching problem with positive restricted weights.

In practice weighted patterns are obtained by rounding log-odd or entropy matrices to integer matrices. Thus the values of the weights depend on how much precision is preserved by this rounding and furthermore practical values of the threshold α depend on the weights. Because of the $\lceil \log \alpha \rceil + 1$ factor in the running time the shift-add algorithm is somewhat sensitive to the precision of this rounding unlike other algorithms.

```
enumerate(p, α)
1.   recurse(1, 0)


string s


recurse(i, score)
1.   if (α > score + max_score(i...m))
2.       return
3.   if (i > m and score ≥ α)
4.       add_string(s)
5.   else
6.       for each c ∈ Σ
7.           s[i] = c
8.           recurse(i + 1, score + p[i, c])
```

Fig. 2. Pseudo code for enumerating all strings that produce a score higher than or equal to the score threshold α

5   Enumeration Algorithms

For short patterns it is possible to enumerate all matching strings which are the strings that produce a score higher than the score threshold when aligned with the weighted pattern. The enumerated strings can then be searched for with an exact multipattern matching algorithm.

The enumeration of matching strings is done with a recursive algorithm. At recursion level i we have constructed a string of length $i - 1$ that is a possible prefix of a matching string and we try to expand that prefix with all characters of the alphabet. This way we have to calculate the score of each prefix only once. The recursion can further be pruned with the lookahead technique. Suppose we have enumerated a prefix of length $i - 1$ with score score_i and the maximum score of a suffix of length $m - i$ is $\text{max\_score}(i...m)$ then if the score threshold $\alpha > \text{score}_i + \text{max\_score}(i...m)$ then at this branch of the recursion no matching strings can be found. The pseudo code for enumerating the matching strings is given in Fig. 2.

Because the number of enumerated strings is often very large, we used the multipattern BNDM with q-grams (BG) [9] algorithm which is especially tuned for large pattern sets. The BG algorithm first builds a filter, which is a pattern of classes of characters. In this filter all characters that appear in any of the single patterns in position i are accepted at that position. The backward nondeterministic DAWG matching (BNDM) [6] algorithm is then used to scan the text with this filter. The returned alignments are verified with a Rabin-Karp [3] style algorithm. When the number of patterns grows the filtering is no longer efficient enough because almost every alignment will match the filter. To boost the filtering efficiency, the BG algorithm uses q-grams instead of single characters in the filtering phase. If matches are sufficiently rare (i.e. the p-value(p, α) is sufficiently low), the BG algorithm has average case running time $O(n \log_{1/d} m/m)$ where $d = 1 - (1 - 1/\sigma^q)^r$ where r is the number of patterns.

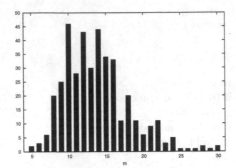

Fig. 3. The length distribution of patterns in the TRANSFAC database

p-value(p, α) gives the probability of a random string to produce a score equal to or greater than α when aligned with the weighted pattern p. If the background model assumes that all characters are chosen independently and uniformly, p-value(p, α) gives the proportion of all possible strings for which the score is at least α. Thus the expected number of enumerated strings is $\sigma^m p$-value(p, α) because there are σ^m different strings of length m.

In practice, it turned out to be reasonably fast to enumerate matching strings up to pattern length 16. With larger patterns we enumerated only 16 characters long prefixes of the matching strings and the algorithm verifies the found matches later.

The enumeration approach is easy to adjust to searching for multiple weighted patterns at once. All we need to do is to enumerate for all of the weighted patterns the strings producing high enough scores and then search for all these enumerated strings.

6   Experimental Results

For all experimental testing we used a computer with a 2.0 GHz AMD Opteron dual-processor and 6 GB of memory. The machine was running the 64-bit version of Linux 2.6.15. The tests were written in C and compiled with the gcc 4.1.0 compiler. The patterns were extracted from the TRANSFAC database [5]. Figure 3 shows the length distribution of the patterns. As can be seen the length of most patterns is between 8 and 22 nucleotides. In particular there are only a few patterns of length over 22 and thus the results concerning these pattern lengths are only tentative. The text we used was a chromosome from the fruitfly genome (20 MB).

Figure 4 shows a runtime comparison of the algorithm by Liefooghe, Touzet and Varré (LTV) [4], shift-add algorithm (sa) and the enumeration algorithm (ebg) for two p-values. The algorithms were run 10 times with each pattern and the average runtime was calculated. The figure shows average runtimes of patterns of same length. The measured runtime excludes the time used for pre-processing.

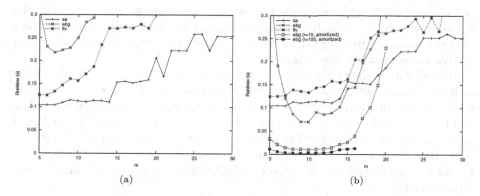

Fig. 4. Runtime comparison of different methods for p-values (a) 10^{-3} and (b) 10^{-5}

For the LTV algorithm we did not count the optimum length of the submatrices as presented in the original paper by Liefooghe et al. [4] because the optimum length calculation does not take into account cache effects and these surely have a significant effect on the runtime. Instead we tried the algorithm with submatrix lengths from 4 to 8 and included the best results in the comparison. With this modification the method is actually the same as the superalphabet algorithm of Pizzi et al. [7].

The optimal value for q in the LTV algorithm is lower for shorter patterns and for higher p-values but it does not affect the runtime of the algorithm very much until it reaches the value 8 when the tables no longer all fit into the cache. We can see that for the p-value 10^{-3} the runtime increases slowly until pattern length 11 and for the p-value 10^{-5} the runtime stays almost constant until pattern length 15. Until that time it is almost always sufficient to calculate the index of the first precalculated score table corresponding to the first submatrix because the lookahead technique then reports that a match at that position is not possible. When the pattern length increases further, more and more accesses are needed to the second precalculated table until at pattern length 14 for the p-value 10^{-3} and at pattern length 19 for the p-value 10^{-5} at almost every position we need to consult both the first and the second precalculated table.

Figure 4 shows that the runtime of the shift-add algorithm increases each time we need more words to represent the state vector. For pattern lengths $\{5-8, 8-14, 15-21, 19-24, 25-30\}$ we need state vectors of size $\{1,2,3,4,5\}$ words, respectively. Between lengths 19 and 21 some patterns need state vectors of 3 words while others need 4 words. Similarly for pattern length 8 some patterns need state vectors of 1 word while others need already 2 words. The number of words needed does not change from the p-value 10^{-3} to the p-value 10^{-5}.

We ran the enumeration algorithm with several different values of q and chose the value that gives the best runtime. For the p-value 10^{-3} and pattern lengths $\{5-7, 8-9, 10, 11, 12-15\}$ the values $\{4,5,6,7,8\}$, respectively, gave the best results and for the p-value 10^{-5} and pattern lengths $\{5-11, 12, 13, 14, 15-20\}$

the values $\{4, 5, 6, 7, 8\}$, respectively, gave the best results. We did not run the enumeration algorithm for longer pattern lengths because the number of enumerated patterns grew too large and already with these pattern lengths the algorithm started to significantly slow down.

Overall Fig. 4 shows that for low significance levels (i.e. high p-values) the shift-add algorithm is the fastest. For higher significance levels (i.e. smaller p-values) the shift-add algorithm is the fastest for pattern lengths smaller than 7. The enumeration algorithm is fastest for patterns lengths 8 to 16. For longer patterns the shift-add algorithm is the fastest at least until pattern length 25. After that the differences between shift-add and LTV are so small that it is hard to say anything conclusive because the TRANSFAC database contained so few long patterns.

The preprocessing of the shift-add algorithm is very fast taking less than 0.01 s regardless of the pattern length. The preprocessing time for the LTV algorithm ranges from less than 0.01 s to 0.09 s. The preprocessing time of the enumeration algorithm is exponential in the length of the pattern. It stays under 0.01 s until pattern length 12 for the p-value 10^{-3} and until pattern length 16 for the p-value 10^{-5}. For longer patterns the preprocessing time increases to 0.93 s for the p-value 10^{-3} and pattern length 15 and to 0.40 s for the p-value 10^{-5} and pattern length 20.

We also ran some experiments with the multiple pattern version of the enumeration algorithm. Because the single pattern algorithm worked well only for high significance levels we ran the multiple pattern version only for the p-value 10^{-5}. To get reliable results, we needed more patterns of each length than is provided by the TRANSFAC database. To increase the number of patterns for each pattern length we took prefixes of longer patterns and added these to our pool of patterns until we had a hundred patterns of each length. This worked up to pattern length 16 after which including prefixes of all longer patterns did not bring the number of patterns to one hundred.

Figure 5 shows how the runtime of the algorithm behaves as a function of pattern length and pattern set size r. As can be seen, the runtime decreases for all pattern sets as pattern length increases until pattern length 8 because the BG algorithm can make longer shifts. After pattern length 12 the filtering efficiency of the BG algorithm starts to deteriorate and we need to make more verifications which increases the runtime. The filtering efficiency could be boosted by increasing the value of parameter q but this would increase the amount of memory needed so that the structures frequently used by the algorithm no longer fit in the data cache and this imposes an even larger penalty on the runtime.

Figure 5b shows that the runtime increases only slightly when the pattern set size is increased for pattern lengths 8 through 14. For shorter pattern lengths the performance of the algorithm deteriorates faster because so many positions match at least one of the patterns. For longer patterns the filtering efficiency is a problem even when searching for a single pattern and this problem is further emphasized by increasing the pattern set size.

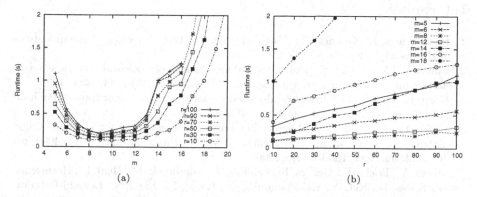

Fig. 5. The runtime of the multipattern enumeration algorithm as a function of (a) pattern length and (b) pattern set size

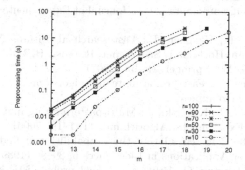

Fig. 6. Preprocessing times for the multiple pattern enumeration algorithm

Preprocessing time of the multipattern algorithm is less than 0.01 s for all pattern set sizes when the pattern length is at most 11. Figure 6 shows the preprocessing times for longer patterns and various pattern set sizes.

The amortized running times (i.e. the running times per pattern) for the multipattern enumeration algorithm are shown also in Fig. 4b for pattern set sizes 10 and 100. As can be seen these times are much lower than the running times of the other algorithms until pattern length 16. After that the runtime starts to increase and after pattern length 20 it is probably faster to match one pattern at a time using either the shift-add or the LTV algorithm.

7   Conclusions

We have presented two efficient algorithms for searching weighted patterns in an unweighted text. We have showed that the algorithms are fast in practice by comparing their performance on real data against the previous algorithm by Liefooghe et al. [4].

References

1. Baeza-Yates, R., Gonnet, G.: A new approach to text searching. Communications of the ACM 35(10), 74–82 (1992)
2. Claverie, J.M., Audic, S.: The statistical significance of nucleotide position-weight matrix matches. Computer Applications in Biosciences 12(5), 431–439 (1996)
3. Karp, R., Rabin, M.: Efficient randomized pattern-matching algorithms. IBM Journal of Research and Development 31, 160–249 (1987)
4. Liefooghe, A., Touzet, H., Varré, J.S.: Large scale matching for position weight matrices. In: Lewenstein, M., Valiente, G. (eds.) CPM 2006. LNCS, vol. 4009, pp. 401–412. Springer, Heidelberg (2006)
5. Matys, V., Fricke, E., Geffers, R., Gößling, E., Haubrock, M., Hehl, R., Hornischer, K., Karas, D., Kel, A., Kel-Margoulis, O., Kloos, D., Land, S., Lewicki-Potapov, B., Michael, H., Münch, R., Reuter, I., Rotert, S., Saxel, H., Scheer, M., Thiele, S., Wingender, E.: TRANSFAC: transcriptional regulation, from patterns to profiles. Nucleic Acids Res. 31, 374–378 (2003)
6. Navarro, G., Raffinot, M.: Fast and flexible string matching by combining bit-parallelism and suffix automata. ACM Journal of Experimental Algorithmics 5(4), 1–36 (2000)
7. Pizzi, C., Rastas, P., Ukkonen, E.: Fast search algorithms for position specific scoring matrices. In: Hochreiter, S., Wagner, R. (eds.) BIRD 2007. LNCS (LNBI), vol. 4414, pp. 239–250. Springer, Berlin (2007)
8. Pizzi, C., Ukkonen, E.: Fast profile matching algorithms – a survey. Theoretical Computer Science (to appear)
9. Salmela, L., Tarhio, J., Kytöjoki, J.: Multi-pattern string matching with q-grams. ACM Journal of Experimental Algorithmics 11, 1–19 (2006)
10. Staden, R.: Methods for calculating the probabilities of finding patterns in sequences. Computer Applications in Biosciences 5, 89–96 (1989)
11. Wu, T., Neville-Manning, C., Brutlag, D.: Fast probabilistic analysis of sequence function using scoring matrices. Bioinformatics 16(3), 233–244 (2000)

Efficient Text Proximity Search

Ralf Schenkel[1], Andreas Broschart[1], Seungwon Hwang[2], Martin Theobald[3], and Gerhard Weikum[1]

[1] Max-Planck-Institut für Informatik, Saarbrücken, Germany
{abrosch,schenkel,weikum}@mpi-inf.mpg.de
[2] POSTECH, Korea
swhwang@postech.ac.kr
[3] Stanford University
theobald@stanford.edu

Abstract. In addition to purely occurrence-based relevance models, term proximity has been frequently used to enhance retrieval quality of keyword-oriented retrieval systems. While there have been approaches on effective scoring functions that incorporate proximity, there has not been much work on algorithms or access methods for their efficient evaluation. This paper presents an efficient evaluation framework including a proximity scoring function integrated within a top-k query engine for text retrieval. We propose precomputed and materialized index structures that boost performance. The increased retrieval effectiveness and efficiency of our framework are demonstrated through extensive experiments on a very large text benchmark collection. In combination with static index pruning for the proximity lists, our algorithm achieves an improvement of two orders of magnitude compared to a term-based top-k evaluation, with a significantly improved result quality.

1 Introduction

Techniques for ranked retrieval of text documents have been intensively studied including relevance scoring models such as tf*idf, Okapi BM25, and statistical language models [13]. Most of the models in these families are based on the (multinomial) bag-of-words representation of documents, with consideration of term frequencies (tf) and inverse document frequencies (idf) but without considering term proximity. However, there are many queries where the best results contain the query terms in a single phrase, or at least in close proximity.

To illustrate the importance of proximity, let us consider the query *"surface area of rectangular pyramids"*. Schemes that do not take proximity into account return general mathematical documents in which all the four terms *surface, area, rectangular* and *pyramid* are individually important, but the document does not necessarily contain information about the surface area of rectangular pyramids (for example, it may discuss the volume of pyramids and the area of rectangular prisms. On the other hand, an exact phrase match *"surface area of rectangular pyramids"* would most certainly ensure that the document retrieved is of

N. Ziviani and R. Baeza-Yates (Eds.): SPIRE 2007, LNCS 4726, pp. 287–299, 2007.

the desired type, but strictly enforcing such phrase matchings in a boolean way would exclude many relevant results. A good proximity-aware scoring scheme should give perfect phrase matches a high score, but reward also high proximity matches such as "*surface area* of a *rectangular*-based *pyramid*" with good scores. There has been a number of proposals in the literature for such proximity-aware scoring schemes [5,6,9,10,16,18,20]; however, none of these proposals considered efficiently finding the best results to queries in a top-k style with dynamic pruning techniques. This paper shows that integrating proximity in the scoring model can not only improve retrieval effectiveness, but also improve retrieval efficiency by up to two orders of magnitude compared to state-of-the-art processing algorithms for purely occurrence-based scoring models.

2   Related Work

Using phrases is a common means in term queries to restrict the results to those that exactly contain the phrase and is often useful for effective query evaluation [7]. A simple way to efficiently evaluate phrases are *word-level indexes*, inverted files that maintain positional information [24]. There have been some proposals for specialized index structures for efficient phrase evaluation that utilize term pair indexes and/or phrase caching, but only in the context of boolean retrieval and hence not optimized for top-k style retrieval with ranked results [8,22,23]. There are proposals to extend phrases to window queries, where users can specify the size of a window that must include the query terms to favor documents containing all terms within such a window [15,17,4]. However, this line of works has treated term proximity only as an afterthought after ranking, i.e., proximity conditions are formulated as a simplistic Boolean condition and optimized as separate post-pruning step after rank evaluation.

More recently, some scoring models were proposed that integrate content and proximity scores for ranking results [5,6,9,10,16,18,20]. These scoring models can be categorized into the following two classes. First, linear combination approaches attempt to reflect proximity in the scoring by linearly combining a proximity score with a text-based content score [5,6,16,18]. Monz quantified the proximity score based on the size of the minimum window containing all query keywords occurring in the document [16]. Rasolofo et al. consider term pairs that occur together in a small window in a document, and use a distance-based proximity score for these term pairs [18]. Büttcher et al. extend on this work by considering adjacent query term occurrences without a limit on the window size and use a proximity score similar to BM25 for text [5,6]. Second, holistic approaches have more tightly integrated proximity metrics and content scoring [9,10,20]. De Kretser and Moffat [10] and Clarke et al. [9] proposed scoring methods that reward the density of query terms in documents, and Song et al. [20] integrate a similar term density score within a BM25-based scoring model. However, none of the proximity proposals we are aware of has been designed to be used within a top-k style evaluation.

3   Processing Model

We consider a document d_i $(i = 1, \ldots, m)$, with which we associate n scores $s_{i1}, \ldots, s_{in}$, each quantifying the relevance of d_i over n different dimensions like terms or term pairs (e.g., tf*idf or BM25-based scores for query terms or, as we will later introduce, proximity scores for term pairs). The scores are aggregated using a *monotonous* function; we will focus on weighted summation as aggregation function for ease of presentation.

Our processing uses algorithms from the family of *Threshold Algorithms* [12], similar to *dynamic pruning* approaches in the IR community [1,2,14]. These algorithms assume that the scores for each dimension j have been precomputed and stored in an inverted list L_j which is sorted by descending score (or, in IR terms, in *frequency* or *impact order*). The algorithms then sequentially scan each list involved in the query execution in an interleaved, round robin manner. As documents are discovered in this process, they are maintained as candidates in an in-memory pool, where each candidate has a current score (aggregated from the scores in dimensions where the document has been encountered so far). Additionally, each candidate d_i has an upper score bound that is computed by setting all unknown scores to the highest possible score $high_j$ corresponding to the score at the current scan positions of the lists:

$$bestscore(d_i) = \sum_{j=1}^{n} \left(\begin{array}{l} s_{ij} \quad \text{if } d_i \text{ seen in } L_j \\ high_j \text{ otherwise} \end{array} \right) \tag{1}$$

For a top-k query, the algorithms maintain a list of the k candidates with the highest current scores. Other candidates whose best score is below the lowest current score of a top-k candidate can be safely pruned from the evaluation. The execution can stop if all but the top-k candidates have been eliminated; this is typically the case long before the lists have been completely read.

To further speed up the execution, some algorithms additionally make random lookups for the scores of promising candidates in dimensions where they have not yet been encountered; as such a random access (RA) is a lot more expensive than a sequential scan (in the order of 100 to 1,000 times for real systems), an intelligent schedule for these RAs has a great influence on efficiency. The different variants within the family of threshold algorithms primarily differ in their RA schedules; the currently most efficient variant [3] schedules all RAs only at the end of the partial scans of inverted lists, namely, when the expected cost for RA is below the cost for all sequential accesses so far.

4   Proximity Scoring

4.1   Proximity Scoring Models

We focus on proximity scoring models that use a linear combination of a content-based score with a proximity score in the form of

$$score(d, q) = \alpha \cdot score_{content}(d, q) + (1 - \alpha) score_{proximity}(d, q)$$

There are several proximity metrics in this category [5,6,16,18]. In preliminary experiments on the TREC Terabyte collection, the scoring model proposed by Büttcher et al. [5,6] (labelled *Büttcher's scoring model* from now on) was the only one to yield significant improvements in result quality over BM25-based scoring, hence we use this model in our proximity-aware evaluation framework.

4.2   Büttcher's Scoring Model

For a document d with length l, we denote the term occurring at position i of d by $p_i(d)$, or p_i when the document is uniquely given by the context. For a term t, we denote by $P_d(t) \subseteq \{1, \ldots, l\}$ the positions in document d where t occurs; or we write $P(t)$. Given a query $q = \{t_1 \ldots t_n\}$, we write $P_d(q) := \cup_{t_i \in q} P_d(t_i)$ for the positions of query terms in document d, or $P(q)$ when d is given by the context. We denote pairs of positions of distinct query terms in document d by

$$Q_d(q) := \{(i,j) \in P_d(q) \times P_d(q) \mid i < j \land p_i \neq p_j\}$$

and pairs of adjacent occurrences of distinct query terms, possibly with non-query terms in between, by

$$A_d(q) := \{(i,j) \in Q_d(q) \mid \forall k \in \{i+1, \ldots, j-1\} : k \notin P_d(q)\}$$

Büttcher's scoring model linearly combines the BM25 scoring function with a proximity score for each query term into a proximity-aware document-level score. Given a query $q = \{t_1, \ldots, t_n\}$ and a document d, they first compute an accumulated interim score acc for each query term that depends on the distance of this term's occurrences to other, adjacent query term occurrences. Formally,

$$acc_d(t_k) = \sum_{(i,j) \in A_d(q) : p_i = t_k} \frac{idf(p_j)}{(i-j)^2} + \sum_{(i,j) \in A_d(q) : p_j = t_k} \frac{idf(p_i)}{(i-j)^2} \qquad (2)$$

where idf is the inverse document frequency. The accumulated proximity score increases the more, the less distant the occurrences of two adjacent terms are and the less frequent the neighboring term is in the collection. The score for a document d is then computed by a linear combination of a standard BM25-based score and a proximity score, which is itself computed by plugging the accumulated proximity scores into a BM25-like scoring function:

$$score_{\text{Büttcher}}(d,q) = score_{\text{BM25}}(d,q) + \sum_{t \in q} min\{1, idf(t)\} \frac{acc_d(t) \cdot (k_1 + 1)}{acc_d(t) + K}$$

where, analogously to the BM25 formula,

$$K = k \cdot [(1-b) + b \cdot \frac{|d|}{avgdl}]$$

and b, k_1, and k are configurable parameters that are set to $b = 0.5$ and $k = k_1 = 1.2$, respectively [5].

4.3   Modified Büttcher Scoring Model

To include Büttcher's proximity score into query processing, it would be intriguing to use a standard word-level inverted list and compute proximity scores on
the fly as a document is encountered. However, this is not feasible in a top-k
style processing as the proximity score is not upper bounded, and hence it is
not possible to compute tight score bounds for candidates which in turn disables
pruning. For an efficient computation of the top-k results, we need to precompute proximity information into index lists that can be sequentially scanned and
compute tight score bounds for early pruning. The main problem with Büttcher's
scoring function in this respect is that $acc_d(t)$ is computed as a sum over adjacent
query term occurrences, which is inherently query dependent, and we cannot precompute query-independent information. An additional, minor issue is that the
scoring function includes the document length which cannot be easily factorized
into a precomputed score contribution.

To solve this, we slightly modify Büttcher's original scoring function; this does
not have much influence on result quality (as can be shown experimentally), but
allows for precomputation. In addition to dropping the document length by
setting $b = 0$ in the formula, we consider *every* query term occurrence, not only
adjacent occurrences. The modified accumulation function acc' is defined as

$$acc'_d(t_k) = \sum_{(i,j) \in Q_d(q):p_i=t_k} \frac{idf(p_j)}{(i-j)^2} + \sum_{(i,j) \in Q_d(q):p_j=t_k} \frac{idf(p_i)}{(i-j)^2} \tag{3}$$

As the value of $acc'_d(t_k)$ does not depend only on d and t_k, but also on the
other query terms, we still cannot precompute this value independently of the
query. However, we can reformulate the definition of $acc'_d(t_k)$ as follows:

$$acc'_d(t_k) = \sum_{t \in q} idf(t) \underbrace{\left(\sum_{\substack{(i,j)\, \in\, Q_d(q): \\ p_i = t_k,\, p_j = t}} \frac{1}{(i-j)^2} + \sum_{\substack{(i,j)\, \in\, Q_d(q): \\ p_i = t,\, p_j = t_k}} \frac{1}{(i-j)^2} \right)}_{:=acc_d(t_k,t)} \tag{4}$$

$$= \sum_{t \in q} idf(t) \cdot acc_d(t_k,t) \tag{5}$$

We have now represented $acc'_d(t_k)$ as a monotonous combination of *query term
pair scores* $acc_d(t_k,t)$. Note that term order does not play a role, i.e., $acc_d(t_k,t) =
acc_d(t,t_k)$. We can precompute these pair scores for all term pairs occurring in
documents and arrange them in index lists that are sorted in descending score
order. Including these lists in the sequential accesses of our processing algorithm,
we can easily compute upper bounds for $acc'_d(t_k)$ analogously to query term
dimensions by plugging in the score at the current scan position in the lists
where d has not yet been encountered. The current score of a document is then
computed by evaluating our modified Büttcher score with the current value of

acc'_d, and the upper bound is computed using the upper bound for acc'_d; this is correct as the modified Büttcher score is monotonous in acc'_d.

5   Indexing and Evaluation Framework

5.1   Precomputed Index Lists and Evaluation Strategies

Our indexing framework consists of the following precomputed and materialized index structures, each primarily used for sequential access, but with an additional option for random access:

- TextIndexList (TL): for each term t_i, a list of the form $(d_k, score_{BM25}(d_k, t_i))$, sorted by descending score.
- ProxIndexList (PXL): for each unordered pair $\{t_i, t_j\}$ of terms with $t_i < t_j$, a list of the form $(d_k, acc_{d_k}(t_i, t_j))$, sorted by descending acc.
- CombinedIndexList (CL): for each unordered pair $\{t_i, t_j\}$ of terms with $t_i < t_j$, a list of the form $(d_k, acc_{d_k}(t_i, t_j), score_{BM25}(d_k, t_i), score_{BM25}(d_k, t_j)))$, sorted by descending acc.

These index structures can be combined into several processing strategies:

- TL: This corresponds to standard, text-based retrieval without proximity.
- PXL: This scans only the proximity lists and uses the proximity part of our modified Büttcher scoring function for ranking.
- TL+PXL: This scans proximity and content lists (which would be the straight-forward implementation of our scoring model with a Threshold algorithm).
- TL+CL: This strategy, which is the main contribution of this paper, exploits the additional content scores in the pair lists to reduce the uncertainty about the score of documents with high proximity scores early in the process, which often allows early termination of the algorithm. We can additionally tighten the bounds when a proximity list for a pair (t_1, t_2) runs empty: If a document was seen in the dimension for t_1, but not in the proximity list, it is certain that it won't appear in the list for t_2 any more.

5.2   Index List Pruning

For large collections, the size of the inverted lists may be too large to completely store them, especially when the index includes proximity lists. As we do not consider only adjacent terms, but *any* terms occurring in the same document, a complete set of proximity lists will be much larger than the original text collection. Lossless index compression techniques (see, e.g., [11]) are one way to solve this problem, but the compression ratio will not be sufficient for really huge collections. We therefore apply *index pruning* (which is a lossy index compression technique) to reduce the size of the index, while at the same time sacrificing as little result quality as possible. Following the literature on inverted lists for text processing, a common way is pruning lists *horizontally*, i.e., dropping entries towards the end of the lists. These entries have low scores and hence will not play a big role when retrieving the best results for queries. Unlike text lists, pair lists contain many entries with very low scores (as the score depends on the distance of term occurrences), so the pruning effect on pair lists should be a lot higher than on text lists.

Our indexing framework provides three different pruning methods, mainly geared towards proximity lists. First, we heuristically limit the distance of term occurrences within a document, as occurrences within a large distance have only a marginal contribution to the proximity score. Second, we heuristically limit the list size to a constant, usually in the order of a few thousand entries. Third, we leverage the seminal work by Soffer et al. [19] for proximity lists. They introduced list pruning with quality guarantees for the scores of query results, assuming top-k style queries with a fixed (or at least bounded) k. For each list l, they consider the score $s_k(l)$ at position k of the list, and drop each entry from that list whose score is below $\epsilon \cdot s_k(l)$, where $0 < \epsilon < 1$ is a tuning parameter.

6   Evaluation

6.1   Setup

We evaluated our algorithms with the Java-based, open-source TopX search engine[1] [21]. Our experiments were run using the TREC Terabyte collection with roughly 25 million documents, corresponding to about 426GB of data. We evaluated our methods with the 100 adhoc topics from the 2004 and 2005 TREC Terabyte tracks. As we are focusing on top-k retrieval, we measured precision at several cutoffs. To evaluate efficiency, we measured the number of sequential (SA) and random (RA) accesses to the index lists and the number of bytes transferred from disk, assuming sizes of 8 bytes for scores and document ids. As random accesses are usually much more expensive than sequential accesses, we additionally compute a byte-based abstract cost $Cost(\gamma) = \#bytesSA + \gamma \cdot \#bytesRA$ for each run, based on the cost ratio $\gamma := c_{RA}/c_{SA}$ of random to sequential accesses. We indexed the documents with the indexer included in the TopX system with stopword removal enabled and computed the proximity lists needed for the queries with an additional tool. For the Okapi BM25 model, we used the parameters $k = k_1 = 1.2$ and $b = 0.5$. We ran the results with TopX configured in RR-LAST mode and a batch size of 5,000, i.e., round-robin sequential accesses in batches of 5,000 items to the index lists and postponing random accesses to the end. Index lists were stored in an Oracle database.

6.2   Results with Unpruned Index Lists

Table 1 shows our experimental results for top-10 retrieval with unpruned index lists and stemming enabled. It is evident that the configuration TL+CL improves precision@10 to 0.6 over the original BM25 setting (which corresponds to TL with a precision of 0.56), with a t-test and a Wilcoxon signed-rank confirming

---

[1] http://topx.sourceforge.net

Table 1. Experimental results for top-10 retrieval of 100 topics on Terabyte

|Configuration|	P@10|	#SA|	#RA|	bytes SA|	bytes RA|	Cost(100)|	Cost(1000)|
TL	0.56	24,175,115	196,174	386,801,840	1,569,392	543,741,040	1,956,193,840
TL+PXL	0.60	24,743,914	149,166	395,902,624	1,193,328	515,235,424	1,589,230,624
TL+CL	0.60	4,362,509	8,663	108,743,568	79,256	116,669,168	187,999,568
PXL	0.40	867,095	2,925	13,873,520	23,400	16,213,520	37,273,520

Table 2. Index sizes (million items) with different length limits, with and without window limit

|index/limit|	500|	1000|	1500|	2000|	2500|	3000|	unpruned|
TL	295	355	402	442	472	496	3,191
PXL/CL (est.)	368,761	435,326	481,949	515,079	542,611	566,277	1,410,238
PXL/CL, window≤ 10 (est.)	23,050	28,855	34,023	38,985	42,085	45,186	87,049

statistical significance. At the same time, it dramatically reduces the number of accesses, bytes transferred, and abstract costs by a factor of 5 to 15 over the BM25 baseline, due to the additional text scores available in CL and the better bounds. The configuration TL+PXL with simple proximity lists achieves the same improvement in precision as it uses the same scoring function, but needs to run longer until it can safely stop. Scanning only the proximity lists exhibits poor result precision, even though it is much faster. We verified by additional experiments (not shown here) that the retrieval quality of our modified Büttcher scoring model was as good as the original Büttcher model.

6.3   Results with Pruned Index Lists

We first study the size of our indexes at different levels of pruning for an index (without stemming as this is an upper bound for the index size with stemming). As the complete set of proximity lists is too large to completely materialize it, we randomly sampled 1,500,000 term pairs with a frequency of at least 10, of which about 1.2% had a nonempty proximity list. Table 2 shows the index sizes (number of list entries) for text (exact) and proximity lists (estimated), for different length limits. They are calculated/estimated according to the kind of data stored in the lists as described in Subsection 5.1. We assume that document identifiers and scores a size of 8 bytes each. Therefore one TL entry or PXL entry (consisting of document identifier and BM25 score or accumulated score

Table 3. Index sizes (disk space) with different length limits, with and without window limit

|index/limit|	500|	1000|	1500|	2000|	2500|	3000|	unpruned|
TL	4.4 GB	5.3 GB	6.0 GB	6.6 GB	7.0 GB	7.4 GB	47.5 GB
PXL (est.)	5.4 TB	6.3 TB	7.0 TB	7.5 TB	7.9 TB	8.2 TB	20.5 TB
PXL, window≤ 10 (est.)	343.5 GB	430 GB	507 GB	580.9 GB	627.1 GB	673.3 GB	1.3 TB
CL (est.)	10.7 TB	12.7 TB	14.0 TB	15.0 TB	15.8 TB	16.5 TB	41.0 TB
CL, window≤ 10 (est.)	686.9 GB	860 GB	1.0 TB	1.1 TB	1.2 TB	1.3 TB	2.5 TB

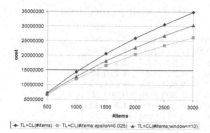

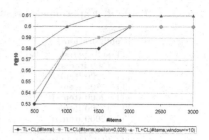

Fig. 1. TL+CL approaches: cost

Fig. 2. TL+CL approaches: P@10

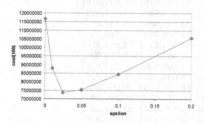

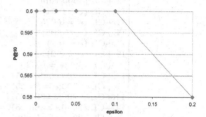

Fig. 3. TL+CL(ϵ varied): cost

Fig. 4. TL+CL(ϵ varied): P@10

respectively) takes a size of 16 bytes whereas one CL entry takes a size of 32 bytes as it stores the document identifier, the accumulated score and two BM25 scores. It is evident that keeping all proximity lists, even with a length limit, is infeasible. However, limiting the window to 10 reduces the size of the index noticably to at most a factor of 8-15 over the unpruned text index, which may be tolerated given the cheap disk space available today. Table 3 shows the index sizes (required disk space) for the very same lists. The sze of TLs is not that big issue as the unpruned TLs only amount to 47.5 GB, and can be further downsized using maximum list lengths. The far more critical indexes are PXLs and CLs that exhibit the prohibitive estimated size of 20.5 TB and 41.0 TB respectively. Limiting the list size helps, although the lists remain too large. Additionally restricting PXLs and CLs by a window size of ten finally leads to tolerable sizes between 343.5 GB and 673.3 GB for PXLs and 686.9GB and 1.3 TB for CLs. As we show later (Table 4), excellent results can be achieved when limiting the index size to 1,500. Hence, we need about 1 TB of disk space to execute TL+CL(1500;window$\leq$10) on a document collection with 426 GB data. Additional lossless compression may further reduce the index sizes.

We then evaluated retrieval quality with pruned (text and proximity) index lists, where we used combinations of window-based pruning with a maximal size of 10, fixed-length index lists and the pruning technique by Soffer et al. [19] for $k = 10$. All measurements were done without random accesses, hence we report only a single cost value based on the number of bytes transferred by sequential accesses. Additional experiments without this constraint showed that TopX only rarely attempts to make RAs in this setting as the pruned lists are often very short.

Table 4. Experimental results for top-10 retrieval with pruned lists

Configuration	P@10	#SA	bytes SA	cost
TL+CL (window≤10)	0.60	5,268,727	111,119,408	111,119,408
TL (500 items)	0.27	148,332	2,373,312	2,373,312
TL (1000 items)	0.30	294,402	4,710,432	4,710,432
TL (1500 items)	0.32	439,470	7,031,520	7,031,520
TL (2000 items)	0.34	581,488	9,303,808	9,303,808
TL (2500 items)	0.36	721,208	11,539,328	11,539,328
TL (3000 items)	0.37	850,708	13,611,328	13,611,328
TL+CL (500 items)	0.53	295,933	7,178,960	7,178,960
TL+CL (1000 items)	0.58	591,402	14,387,904	14,387,904
TL+CL (1500 items)	0.58	847,730	20,605,312	20,605,312
TL+CL (2000 items)	0.60	1,065,913	25,971,904	25,971,904
TL+CL (2500 tuples)	0.60	1,253,681	30,648,064	30,648,064
TL+CL (3000 tuples)	0.60	1,424,363	34,904,576	34,904,576
TL+CL ($\epsilon = 0.01$)	0.60	4,498,890	87,877,520	87,877,520
TL+CL ($\epsilon = 0.025$)	0.60	3,984,801	73,744,304	73,744,304
TL+CL ($\epsilon = 0.05$)	0.60	4,337,853	75,312,336	75,312,336
TL+CL ($\epsilon = 0.1$)	0.60	5,103,970	84,484,976	84,484,976
TL+CL ($\epsilon = 0.2$)	0.58	6,529,397	105,584,992	105,584,992
TL+CL (500;$\epsilon = 0.025$)	0.54	281,305	6,628,528	6,628,528
TL+CL (1000;$\epsilon = 0.025$)	0.58	521,519	12,034,320	12,034,320
TL+CL (1500;$\epsilon = 0.025$)	0.59	732,919	16,606,064	16,606,064
TL+CL (2000;$\epsilon = 0.025$)	0.60	910,721	20,377,904	20,377,904
TL+CL (2500;$\epsilon = 0.025$)	0.60	1,060,994	23,519,296	23,519,296
TL+CL (3000;$\epsilon = 0.025$)	0.60	1,191,956	26,211,376	26,211,376
TL+CL (500;window≤10)	0.58	290,788	6,931,904	6,931,904
TL+CL (1000;window≤10)	0.60	543,805	12,763,376	12,763,376
TL+CL (1500;window≤10)	0.61	780,157	18,117,552	18,117,552
TL+CL (2000;window≤10)	0.61	984,182	22,734,544	22,734,544
TL+CL (2500;window≤10)	0.61	1,166,144	26,854,608	26,854,608
TL+CL (3000;window≤10)	0.61	1,325,250	30,466,512	30,466,512

Table 5. Experimental results for top-100 retrieval with unpruned and pruned lists

Configuration	P@100	MAP@100	#SA	#RA	bytes SA	bytes RA
TL	0.37	0.13	42,584,605	434,233	681,353,680	3,473,864
TL+PXL	0.39	0.14	44,450,513	394,498	711,208,208	3,155,984
TL+CL	0.39	0.14	12,175,316	32,357	302,386,896	380,552
PXL	0.27	0.09	867,095	2,925	13,873,520	23,400
TL+CL (window≤ 10)	0.39	0.14	17,714,952	0	346,997,712	0
TL+CL (500 items)	0.34	0.11	310,469	0	7,558,816	0
TL+CL (1000 items)	0.37	0.13	610,983	0	14,838,144	0
TL+CL (1500 items)	0.38	0.13	904,910	0	21,911,520	0
TL+CL (2000 items)	0.38	0.14	1,184,658	0	28,615,776	0
TL+CL (2500 items)	0.39	0.14	1,457,093	0	35,138,176	0
TL+CL (3000 items)	0.39	0.14	1,723,204	0	41,493,728	0
TL+CL (500;$\epsilon = 0.025$)	0.33	0.11	281,485	0	6,631,408	0
TL+CL (1000;$\epsilon = 0.025$)	0.36	0.12	527,171	0	12,156,256	0
TL+CL (1500;$\epsilon = 0.025$)	0.37	0.13	753,012	0	17,054,112	0
TL+CL (2000;$\epsilon = 0.025$)	0.37	0.13	957,593	0	21,371,376	0
TL+CL (500;window≤10)	0.34	0.12	290,968	0	6,934,784	0
TL+CL (1000;window≤ 10)	0.37	0.13	551,684	0	12,940,576	0
TL+CL (1500;window≤ 10)	0.38	0.13	802,538	0	18,638,752	0
TL+CL (2000;window≤ 10)	0.38	0.13	1,039,466	0	23,969,632	0
TL+CL (2500;window≤ 10)	0.38	0.13	1,261,124	0	28,907,200	0
TL+CL (3000;window≤ 10)	0.38	0.13	1,483,154	0	33,856,144	0

Table 4 shows the experimental results for top-10 queries in this setup, again with stemming enabled. It is evident that TL+CL with length-limited lists and a lim-

Table 6. Costs for top-100 retrieval with unpruned and pruned lists

Configuration	Cost(100)	Cost(1000)
TL	1,028,740,080	4,155,217,680
TL+PXL	1,026,806,608	3,867,192,208
TL+CL	340,442,096	682,938,896
PXL	16,213,520	37,273,520
TL+CL (window≤ 10)	346,997,712	346,997,712
TL+CL (500 items)	7,558,816	7,558,816
TL+CL (1000 items)	14,838,144	14,838,144
TL+CL (1500 items)	21,911,520	21,911,520
TL+CL (2000 items)	28,615,776	28,615,776
TL+CL (2500 items)	35,138,176	35,138,176
TL+CL (3000 items)	41,493,728	41,493,728
TL+CL (500;$\epsilon = 0.025$)	6,631,408	6,631,408
TL+CL (1000;$\epsilon = 0.025$)	12,156,256	12,156,256
TL+CL (1500;$\epsilon = 0.025$)	17,054,112	17,054,112
TL+CL (2000;$\epsilon = 0.025$)	21,371,376	21,371,377
TL+CL (2500;$\epsilon = 0.025$)	25,288,646	25,288,646
TL+CL (3000;$\epsilon = 0.025$)	28,924,720	26,211,376
TL+CL (500;window≤10)	6,934,784	6,934,784
TL+CL (1000;window≤ 10)	12,940,576	12,940,576
TL+CL (1500;window≤ 10)	18,638,752	18,638,752
TL+CL (2000;window≤ 10)	23,969,632	23,969,632
TL+CL (2500;window≤ 10)	28,907,200	28,907,200
TL+CL (3000;window≤ 10)	33,856,144	33,856,144

ited window size gives a factor of 50-150 over the unpruned TL baseline in terms of saved cost, while yielding the same result quality (TL+CL (1000;window≤10)). Using TL with text lists of limited length is a lot worse in effectiveness. Pruning with ϵ is not as efficient, and large values for ϵ in fact *increase* cost: Many entries from the proximity lists are pruned away, but at the same time the additional content scores available from these entries are not available any more. In combination with length limiting, results are comparable to our best configuration, but with slightly longer lists. Figures 1 to 4 illustrate some of these experimental results. We obtain the best precision values when for limiting the list size to 1,500 or more elements. Out of the approaches depicted in Figures 1 and 2, TL+CL(#items) is the approach with the worst precision values at the highest cost. TL+CL(#items, window≤10) provides the best precision values at a medium cost, whereas TL+CL(#items, $\epsilon = 0.025$) only comes up with a slightly better precision than TL+CL(#items), however at the best costs. For mere static index list pruning, precision values are most favorable for choices of ϵ below 0.1.

As especially pruning along the lines of Soffer et al. [19] is done for a specific value of k, it is interesting to see how good results using the index pruned with $k = 10$ are for larger values of k. For space reasons, we limit the presentation to $k = 100$; Tables 5 and 6 shows the results for pruned and unpruned lists. Even though proximity awareness cannot improve much on result quality, most runs with pruning are at least as effective as the unpruned runs, while saving one or two orders of magnitude in accesses, bytes transferred, and cost. The combination of length-limited lists and limited window size is again best, with a peak factor of 350 over the unpruned TL baseline at the same quality (TL+CL (1000;window≤10)).

7   Conclusion

This paper presented novel algorithms and implementation techniques for efficient evaluation of top-k queries on text data with proximity-aware scoring. We have shown that our techniques can speed up evaluation by one or two orders of magnitude, trading in runtime for cheap disk space and maintaining the very high result quality (effectiveness) of proximity-aware scoring models. Our future work will focus on smarter data structures for indexes and applying index compression techniques.

References

1. Anh, V.N., de Kretser, O., Moffat, A.: Vector-space ranking with effective early termination. In: SIGIR, pp. 35–42 (2001)
2. Anh, V.N., Moffat, A.: Pruned query evaluation using pre-computed impacts. In: SIGIR, pp. 372–379 (2006)
3. Bast, H., et al.: Io-top-k: Index-access optimized top-k query processing. In: VLDB, pp. 475–486 (2006)
4. Botev, C., et al.: Expressiveness and performance of full-text search languages. In: Ioannidis, Y., Scholl, M.H., Schmidt, J.W., Matthes, F., Hatzopoulos, M., Boehm, K., Kemper, A., Grust, T., Boehm, C. (eds.) EDBT 2006. LNCS, vol. 3896, pp. 349–367. Springer, Heidelberg (2006)
5. Büttcher, S., Clarke, C.L.A.: Efficiency vs. effectiveness in terabyte-scale information retrieval. In: TREC (2005)
6. Büttcher, S., Clarke, C.L.A., Lushman, B.: Term proximity scoring for ad-hoc retrieval on very large text collections. In: SIGIR, pp. 621–622 (2006)
7. Callan, J.P., Croft, W.B., Broglio, J.: Trec and tipster experiments with inquery. Inf. Process. Manage. 31(3), 327–343 (1995)
8. Chang, M., Poon, C.K.: Efficient phrase querying with common phrase index. In: Lalmas, M., MacFarlane, A., Rüger, S., Tombros, A., Tsikrika, T., Yavlinsky, A. (eds.) ECIR 2006. LNCS, vol. 3936, pp. 61–71. Springer, Heidelberg (2006)
9. Clarke, C.L.A., Cormack, G.V., Tudhope, E.A.: Relevance ranking for one to three term queries. Inf. Process. Manage. 36(2), 291–311 (2000)
10. de Kretser, O., Moffat, A.: Effective document presentation with a locality-based similarity heuristic. In: SIGIR, pp. 113–120 (1999)
11. de Moura, E.S., et al.: Fast and flexible word searching on compressed text. ACM Trans. Inf. Syst. 18(2), 113–139 (2000)
12. Fagin, R., Lotem, A., Naor, M.: Optimal aggregation algorithms for middleware. J. Comput. Syst. Sci. 66(4), 614–656 (2003)
13. Grossman, D.A., Frieder, O.: Information Retrieval. Springer, Heidelberg (2005)
14. Hawking, D.: Efficiency/effectiveness trade-offs in query processing. SIGIR Forum 32(2), 16–22 (1998)
15. Metzler, D., et al.: Indri at TREC 2004: Terabyte track. In: TREC (2004)
16. Monz, C.: Minimal span weighting retrieval for question answering. In: IR4QA (2004)
17. Papka, R., Allan, J.: Why bigger windows are better than small ones. Technical report, CIIR (1997)

18. Rasolofo, Y., Savoy, J.: Term proximity scoring for keyword-based retrieval systems. In: Sebastiani, F. (ed.) ECIR 2003. LNCS, vol. 2633, pp. 207–218. Springer, Heidelberg (2003)
19. Soffer, A., et al.: Static index pruning for information retrieval systems. In: SIGIR, pp. 43–50 (2001)
20. Song, R., et al.: Viewing term proximity from a different perspective. Technical Report MSR-TR-2005-69, Microsoft Research Asia (May 2005)
21. Theobald, M., Schenkel, R., Weikum, G.: An efficient and versatile query engine for TopX search. In: VLDB, pp. 625–636 (2005)
22. Williams, H.E., et al.: What's next? index structures for efficient phrase querying. In: Australasian Database Conference, pp. 141–152 (1999)
23. Williams, H.E., et al.: Fast phrase querying with combined indexes. ACM Trans. Inf. Syst. 22(4), 573–594 (2004)
24. Witten, I.H., Moffat, A., Bell, T.: Managing Gigabytes. Morgan Kaufman, San Francisco (1999)

Prefix-Shuffled Geometric Suffix Tree

Tetsuo Shibuya

Human Genome Center, Institute of Medical Science, University of Tokyo
4-6-1 Shirokanedai, Minato-ku, Tokyo 108-8639, Japan
tshibuya@hgc.jp

Abstract. Protein structure analysis is one of the most important research issues in the post-genomic era, and faster and more accurate index data structures for such 3-D structures are highly desired for research on proteins. The geometric suffix tree is a very sophisticated index structure that enables fast and accurate search on protein 3-D structures. By using it, we can search from 3-D structure databases for all the substructures whose RMSDs (root mean square deviations) to a given query 3-D structure are not larger than a given bound. In this paper, we propose a new data structure based on the geometric suffix tree whose query performance is much better than the original geometric suffix tree. We call the modified data structure the prefix-shuffled geometric suffix tree (or PSGST for short). According to our experiments, the PSGST outperforms the geometric suffix tree in most cases. The PSGST shows its best performance when the database does not have many substructures similar to the query. The query is sometimes 100 times faster than the original geometric suffix trees in such cases.

1 Introduction

Protein 3-D structure analysis is one of the most important post-genomic research topics in molecular biology. Recently, more and more protein structures are solved by state-of-the-art technologies such as NMR (nuclear magnetic resonance), and the size of the protein 3-D structure database increases larger and larger. Now, there are more than 40,000 entries in the PDB database [2] and it is still increasing. The protein structures are said to have similar functions if their 3-D structures are similar. Thus, to analyze the functions of a protein whose structure is newly determined, it is very important to search for similar (sub)structures from the growing database. There are many comparison algorithms for protein structures [5], and the results could be very accurate, but it will require enormous amount of time to apply them against the very large databases. Hence, indexing techniques for protein structure databases are highly desired to avoid the large computation time.

The similarity of two protein structures is often measured by the RMSD (root mean square deviation) [1,4,11]. The geometric suffix tree [12] is an indexing data structure that enables efficient search from a 3-D structure database for all the substructures whose RMSDs to a given query are not larger than some given bound. It also has many potential applications, such as 3-D motif finding and

N. Ziviani and R. Baeza-Yates (Eds.): SPIRE 2007, LNCS 4726, pp. 300–309, 2007.

functional prediction. The geometric suffix tree is based on the famous suffix trees for alphabet strings [6,8,10,13,14], but it deals with 3-D coordinates instead of alphabet characters. In this paper, we propose a new data structure based on the geometric suffix tree, which we call the prefix-shuffled geometric suffix tree, or PSGST for short. It improves the query performance of the geometric suffix tree by changing the order of atoms in each substructure. We will demonstrate the PSGSTs' performance through experiments.

This paper is organized as follows. In section 2, we explain the preliminaries. In section 3, we explain a new notion called the 'prefix-shuffled structure' that would help us to improve the query performance of the geometric suffix trees. Then, in section 4, we explain the newly proposed data structure, the prefix-shuffled geometric suffix tree. In section 5, we demonstrate the performance of it through experiments. Finally in section 6, we conclude our results and discuss future work.

2   Preliminaries

2.1   RMSD: The Root Mean Square Deviation

A protein is a chain of amino acids. Each amino acid has one unique carbon atom named C_α, and the set of all the C_α atoms in a protein is called the backbone of the protein. The backbone is topologically linear, but it forms a geometrically very complex structure in the 3-D space. Most previous work on protein 3-D structures deals with the coordinates of the backbone atoms. Thus, we also consider the coordinates of the backbone atoms as the target to index. The most popular and basic measure to determine geometric similarity between two sets of points in 3-D, like the positions of backbone atoms, is the RMSD (root mean square deviation) [1,4,11].

Before defining the RMSD, let us define the measures that we call the MSSD (minimum sum squared distance) and the RSSD (Root Sum Square Distance). Let the two sets of points (*i.e.*, structures) to be compared be $P=\{p_1, p_2, \ldots, p_n\}$ and $Q = \{q_1, q_2, \ldots, q_n\}$, where p_i and q_j are 3-D coordinates. To compute the MSSD/RSSD/RMSD between two sets of 3-D coordinates, we must know which atom in one structure corresponds to which atom in the other. Here we consider p_i corresponds to q_i for all i. Let $mssd(P,Q)$ be the minimum value of $\sum_{i=1}^{n} \|p_i - (R \cdot q_i + v)\|^2$ over all the possible rotation matrices R and translation vectors v, where $\| \cdot \|$ denotes the norm. Let $\hat{R}(P,Q)$ and $\hat{v}(P,Q)$ be the rotation matrix and the translation vector that satisfies $\sum_{i=1}^{n} \|p_i - (\hat{R}(P,Q) \cdot q_i + \hat{v}(P,Q))\|^2 = mssd(P,Q)$. Then the RSSD is defined as the squared root of it: $rssd(P,Q) = \sqrt{mssd(P,Q)}$, and the RMSD is finally defined as $rssd(P,Q)/\sqrt{n}$.

It is known that $\hat{v}(P,Q) = \sum_{i=1}^{n} (p_i - \hat{R}(P,Q) \cdot q_i)/n$. It means that the centroids of the two point sets must be translated to the same point by $\hat{v}(P,Q)$. Hence, if both of the point sets are translated so that their centroids are located at the origin of the coordinates, the RMSD problem is reduced to a problem of finding R that minimizes $f(R) = \sum_{i=1}^{n} \|p_i - R \cdot q_i\|^2$. We can solve this problem

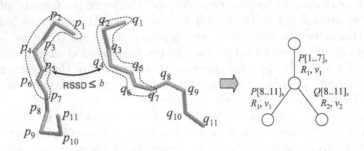

Fig. 1. A geometric trie for two protein 3-D structures. A node is constructed for $P[1..7]$ and $Q[1..7]$, as the RSSD between $P[1..7]$ and $Q[1..7]$ is smaller than the threshold b_{RSSD}. The combined edge is represented by arbitrary one of the two substructures — $P[1..7]$ is chosen in this example.

in linear time by using the singular value decomposition (SVD) [1] as follows. Let $H = \sum_{i=1}^{n} \boldsymbol{p}_i \cdot \boldsymbol{q}_i^t$, where $\boldsymbol{v}^t$ means the transpose of vector $\boldsymbol{v}$. Then $f(R)$ can be described as $\sum_{i=1}^{n} (\boldsymbol{p}_i^t \boldsymbol{p}_i + \boldsymbol{q}_i^t \boldsymbol{q}_i) - trace(R \cdot H)$, and $trace(RH)$ is maximized when $R = VU^T$, where $U \Lambda V$ is the SVD of H, and U^T denotes the transpose of matrix U. The SVD of H can be done in constant time as H is a fixed-size 3×3 matrix (see [7] for SVD algorithms). Hence the optimal rotation matrix can be obtained in constant time from H. In this way, we can compute the RMSD in $O(n)$ time. Note that there are rare degenerate cases where $det(VU^T) = -1$, which means that VU^T is a reflection matrix. We ignore the degenerate cases in this paper.

According to [12], the RMSD value, the optimal rotation matrix $\hat{R}(P, Q)$, and the optimal translation vector $\hat{v}(P, Q)$ can be computed incrementally by keeping some additional values for computation, *i.e.*, we can compute $\hat{R}(P_i, Q_i)$ and $\hat{v}(P_i, Q_i)$ for $P_i = \{\boldsymbol{p}_1, \boldsymbol{p}_2, \ldots, \boldsymbol{p}_i\}$ and $Q_i = \{\boldsymbol{q}_1, \boldsymbol{q}_2, \ldots, \boldsymbol{q}_i\}$ in $O(1)$ time after the computation of $\hat{R}(P_{i-1}, Q_{i-1})$ and $\hat{v}(P_{i-1}, Q_{i-1})$ for $P_{i-1} = \{\boldsymbol{p}_1, \boldsymbol{p}_2, \ldots, \boldsymbol{p}_{i-1}\}$ and $Q_{i-1} = \{\boldsymbol{q}_1, \boldsymbol{q}_2, \ldots, \boldsymbol{q}_{i-1}\}$, for any i (see [12] for more details).

2.2   Geometric Suffix Trees

The suffix tree of a string is the compacted trie of all its suffixes. Likewise, the geometric suffix tree [12] is based on a data structure called the *geometric trie*, which is defined as follows.

Consider a set of n 3-D structures $\mathbf{W} = \{W_1, W_2, \ldots, W_n\}$, and let ℓ_i be the length of W_i. Let $\boldsymbol{w}_j^{(i)}$ denote the coordinates of the j-th atom of W_i. Let $W_i[j..k]$ denote $\{\boldsymbol{w}_j^{(i)}, \boldsymbol{w}_{j+1}^{(i)}, \ldots, \boldsymbol{w}_k^{(i)}\}$, which means a structure formed by the $(k - j + 1)$ atoms from the j-th atom to the k-th atom in W_i. We call it a substructure of W_i. Furthermore, we call $W_i[1..j]$ $(1 \leq j \leq \ell_i)$ a prefix substructure of W_i. Conversely, $W_i[j..\ell_i]$ is called a suffix substructure. Then, the geometric trie for $\mathbf{W}$ is a rooted tree data structure that has the following features (Figure 1):

1. All the internal nodes (nodes other than the root and the leaves) have more than one child.
2. The tree has n leaves, each of which corresponds to one protein structure in **W**, and no two leaves correspond to the same structure. Let $leaf(i)$ denote the leaf that corresponds to W_i.
3. All the edges e except for some of edges that end at leaves correspond to a substructure $P(e) = W_i[j..k]$, and they have information of some 3-D rotation matrix $R(e)$ and some 3-D translation vector $v(e)$ for each.
4. Let $S(e)$ be $P(e)$ rotated by $R(e)$ and translated by $v(e)$, which is called the 'edge structure' of e. For a node x in the tree, let $S(x)$ be a structure that is constructed by concatenating all the edge structures of the edges on the path from the root to x, which we call the 'node structure' of x. For any leaf $v = leaf(i)$ and its node structure $S(v)$, the RSSD between any prefix substructure of $S(v)$ and the prefix substructure of W_i (of the same length) must not be larger than some given fixed bound b_{RSSD}.
5. For an edge $e = (v, w)$ with some corresponding substructure $P(e)$, the 'branching structure' $str(e)$ is defined as a structure that is obtained by adding the coordinates of the first atom of $S(e)$ (i.e., $S(e)[1]$) after $S(v)$. For any internal node v with more than one outgoing edge with corresponding substructures, the RSSD between $str(e_1)$ and $str(e_2)$ must be larger than b_{RSSD}, where e_1 and e_2 are arbitrary two of the edges.

Then the *geometric suffix tree* of a structure $P = \{p_1, p_2, \ldots, p_n\}$ is defined as the geometric trie of all the suffix substructures of P. The geometric suffix tree can be stored in $O(n)$ memory, though there are $O(n^2)$ substructures in the target structure. It can be built in $O(n^2)$ time by just adding suffix substructures into the tree one by one, using the incremental RMSD computation technique. The geometric suffix tree can be easily extended to deal with all the suffix substructures of a set of structures, like the generalized suffix trees for ordinary alphabet strings [8].

A prefix substructure of a node structure is called a 'representative structure'. To search for a substructure similar (*i.e.*, RMSD is within some bound b_{RMSD}) to a query $Q[1..m]$ using the geometric suffix tree, we first search for all the representative structures of length m whose RMSD to Q is within $b_{RMSD} + (b_{RSSD}/\sqrt{m})$. There always exist (one or more) original substructures that correspond to each representative structure. Finally, if the RMSDs between the query and the enumerated original substructures are actually within b_{RMSD}, we output them as the answers.

3 Prefix-Shuffled Structures

When we search for similar substructures from the geometric suffix trees, we incrementally compare RSSDs between the prefix substructures of the query structure and representative structures. In Figure 2, the line noted as 'Normal' shows the RSSDs of prefix substructures (of various lengths) of two very different

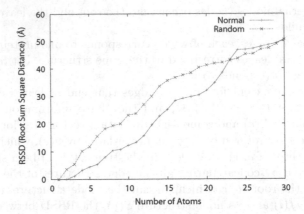

Fig. 2. Prefix RSSDs and Shuffled Prefix RSSDs. The RSSD goes up faster if the order of the atoms are shuffled.

proteins (a myoglobin and a rhodopsin taken from the set of structures used in section 5). In this example, the RMSD between two prefix substructures of length 30 is 9.40Å (*i.e.*, RSSD is 50.62Å), which means that the two structures are not at all similar to each other.

Consider the case that the myoglobin structure above is stored in the geometric suffix tree as a representative structure, and we want to find all the representative structures whose RSSDs to the rhodopsin structure above is within 20.0Å. Then we must incrementally compare these prefix structures up to 12 atoms. It means that we have to meaninglessly compute RSSDs 12 times, though these two structures are not at all similar to each other.

Let $\pi = \{\pi_1, \pi_2, \ldots, \pi_k\}$ be some permutation of length k. For a structure $P = \{\boldsymbol{p}_1, \boldsymbol{p}_2, \ldots, \boldsymbol{p}_n\}$ such that $n \geq k$, consider a new structure $H_\pi(P) = \{\boldsymbol{p}_{\pi_1}, \boldsymbol{p}_{\pi_2}, \ldots, \boldsymbol{p}_{\pi_k}, \boldsymbol{p}_{k+1}, \boldsymbol{p}_{k+2}, \ldots, \boldsymbol{p}_n\}$, which we call the prefix-shuffled structure of P by π.

In Figure 2, the dotted line noted as 'Random' shows the RSSDs between the prefix substructures of the prefix-shuffled structures of the same two structures (a myoglobin and a rhodopsin) with a randomly-generated permutation of length 30.[1] In other words, we compare $H_\pi(P)$ and $H_\pi(Q)$ instead of P and Q. According to the figure, the RSSD exceeds 20.0Å if the prefix substructure length becomes larger than 7, which is much smaller than the '12' in the previous 'Normal' case. It is a very reasonable result, because the distances between two adjacent atoms in the prefix-shuffled structure is often much larger than those in the original structure. Based on these observations, we consider that we may be able to improve the query performance by shuffling the structures with some appropriate permutation (both for the database and the query). The new data structure proposed in the next section is based on this intuition.

---

[1] The permutation we used here is $\{3, 25, 12, 29, 2, 13, 19, 16, 17, 10, 11, 9, 7, 1, 8, 18, 26, 27, 23, 5, 28, 15, 21, 20, 24, 14, 30, 22, 4, 6\}$.

4 Prefix-Shuffled Geometric Suffix Trees

We define the *prefix-shuffled geometric suffix tree* (*PSGST* for short) for a structure P as the geometric trie over all the prefix-shuffled suffix substructures of P by some permutation π (*i.e.*, $\{H_\pi(P[i..n])|1 \leq i \leq n - |\pi| + 1\}$). The memory requirement for storing the PSGST is $O(n)$ (same as the geometric suffix tree). Recall that the geometric suffix tree is built by just adding each suffix substructures one by one. The PSGSTs can also be built in the same way as the geometric suffix trees, which requires $O(n^2)$ time. Moreover, we can search for substructures that is similar to Q by just searching for representative structures that is similar to the prefix-shuffled query $H_\pi(Q)$ on the PSGST, if the length of Q is not smaller than the length of π. In this paper, we do not deal with queries which are shorter than the permutation π.

To construct the PSGSTs, we need some appropriate permutation of a given length. A random permutation can be used for this purpose. A uniform random permutation of length k can be generated in $O(k)$ time by iteratively swapping each position i ($1 \leq i \leq k$, in increasing order) of a list $\{1, 2, ..., k\}$ with a randomly chosen position among positions j such that $j \geq i$ (see [3] for details).

Other than the random permutations, the following permutation can also be used. A permutation $\pi = \{\pi_1, \pi_2, ..., \pi_k\}$ is called a *furthest permutation* if it satisfies $\pi_1 = 1$, $\pi_2 = k$, and $\min_{\ell < i} |\pi_i - \pi_\ell| \geq \min_{\ell < i} |\pi_j - \pi_\ell|$ for any i and j such that $i < j$. We call it 'furthest' because π_i is furthest from $\{\pi_1, ..., \pi_{i-1}\}$ among $\pi_i, \pi_{i+1}, ..., \pi_k$. For example, $\{1, 9, 5, 3, 7, 2, 4, 6, 8\}$ is a furthest permutation. We can assume that the distance between two atoms $\boldsymbol{p}_{\pi_i}$ and $\boldsymbol{p}_{\pi_j}$ would be large if $|\pi_i - \pi_j|$ is large. Therefore we consider the furthest permutation might be suitable for the PSGSTs.

The furthest permutation of length k can also be computed in $O(k)$ time with a bit operation technique as follows. To ease discussion, we first assume that $k - 1$ is a power of 2 and let $d = \log_2(k - 1)$. Let $rev_d(x)$ be a function that reverses the last d bit of x. For example, $rev_3(3) = rev_3(011(2)) = 110(2) = 6$. The function $rev_d(x)$ can be computed in $O(1)$ time.[2] Then consider a permutation $\pi^{(k)}$ of length k where $\pi_1^{(k)} = 1$, $\pi_2^{(k)} = k$, and $\pi_i^{(k)} = rev_d(i - 2) + 1$ for $i \geq 3$. It is the furthest permutation of length k, and it can be computed in $O(k)$ time. In case that $k - 1$ is not a power of 2, let k' be the smallest power of 2 that is not smaller than k, and construct the furthest permutation $\pi^{(k'+1)}$ with the above method. Then the furthest permutation of length k can be obtained by just removing numbers larger than k from $\pi^{(k'+1)}$, which requires only $O(k)$ time.

---

[2] We can compute $rev_d(x)$ by using a pre-computed table of the values of $rev_{\lceil d/c \rceil}(x)$ for $0 \leq x < 2^{\lceil d/c \rceil}$, where c is an appropriate constant positive integer. If we use appropriate c, the table size must be reasonably small, even if k (*i.e.*, $2^d + 1$) is very large. But even without such table, it takes only $O(\log d)$ time to compute $rev_d(x)$ with the basic bit operations of AND, OR, and SHIFT, and consequently the total computing time is still $O(k \log \log k)$ time. If k is a 32-bit (or even a 64-bit) integer, we can assume it as linear time. Note that it is very easy to design a digital circuit that computes $rev_d(x)$ in constant time.

Table 1. Time (in second) for constructing a geometric suffix tree and PSGSTs. The 'Random' columns shows the average/minimum/maximum construction time of 100 PSGSTs constructed with different permutations. The 'Furthest' column shows the construction time for the PSGST constructed with the furthest permutation.

| | GST | PSGST | | | |
| | | Random | | | Furthest |
		Average	Minimum	Maximum	
Time (sec)	39.10	37.26	35.87	38.65	37.89

Thus we conclude that the total computation time for constructing the furthest permutation of length k is $O(k)$.

In the next section, we will show through experiments how well our simple strategy works for 3-D substructure search.

5   Experiments

In this section, we demonstrate the performance of the PSGSTs. All the experiments are done on a Sun Fire 15K super computer with 288 GB memory and 96 UltraSPARC III Cu CPUs running at 1.2GHz.[3] As a data for experiments, we used a set of 228 myoglobin or myoglobin-related PDB data files containing 275 protein structures, which is same as the set used in the experiments by [12]. The total number of amino acids in the protein set is 41,719.

At first, we compared the construction time of PSGSTs against the construction time of the geometric suffix trees, by setting the RSSD bound $b_{RSSD} = 20.0$Å (Table 1). In the table, the 'GST' column shows the construction time of the geometric suffix tree against the myoglobin database. Next, we constructed 100 PSGSTs with different random permutations of length 50.[4] The 'Random' column shows the average, minimum, and maximum construction time among these 100 experiments. They are a little faster than the case of the geometric suffix tree, but it is not much different. We also did experiments by using the furthest permutation of length 50. The 'Furthest' column shows the result. The result is almost the same as the average of the results of random permutations. We assume these results are very reasonable, as there is no difference between the algorithms for the PSGSTs and the geometric suffix trees except for the prefix shuffling.

We next examined the query speed of the above 101 PSGSTs (*i.e.*, the 100 PSGSTs constructed with different random permutations, and the one constructed with the furthest permutation) and the geometric suffix tree (Table 2). We used two protein substructures as queries: (a) A substructure from the 20th amino acid to the 69th amino acid of the backbone structure of a rhodopsin[5]

---

[3] We used only one CPU for each experiment.

[4] We used the Mersenne-Twister [9] for generating random numbers.

[5] As seen in section 3, rhodopsins have nothing to do with myoglobins, and their structures are totally different.

Table 2. Time (in second) for queries on the geometric suffix trees and the PSGSTs. In (a), we used as a query a protein structure unrelated to any of the structures in a myoglobin structure database. In (b), by contrast, we used a myoglobin structure that is included in the same database.

(a) A rhodopsin query against the myoglobin database.

b_{RMSD} (Å)		0.2	0.4	0.6	0.8	1.0	1.2	1.4	1.6	1.8	2.0
#hits		0	0	0	0	0	0	0	0	0	0
GST		0.0207	0.0825	0.1736	0.2567	0.3306	0.3960	0.4554	0.5146	0.5726	0.6321
PS-GST	Random avg	0.0028	0.0063	0.0127	0.0241	0.0428	0.0716	0.1130	0.1681	0.2379	0.3244
	Random min	0.0002	0.0008	0.0018	0.0037	0.0080	0.0185	0.0372	0.0679	0.1053	0.1607
	Random max	0.0167	0.0461	0.0866	0.1241	0.1621	0.1986	0.2350	0.3020	0.4008	0.5130
	Furthest	0.0013	0.0022	0.0044	0.0081	0.0332	0.0576	0.1012	0.1605	0.2423	0.0163

(b) A myoglobin query against the myoglobin database.

b_{RMSD} (Å)		0.2	0.4	0.6	0.8	1.0	1.2	1.4	1.6	1.8	2.0
#hits		1	1	4	9	19	26	33	44	86	142
GST		0.0654	0.1065	0.1547	0.2333	0.3145	0.3994	0.4662	0.5473	0.7101	0.7859
PS-GST	Random avg	0.0583	0.0887	0.1219	0.1688	0.2372	0.2960	0.3519	0.4287	0.5443	0.6590
	Random min	0.0472	0.0712	0.0986	0.1243	0.1889	0.2445	0.2958	0.3476	0.4480	0.5608
	Random max	0.0778	0.1143	0.1580	0.2180	0.2984	0.3624	0.4359	0.5174	0.6835	0.7871
	Furthest	0.0615	0.0930	0.1637	0.2122	0.2633	0.3164	0.3775	0.4829	0.5796	0.7463

(named 1F88) obtained from the PDB, and (b) A substructure from the 20th amino acid to the 69th amino acid of the backbone structure of a myoglobin (named 103M), which is contained in the myoglobin database we used for constructing the geometric suffix trees and the PSGSTs. Note that these queries are same as those used in [12]. For each query, we searched for similar substructures with 10 different settings of the RMSD bound (b_{RMSD}). In the table, the '#hits' rows show the numbers of similar structures obtained with the designated b_{RMSD} settings, the 'GST' rows show the query time (in second) on the geometric suffix tree, and the 'PSGST' rows show the query time (in second) on the PSGSTs. In the 'PSGST' rows, the 'Random' rows show the average/minimum/maximum query time among the 100 PSGSTs constructed with different random permutations, while the 'Furthest' rows show the query time on the PSGST constructed with the furthest permutation.

In the experiment (a), the PSGST outperforms the geometric suffix tree in all the 101 cases. The PSGSTs constructed with random permutations perform about 1.9–13 times better than the geometric suffix tree in average. Moreover, the PSGSTs perform more than 100 times better than the geometric suffix tree in the best case. If we use the furthest permutation, the PSGST performs about 2.6–37.5 times better than the geometric suffix tree. The results by the furthest permutation is better than the average of results by random permutations, but it is not the best one among the 101 permutations we tried.

Consider a Figure 2-like graph for two similar structures. In this case, the RSSD will not go up until the end of the structure. Thus, we can easily imagine that the PSGSTs are not so efficient if the database has many structures similar to the query, which can be seen in the experiment (b). But, according to the table, the PSGST outperforms the geometric suffix tree in most cases. If we use a random permutation, the PSGST performs about 1.5 times better than the geometric suffix tree in average. If we use the furthest permutation, the PSGST outperforms the geometric suffix tree in all the cases but 1 case. All in all, we can conclude that the PSGST outperforms the geometric suffix tree.

6 Discussion

We proposed a new data structure based on the geometric suffix tree, which we call the prefix-shuffled geometric suffix tree (PSGST). The PSGSTs show higher query performance than the geometric suffix trees in most cases, though the construction time is almost the same. In the best case, a query on a PSGST is more than 100 times faster than the same query on the geometric suffix tree.

Several tasks remain as future work. The PSGST performs well especially when there are not many substructures similar to the query in the database. It means that the PSGST can be used as a very powerful filtering tool for some other more flexible similarity search algorithms on 3-D structures, which is one of the future tasks. Another future task is finding gapped 3-D motifs of proteins by using the PSGST. We do not know how to get the optimal permutation for the PSGST, which is an open problem. On PSGSTs, we cannot search for queries shorter than the permutation used for constructing the PSGST. It is also an open problem how to smartly deal with such short queries on PSGSTs.

Acknowledgement

All the computational experiments in this research were done on the Super Computer System, Human Genome Center, Institute of Medical Science, University of Tokyo.

References

1. Arun, K.S., Huang, T.S., Blostein, S.D.: Least-squares fitting of two 3-D point sets. IEEE Trans Pattern Anal. Machine Intell. 9, 698–700 (1987)
2. Berman, H.M., Westbrook, J., Feng, Z., Gilliland, G., Bhat, T.N., Weissig, H., Shindyalov, I.N., Bourne, P.E.: The protein data bank. Nucl. Acids Res. 28, 235–242 (2000)
3. Cormen, T.H., Leiserson, C.E., Rivest, R.L.: Introduction to Algorithms, 2nd edn. MIT Press, Cambridge (2001)
4. Eggert, D.W., Lorusso, A., Fisher, R.B.: Estimating 3-D rigid body transformations: a comparison of four major algorithms. Machine Vision and Applications 9, 272–290 (1997)
5. Eidhammer, I., Jonassen, I., Taylor, W.R.: Structure Comparison and Structure Patterns. J. Computational Biology 7(5), 685–716 (2000)

6. Farach, M.: Optimal suffix tree construction with large alphabets. In: Proc. 38th IEEE Symp. Foundations of Computer Science, pp. 137–143. IEEE Computer Society Press, Los Alamitos (1997)
7. Golub, G.H., Van Loan, C.F.: Matrix Computation, 3rd edn. John Hopkins University Press (1996)
8. Gusfield, D.: Algorithms on strings, trees, and sequences: computer science and computational biology. Cambridge University Press, Cambridge (1997)
9. Matsumoto, M., Nishimura, T.: A nonempirical test on the weight of pseudorandom number generators. In: Fang, K.T., et al. (eds.) Monte Carrlo and Quasi-Monte Carlo Methods 2000, pp. 381–395. Springer, Heidelberg (2002)
10. McCreight, E.M.: A space-economical suffix tree construction algorithm. J. ACM. 23, 262–272 (1976)
11. Schwartz, J.T., Sharir, M.: Identification of partially obscured objects in two and three dimensions by matching noisy characteristic curves. Intl. J. of Robotics Res. 6, 29–44 (1987)
12. Shibuya, T.: Geometric Suffix Tree: A New Index Structure for Protein 3-D Structures. In: Lewenstein, M., Valiente, G. (eds.) CPM 2006. LNCS, vol. 4009, pp. 84–93. Springer, Heidelberg (2006)
13. Ukkonen, E.: On-line construction of suffix-trees. Algorithmica 14, 249–260 (1995)
14. Weiner, P.: Linear pattern matching algorithms. In: Proc. 14th Symposium on Switching and Automata Theory, pp. 1–11 (1973)

Author Index

Lecture Notes in Computer Science

Sublibrary 1: Theoretical Computer Science and General Issues

For information about Vols. 1– 4490
please contact your bookseller or Springer

Vol. 4646: J. Duparc, T.A. Henzinger (Eds.), Computer Science Logic. XIV, 600 pages. 2007.

Vol. 4644: N. Azémard, L. Svensson (Eds.), Integrated Circuit and System Design. XIV, 583 pages. 2007.

Vol. 4641: A.-M. Kermarrec, L. Bougé, T. Priol (Eds.), Euro-Par 2007 Parallel Processing. XXVII, 974 pages. 2007.

Vol. 4639: E. Csuhaj-Varjú, Z. Ésik (Eds.), Fundamentals of Computation Theory. XIV, 508 pages. 2007.

Vol. 4638: T. Stützle, M. Birattari, H. H. Hoos (Eds.), Engineering Stochastic Local Search Algorithms. X, 223 pages. 2007.

Vol. 4630: H.J. van den Herik, P. Ciancarini, J. Donkers (Eds.), Computers and Games. XII, 283 pages. 2007.

Vol. 4628: L.N. de Castro, F.J. Von Zuben, H. Knidel (Eds.), Artificial Immune Systems. XII, 438 pages. 2007.

Vol. 4627: M. Charikar, K. Jansen, O. Reingold, J.D.P. Rolim (Eds.), Approximation, Randomization, and Combinatorial Optimization. XII, 626 pages. 2007.

Vol. 4624: T. Mossakowski, U. Montanari, M. Haveraaen (Eds.), Algebra and Coalgebra in Computer Science. XI, 463 pages. 2007.

Vol. 4623: M. Collard (Ed.), Ontologies-Based Databases and Information Systems. X, 153 pages. 2007.

Vol. 4621: D. Wagner, R. Wattenhofer (Eds.), Algorithms for Sensor and Ad Hoc Networks. XIII, 415 pages. 2007.

Vol. 4619: F. Dehne, J.-R. Sack, N. Zeh (Eds.), Algorithms and Data Structures. XVI, 662 pages. 2007.

Vol. 4618: S.G. Akl, C.S. Calude, M.J. Dinneen, G. Rozenberg, H.T. Wareham (Eds.), Unconventional Computation. X, 243 pages. 2007.

Vol. 4616: A.W.M. Dress, Y. Xu, B. Zhu (Eds.), Combinatorial Optimization and Applications. XI, 390 pages. 2007.

Vol. 4614: B. Chen, M. Paterson, G. Zhang (Eds.), Combinatorics, Algorithms, Probabilistic and Experimental Methodologies. XII, 530 pages. 2007.

Vol. 4613: F.P. Preparata, Q. Fang (Eds.), Frontiers in Algorithmics. XI, 348 pages. 2007.

Vol. 4600: H. Comon-Lundh, C. Kirchner, H. Kirchner (Eds.), Rewriting, Computation and Proof. XVI, 273 pages. 2007.

Vol. 4599: S. Vassiliadis, M. Bereković, T.D. Hämäläinen (Eds.), Embedded Computer Systems: Architectures, Modeling, and Simulation. XVIII, 466 pages. 2007.

Vol. 4598: G. Lin (Ed.), Computing and Combinatorics. XII, 570 pages. 2007.

Vol. 4596: L. Arge, C. Cachin, T. Jurdziński, A. Tarlecki (Eds.), Automata, Languages and Programming. XVII, 953 pages. 2007.

Vol. 4595: D. Bošnački, S. Edelkamp (Eds.), Model Checking Software. X, 285 pages. 2007.

Vol. 4590: W. Damm, H. Hermanns (Eds.), Computer Aided Verification. XV, 562 pages. 2007.

Vol. 4588: T. Harju, J. Karhumäki, A. Lepistö (Eds.), Developments in Language Theory. XI, 423 pages. 2007.

Vol. 4583: S.R. Della Rocca (Ed.), Typed Lambda Calculi and Applications. X, 397 pages. 2007.

Vol. 4580: B. Ma, K. Zhang (Eds.), Combinatorial Pattern Matching. XII, 366 pages. 2007.

Vol. 4576: D. Leivant, R. de Queiroz (Eds.), Logic, Language, Information and Computation. X, 363 pages. 2007.

Vol. 4547: C. Carlet, B. Sunar (Eds.), Arithmetic of Finite Fields. XI, 355 pages. 2007.

Vol. 4546: J. Kleijn, A. Yakovlev (Eds.), Petri Nets and Other Models of Concurrency – ICATPN 2007. XI, 515 pages. 2007.

Vol. 4545: H. Anai, K. Horimoto, T. Kutsia (Eds.), Algebraic Biology. XIII, 379 pages. 2007.

Vol. 4533: F. Baader (Ed.), Term Rewriting and Applications. XII, 419 pages. 2007.

Vol. 4528: J. Mira, J.R. Álvarez (Eds.), Nature Inspired Problem-Solving Methods in Knowledge Engineering, Part II. XXII, 650 pages. 2007.

Vol. 4527: J. Mira, J.R. Álvarez (Eds.), Bio-inspired Modeling of Cognitive Tasks, Part I. XXII, 630 pages. 2007.

Vol. 4525: C. Demetrescu (Ed.), Experimental Algorithms. XIII, 448 pages. 2007.

Vol. 4514: S.N. Artemov, A. Nerode (Eds.), Logical Foundations of Computer Science. XI, 513 pages. 2007.

Vol. 4513: M. Fischetti, D.P. Williamson (Eds.), Integer Programming and Combinatorial Optimization. IX, 500 pages. 2007.

Vol. 4510: P. Van Hentenryck, L.A. Wolsey (Eds.), Integration of AI and OR Techniques in Constraint Programming for Combinatorial Optimization Problems. X, 391 pages. 2007.

Vol. 4507: F. Sandoval, A.G. Prieto, J. Cabestany, M. Graña (Eds.), Computational and Ambient Intelligence. XXVI, 1167 pages. 2007.

Vol. 4502: T. Altenkirch, C. McBride (Eds.), Types for Proofs and Programs. VIII, 269 pages. 2007.

Vol. 4501: J. Marques-Silva, K.A. Sakallah (Eds.), Theory and Applications of Satisfiability Testing – SAT 2007. XI, 384 pages. 2007.

Vol. 4497: S.B. Cooper, B. Löwe, A. Sorbi (Eds.), Computation and Logic in the Real World. XVIII, 826 pages. 2007.

Vol. 4494: H. Jin, O.F. Rana, Y. Pan, V.K. Prasanna (Eds.), Algorithms and Architectures for Parallel Processing. XIV, 508 pages. 2007.

Vol. 4493: D. Liu, S. Fei, Z. Hou, H. Zhang, C. Sun (Eds.), Advances in Neural Networks – ISNN 2007, Part III. XXVI, 1215 pages. 2007.

Vol. 4492: D. Liu, S. Fei, Z. Hou, H. Zhang, C. Sun (Eds.), Advances in Neural Networks – ISNN 2007, Part II. XXVII, 1321 pages. 2007.

Vol. 4491: D. Liu, S. Fei, Z.-G. Hou, H. Zhang, C. Sun (Eds.), Advances in Neural Networks – ISNN 2007, Part I. LIV, 1365 pages. 2007.